006.32

Discrete Neural Computation
A Theoretical Foundation

**Prentice Hall Information
and System Sciences Series**
Thomas Kailath, Editor

Anderson & Moore	Optimal Control: Linear Quadratic Methods
Anderson & Moore	Optimal Filtering
Åström & Wittenmark	Computer-Controlled Systems: Theory and Design, second edition
Basseville & Nikiforov	Detection of Abrupt Changes: Theory and Application
Dickinson	Systems: Analysis, Design & Computation
Gardner	Statistical Spectral Analysis: A Nonprobabilistic Theory
Goodwin & Sin	Adaptive Filtering, Prediction, and Control
Gray & Davisson	Random Processes: A Mathematical Approach for Engineers
Grewal & Andrews	Kalman Filtering: Theory and Practice
Haykin	Adaptive Filter Theory
Haykin, ed.	Blind Deconvolution
Jain	Fundamentals of Digital Image Processing
Jamshidi, Tarokh & Shafai	Computer-Aided Analysis and Design of Linear Control Systems
Johansson	System Modeling & Identification
Kailath	Linear Systems
Kumar & Varaiya	Stochastic Systems
Kung	Digital Neural Networks
Kung	VLSI Array Processors
Kung, Whitehouse & Kailath, eds.	VLSI and Modern Signal Processing
Kwakernaak & Sivan	Signals & Systems
Landau	System Identification and Control Design Using P.I.M. + Software
Ljung	System Identification: Theory for the User
Ljung & Glad	Modeling of Dynamic Systems
Macovski	Medical Imaging Systems
Melsa & Sage	An Introduction to Probability and Stochastic Processes
Middleton & Goodwin	Digital Control & Estimation
Narendra & Annaswamy	Stable Adaptive Systems
Porat	Digital Processing of Random Signals: Theory & Methods
Rugh	Linear System Theory
Sastry & Bodson	Adaptive Control: Stability, Convergence, and Robustness
Siu, Roychowdhury & Kailath	Discrete Neural Computation: A Theoretical Foundation
Soliman & Srinath	Continuous and Discrete Signals and Systems
Spilker	Digital Communications by Satellite
Williams	Designing Digital Filters

Discrete Neural Computation
A Theoretical Foundation

Kai-Yeung Siu

University of California, Irvine

Vwani Roychowdhury

Purdue University

Thomas Kailath

Stanford University

Prentice Hall PTR
Englewood Cliffs, New Jersey 07632

Editorial/production supervision: *Camille Trentacoste*
Manufacturing manager: *Alexis R. Heydt*
Acquisitions editor: *Karen Gettman*
Editorial assistant: *Barbara Alfieri*
Cover design: *Maureen Eide*
Series logo design: *A. M. Bruckstein*

ISBN 0-13-300708-1

Prentice-Hall International (UK) Limited, *London*
Prentice-Hall of Australia Pty. Limited, *Sydney*
Prentice-Hall Canada Inc., *Toronto*
Prentice-Hall Hispanoamericana, S.A., *Mexico*
Prentice-Hall of India Private Limited, *New Delhi*
Prentice-Hall of Japan, Inc., *Tokyo*
Simon & Schuster Asia Pte. Ltd., *Singapore*
Editora Prentice-Hall do Brasil, Ltda., *Rio de Janeiro*

To my family and to my students

Thomas Kailath

To my parents and Helen

Kai-Yeung Siu

To Ma and Bapi

Vwani Roychowdhury

Contents

Foreword

This is a wonderful book of stories about the kinds of parallel computers called neural networks. They tell about how much more hardware and time those machines need for doing increasingly larger jobs.

Of course, those kinds of stories wouldn't interest most people. Less, I'd guess, than one in ten thousand. This is a pity because this is also a book about people – because we ourselves are instances of the most advanced parallel machines. Despite that, the theory of machines does not excite the public imagination the ways that sports contests do, or game shows, rock concerts, or the scandals of the private lives of celebrities. No matter. When we're doing mathematics, we don't have to consider the weights of the opinions of majorities.

The book's subject is the computational complexity of problems being solved by parallel machines – and it comes at a timely moment of history. The power of serial computers has grown exponentially for several decades, but in recent years that exponent has started to decrease. Fortunately, at the very same time, the exponent of parallelism has started to grow – while the exponent of memory cost has held its own. This combination of historical trends means that the exponential growth of the past 50 years may well continue long enough to supply us with another trillion fold more of computer power-per-dollar. That log-log curve might even turn up, when we come to the Age of Nanotechnology.

It is now almost exactly 25 years since Seymour Papert and I published *Perceptrons: An Introduction to Computational Geometry.* I see *Discrete Neural Computation: A Theoretical Foundation* as the natural successor to our book. But why did so relatively little happen within that quarter-century interval?

First, we ought to explain why we feel that these machines are so important. One reason is the conviction that our very own brains are composed of neural network machines. It would not be very useful to say that a brain *is* a neural network – because that expression is normally used to mean a network of components with a highly uniform architecture. It would be better to describe a brain as a highly evolved assembly composed by interconnecting, in special ways, many different kinds of neural networks. From this point of view, the brain can be seen as like a huge computer with quite a few different and specialized processors, memory systems, and data paths. This means that if we're ever to understand how we work, we'll need good theories both about what each of those networks is able to do, and about how they interact on the larger scale. In the brain, that probably means procedures that are less parallel and more serial.

(Of course, some might argue that this is wrong, that brains are continuous, not discrete. Others might claim that they're infinite. But we need not consider such questions here. When dealing with big mysteries, we should consider the simplest theories first. Philosophical beggars cannot be choosers.)

Neural networks also have practical uses – especially when equipped with procedures for learning from experience. Machines that can learn do not have to be programmed! Of course, we could say the same about any other kind of computer, so why do people emphasize that possibility when discussing neural net machines. Simply, I think, because the knowledge embedded in neural nets is represented in such an exceedingly opaque form. Those collections of numerical coefficients seem, at least on the surface, to be so semantically vacuous that we can scarcely imagine any other practical way to program them!

Now what can such network computers do? In the case of conventional programming for serial computers, we know that everything is possible because, given enough time and memory, those machines can compute any computable function. However, this is not usually the case for neural network machines. Many such systems (and feedforward networks in particular) are not universal computers; what they can potentially compute depends on their specific architectures. Here are some things that we'd like to know for each class C of network machines:

> Competence – What kinds of computations can C-nets be programmed to do?

Learnability – What kinds of computations can C-nets be made to learn?

Learning Time – How long will that learning take to do?

Scalability – How must the size and complexity of C-nets grow in scale, with each type of increase in problem complexity?

Both the present book and the older *Perceptrons* are mainly concerned with Scalability questions. This is because, in the case of networks with a fixed number N of input bits, we can always make a two-layer network machine to recognize any particular subset of such inputs – simply because any such function can be expressed as a disjunction of Boolean conjunctions. Furthermore, if any such recognition is possible at all, it could be accomplished by a trivial kind of "learning" machine that simply tries all possible assignments of coefficients to such a network. Thus Competence and Learnability questions are not interesting unless we ask for quantitative answers about how those answers "scale up" with increasing N. And although Learning-Time questions can be interesting indeed, they make sense only in the case of functions for which the network is Competent.

Earlier I remarked that the opinions of majorities lack weight in mathematical matters. When *Perceptrons* appeared, most theorists acclaimed it. But later there came strong complaints from other neural net researchers, who made remarks like these:

"Those Competence results apply only to 2-layer nets. Multilayer networks can escape those limitations."

Generally, this objection was unjustified, because in *Perceptrons* we almost never considered Competence questions at all, but only Scalability questions about networks under various conditions of bounded fan-in (see Chapter 1). In such cases, the numbers of units and connections still tend to grow exponentially, although having more layers reduces the sizes of exponents. This is often not the case for unbounded fan-ins, and the present book presents many surprising results of this kind.

"Those results may apply to the original Perceptron Learning Algorithm, but things are different now. Today we know how to use Back-Propagation, or Simulated Annealing, or other techniques that transcend those limitations."

This is related to another objection that we frequently heard:

> "They claimed that feedforward networks can't ever learn to compute function X. However, this did not turn out to be true because, in our experiments, the machines learned to do X quite satisfactorily."

The problem word here is "satisfactorily." It has no meaning in the world of pure theory. When we move to the world of practical matters, then we may agree to accept some errors. In pure mathematics we always agree to insist on obeying the rules that we've made – whereas in applied mathematics we agree to consider, instead, systems that work only as well as we need. This in turn raises some of the most difficult – and most important issues in all of science: how to design, and how to test, theoretical models of real-world problems. For example, in practical life, it might be reasonable to ask if a certain kind of network would be good for recognizing faces, or for reading handwriting. There usually is no practical way, then, to agree on a formal definition of such a problem that will always work "perfectly." The trouble, of course, is that those portraits and scrawls are not generated by formal systems with rules that we have access to. For example, how would you start to make a theory of how a machine could distinguish between pictures of dogs and cats? Unless you can set down a practical model that serves to define the boundaries, you cannot begin to prove theorems. The trouble is that terms like "good enough" are never good enough for that.

Once we appreciate this, we may be able to understand what happened to progress in making theories about these machines. It seems to be true that not much happened in the decade immediately after the publication of *Perceptrons*. Thus, the bibliography of the present book has virtually no citations in the 1970s, save for Paul Werbos' prescient thesis. How can we explain those quiet years? Many reporters have repeated the tale that this was because of some gloomy remarks that Papert and I made in our book – which persuaded all those researchers to abandon their work. But this story simply does not ring true. There is no incentive for scientists so strong as the wish to prove that the others are wrong. Another oft repeated tale is that we proved those (still-true) theorems only to get the foundations and sponsors to take those researchers' moneys away. But that compliment is too grand to accept! Mathematicians have no such great powers.

The truth is that it's quite usual for a new mathematical approach to lie fallow for a decade or so. But I think that we did make a different mistake. Our book included, all at once, proofs of almost every easy theorem – and that left almost nothing for the next generation of students to teethe on, so to speak. It was not until the 1980s that the new field of computational complexity had matured enough to supplement, with other techniques, the group-theoretic exploitations of pattern symmetries that we had developed. And it was not until now that, finally, young Kai-Yeung Siu and Vwani Roychowdury, working together with our long-time friend Tom Kailath, managed to bring all those ideas together, to build the almost-unified theory of the complexity of symmetric functions that is displayed in this wonderful book.

There still remains the question of why, in the 1980s, we saw no such theoretical progress coming from the many tens of thousands of enthusiasts who were working with practical neural networks. Scott Fahlman has suggested that although there still was a lack of certain key ideas, this was also due in part to inadequate computer power. To support this, he observes that it was not until Geoffrey Hinton got the backpropagation idea from David Rumelhart and started running examples on his Symbolics machine that it was recognized that the problem of local minima need not be fatal in practice.

It seems to me that another, rather paradoxical factor extended this long delay. It turned out that, once those computations became practical, the feedforward learning machines excelled in many domains in which statistical methods had had little success. This led quickly to useful results, especially in the domain of practical pattern recognition applications. The trouble was that this very success made theory seem less attractive. If your machine failed to work on a certain problem, well, who cares? There are plenty of other fish to fry. So the journals filled up with reports of success – and the failures were rarely mentioned at all. It was exciting to make such discoveries, and rewarding when others found uses for them. Almost no one found time to do theory.

But now we've come to another high peak. What will we next be able to see?

Marvin Minsky
Cambridge, Massachusetts

Preface

Interest in artificial neural networks has sparked research efforts in many disciplines, including neurobiology, physics, mathematics, computer science, and engineering. The results of such studies offer the promise of a new generation of computing devices with computational power beyond the reach of current computers. Although such promise remains to be fulfilled, progress in understanding the capabilities and limitations of various models of neural networks has been made.

This book brings together in one volume the recent developments in discrete neural computation. The vast and diversified research literature makes it impossible to emphasize every important topic in neural networks. As a result, only the computational issues in neural networks with discrete inputs and outputs are addressed. We believe that the topics covered in this book constitute an important aspect of neural computation, which should be made more accessible to the research community at large. Moreover, most of the material has not been previously presented in an integrated fashion in other books. Our aim is to integrate a variety of important ideas and analytical techniques and establish a theoretical foundation for discrete neural computation.

Given any model of neural computation, some issues that should be addressed are:

1. How does the neural network model compare with conventional models?

2. What are the systematic procedures for designing neural networks in the given model?

3. What are the computational limitations of such a model?

This book addresses the above issues in the context of discrete models of neural networks such as threshold circuits and discrete time Hopfield networks. For example, to address the first issue, we show that arithmetic functions admit more efficient realizations in the threshold circuit model than in conventional models for digital computation. As one approach to examining the second issue concerning the design of networks, we present systematic methodologies for implementing functions with well-defined structures such as symmetric functions and their generalizations. To understand the computational limitations, one method we employ is to study the minimum amount of resources needed in neural networks by establishing theoretical lower bounds.

We have adopted a computational complexity approach to establish a theoretical framework for studying the above issues in discrete neural computation. Within this approach, the efficiency of the networks is analyzed in terms of complexity measures such as size and depth. The size of a network, which refers to the number of neural elements or the number of weights, gives a natural measure of the amount of hardware required for implementation. Similarly, the depth of a network, which is the number of layers in a feedforward model, determines the computational delay and, hence, the speed of computation. Other complexity measures, which quantify the network connectivity and the precision of the parameters in the networks, are also considered. As much of computational complexity theory is concerned with the questions of scale, the results presented in this book describe how the costs of networks (measured in terms of number of neural elements, number of layers, and magnitudes of weights) grow with the problem sizes.

This book studies neural networks from the perspective of parallel computation. The specific models of threshold circuits and their variants are chosen because of their amenability to analytical techniques and the rigorous results that are obtained. We believe that the results presented for these special cases are suggestive and can be projected to other related networks. Thus, this book represents another step towards understanding the principles of the theory of parallel computation. Finally, although the results in this book are not concerned with hardware implementations, it is the promise of practical realizations of neural networks that makes the theory developed in this book more appealing.

Outline and Scope of the Book

The contents of this book can be broadly classified into three parts. In the first part, basic models for discrete neural computation and fundamental concepts in computational complexity are discussed. In the second part, efficient designs of threshold circuits for computing various functions are established. Finally in the third part, techniques for analyzing the computational power of neural models are developed.

Chapter 1 presents generic issues in neural computation and specific models for discrete neural computation. Complexity notions such as depth, size, and precision of weights are introduced. This is followed by an exposition of the problems examined in the rest of the book. Examples are presented to elicit relevant computational issues and techniques. The chapter concludes with brief summaries of the topics discussed in later chapters.

Chapter 2 deals with the model of linear threshold element, which is a common processing unit in a discrete neural network. Concepts such as linear separability/nonseparability, capacity and learning are introduced. We present the perceptron learning algorithm and study its limitations. Behavior of the perceptron learning algorithm for linearly nonseparable training sets is also addressed. Apart from this discussion, the rest of the book does not address any other aspect of learning in neural networks. We feel that the general topic of learning deserves separate treatment.

The next three chapters discuss efficient implementations of several classes of functions. Chapter 3 is devoted to the computation of symmetric functions and their generalizations. The theory developed in this chapter is useful for studying more general classes of functions. For example, the last section of Chapter 3 presents almost optimal realizations of general Boolean functions.

Chapter 4 deals with computations of common arithmetic functions such as comparison, sorting, addition, multiplication, exponentiation, and division. The emphasis is on designing depth-efficient networks. Techniques of spectral approximation, theory of error-correcting codes, and number theory are employed.

Chapter 5 addresses depth-size trade-off issues in computing periodic symmetric functions and the arithmetic functions discussed in Chapter 4. Ideas such as the 'telescopic' technique and the 'block-save' method are developed. Trade-off results for node and edge complexities are established, and restricted fan-in/fan-out networks are considered.

Chapter 6 addresses issues related to the precision necessary for the weights in threshold circuits. The computational power of threshold circuits with weights of limited precision is studied. In particular, this chapter demonstrates that networks requiring weights of high precision can be simulated by networks using weights with limited precision at the expense of a small increase in depth.

The next four chapters explore the computing power and limitations of discrete neural networks. Toward this end, a variety of techniques are developed to derive lower bound results on several complexity measures. The related results show that with respect to these complexity measures, many of the networks described in earlier chapters are optimal or almost optimal.

Chapter 7 develops the theory of rational approximation and demonstrates how the output of a multilayer neural network can be approximated by a single rational function of appropriate degree. From the properties of the resultant rational function, effective lower bounds are established.

Chapter 8 uses vector space concepts and linear algebraic tools for studying threshold circuit complexity. This chapter provides a geometric framework that generalizes many of the concepts and results of harmonic analysis for characterizing the limitations of threshold circuits.

Chapter 9 compares the computing power of conventional models of computation with that of discrete neural models. For example, with the use of probabilistic methods and polynomial approximation techniques, exponential lower bounds on the size of bounded depth AND-OR circuits are derived.

Chapter 10, on the other hand, addresses complexity issues in unrestricted depth circuits by relating the communication complexity of a function to the size of any circuit computing it. Concepts such as communication matrix, monochromatic rectangle, and decomposition number are used to derive tight lower bounds.

Chapter 11 introduces discrete time Hopfield networks and discusses applications such as associative memory and combinatorial optimization. Convergence to stable states under various modes of iteration, transient period, limit cycles, and spurious memories are also addressed. Mappings of several combinatorial optimization problems onto the Hopfield model are presented.

This book can serve as a reference text for researchers in neural computation and related disciplines. It can also be used in graduate courses in engineering, computer science, and discrete mathematics. In fact, the material in this book has been taught by the authors in graduate courses at Stanford University, Purdue University, and the University of California at Irvine. We expect the students to have elementary knowledge of Boolean algebra, probability theory, and linear algebra from their undergraduate studies in engineering and computer science. Those with some background in complexity theory will have a deeper appreciation of the results. More than a hundred exercises have been integrated into the chapters. These exercises supplement the text and in most cases represent contemporary research. Without undermining the challenge for the readers, we have outlined key ideas for the solutions of the exercises.

We take the liberty here of mentioning that the material in this book is primarily based on our research over the past few years. Wherever appropriate, we have incorporated results obtained by several other researchers. To preserve a coherent presentation, however, we have derived most of these results from our perspective. We have made our best efforts to include all the relevant references in the bibliographic notes at the end of each chapter.

This book has left unresolved many intriguing questions in discrete neural computation. Nevertheless, we hope that our efforts in presenting an integrated approach will promote a better understanding of the foundations of neural computation and stimulate further development.

Acknowledgments

Kai-Yeung Siu and Vwani Roychowdhury would like to thank Professor Bernard Widrow for being a constant source of inspiration in neural network research. The authors would also like to thank Professor Thomas Cover for his invaluable comments. Alon Orlitsky has contributed significantly to many of the results presented here and has generously given his time toward the enhancement of this book; we feel greatly indebted to him.

The funding for our research at Stanford University was provided in part by the Joint Services Program at Stanford University (United States Army, United States Navy, United States Air Force) under Contract DAAL03-88-C-0011, by the SDIO/IST, managed by the Army Research Office under

Contract DAAL03-90-G-0108, and by the Department of the Navy, NASA Headquarters, Center for Aeronautics and Space Information Sciences under Grant NAGW-419-S6. Kai-Yeung Siu has also received funding for this work from the National Science Foundation (NSF) Young Investigator Awards Program and from the School of Engineering at the University of California at Irvine. Vwani Roychowdhury has also received funding provided by the NSF's Research Initiation Awards Program and by the General Motors Faculty Fellowship from the Schools of Engineering at Purdue University.

Kai-Yeung Siu
Vwani Roychowdhury
Thomas Kailath

Chapter 1

Introduction

The objective of this book is to introduce important computational aspects of neural networks and present fundamental results addressing the key issues. Of the many areas in neural networks, we focus on the study of discrete models of neural computation, in which discrete sets of inputs and outputs are related by the network parameters.

The idea of modeling neurons by discrete units was first introduced in a seminal paper by McCulloch and Pitts [77] in 1943. Each neuron was modeled as a *linear threshold element* with a binary output. This model was inspired by early neurobiological evidence showing that neurons exhibit all-or-none firing patterns. Since their inception, discrete models of neural networks have been an integral part of studies on neural computation.

Undoubtedly, such a binary model cannot be expected to capture all the complexities inherent in a biological neural system. However, the appeal of the binary model lies in its very simplicity and its susceptibility to rigorous analytical studies. The popularity of this model is evident from the vast research literature that it has spawned over half a century.

The binary model of McCulloch and Pitts has not only stimulated research in neural networks, but it has also influenced the early development of digital computers. While digital computation has since evolved independently and reached considerable maturity, the early promise of neuromorphic computing systems remains unfulfilled. Recent research on neural networks represents a continuing effort toward developing neurobiology-based computing principles. Such efforts have in part been driven by an increased understanding of the limitations of conventional digital computers.

This book represents accumulated research aimed at understanding the discrete models of neural networks from a computational perspective. Concepts and constructs from modern developments in the theory of computation are employed to address the computational power of discrete neural models. Such studies, one hopes, will lead to the design of a new generation of more efficient computers.

We now present an informal description of the discrete models considered in this book. Precise definitions appear in Section 1.2. The models can be classified into two general categories, depending on whether or not the network incorporates feedback connections. The first category consists of *feedforward* networks such as threshold circuits and their variants. The second category comprises networks with feedback connections such as the discrete time Hopfield model. The parameters in a neural network of either category reside in the connection strengths among the neurons.

In a feedforward network, the neurons can be arranged in layers such that the outputs of the neurons in one layer depend only on those in previous layers. An example of such networks is a threshold circuit. Each formal neuron in a threshold circuit is a linear threshold element as introduced by McCulloch and Pitts. The computation in a threshold circuit can be considered as proceeding in a feedforward manner: The outputs of the neurons are updated layer by layer, with the layer closest to the inputs (i.e., the first layer) being updated first. The outputs of the circuit are taken as the outputs of some prespecified neurons in the last layer (and possibly other layers as well). In this sense, the number of layers corresponds to the time of parallel computation in the circuit.

If the inputs are binary, then a threshold circuit becomes a combinational Boolean circuit. A substantial portion of this book is devoted to the analysis of the threshold circuit model with binary inputs. A variant of the threshold circuit model is a feedforward network of sigmoidal elements. Sigmoidal elements compute continuous functions that approximate step output characteristics of linear threshold elements. Thus, the class of feedforward networks of sigmoidal elements can be considered as a generalization of threshold circuits. Computational power of such networks will be studied and compared with that of threshold circuits.

Another model that we shall examine in some detail is the discrete time Hopfield network. A Hopfield network comprises a fully connected set of linear threshold elements, where the output of each element depends on the outputs of all other elements as determined by the connection strengths. It can be considered as a sequential Boolean circuit and thus differs from the combinational nature of computation in a threshold circuit. Since each element has a binary output, it can be in one of two possible states. The inputs to the network are applied as the network's *initial states*. At every discrete time step of the computation, the states of a subset of the elements are updated. For appropriate choices of network parameters, the network goes through a succession of states until it reaches a *stable state*, when the outputs of the elements no longer change. The outputs of the network are then taken to be the steady-state outputs of the elements. Thus, the computation performed by the network can be viewed as an evolution process of a dynamical system. We shall explore possible applications of the Hopfield model for associative memory and combinatorial optimization, as well as its limitations for such applications.

In the remainder of this introductory chapter, we first discuss two major aspects of neural computation, namely, *learning* and *representation*, in Section 1.1. The scope and objectives of this book will be further clarified in that section. In Section 1.2, we present precise definitions of the models and the relevant concepts in discrete neural computation. In Section 1.3, we briefly discuss the various complexity issues in discrete neural computation, which are to be examined in detail in later chapters. This will be followed by several examples in Section 1.4 that illustrate some of the important issues addressed in this book. Then, in Section 1.5, we highlight several analytical techniques, which constitute the set of mathematical tools to be used in addressing the complexity issues. Some historical remarks on the key developments in discrete neural computation are given in Section 1.6. In Section 1.7, we give an overview of the results presented in the rest of the chapters. Finally, in Section 1.8, we discuss some of the terminologies that will be used frequently in this book.

1.1   Aspects of Neural Computation

Current research in the area of neural networks can be broadly classified into two major categories: *learning* and *representation*. Next, we provide brief discussions of these two topics, and use simple examples to illustrate the underlying concepts.

1.1.1   Learning

An important aspect of neural computation is the adaptive mechanism involved in the functioning of the networks. This adaptive mechanism is often referred to as learning or training. There is usually an assumed architecture of the network and a desired mapping. Learning/training can be viewed as a process of adapting the parameters so that the resultant network will execute the desired mapping. Very often, only a partial description of the mapping is given and the outputs of the mapping are specified only for some subset of the possible inputs, called the training set. One of the main issues in learning is the design of efficient learning algorithms with good generalization properties. Although used frequently in the literature, generalization is not a mathematically well-defined term; it usually refers to the capability of the networks to produce desirable responses to inputs that are not included in the training set. Other objectives in the design of learning algorithms include the minimization of the execution time and resources required in implementing the algorithms. The required resources could include, for instance, memory space for storing parameters and specifications of the desired mapping.

Example 1.1: Consider the perceptron learning algorithm proposed by Rosenblatt [109]. The architecture comprises a linear threshold element (LTE). It has a binary-valued output, 0 or 1, a weight vector $\mathbf{w}$ and a threshold value t as its parameters. Given an input vector X_i, the perceptron outputs 1 if the inner product $X_i^T \mathbf{w} \geq t$ and outputs 0 otherwise. During the learning phase of the perceptron, a set of real pattern vectors (the training set), $P = \{X_1, \ldots, X_k, Y_1, \ldots, Y_l\}$, and a dichotomy of the pattern vectors, $X = \{X_1, \ldots, X_k\}$ and $Y = \{Y_1, \ldots, Y_l\}$, are given. The perceptron learning algorithm is an iterative algorithm for finding the appropriate values of the parameters $\mathbf{w}$ and t such that the output will be 1 or 0 accordingly as the input pattern vector belongs to X or Y. If there exist a weight vector $\mathbf{w}$ and a threshold t such that the patterns can be classified correctly by the perceptron, then the patterns associated with the specified dichotomy are said to be *linearly separable*. Rosenblatt showed that if the patterns are linearly separable, then the perceptron learning algorithm will terminate with the desired parameters.   □

A closely related model is the Adaline proposed by Widrow and Hoff [144], which is also a linear threshold classifier. The major difference is in the associated learning algorithm. The Adaline uses a Least Mean Square (LMS) algorithm to adjust the weights to minimize the difference between the actual and the desired outputs. The idea behind Adaline has developed into powerful techniques for adaptive signal processing such as equalization and adaptive beam forming [146]. Other examples of learning algorithms include Backpropagation [116], the Madaline rule and its variants [145], Adaptive Resonance Theory (ART) [24], and Self-Organizing Feature Maps [66].

1.1.2   Representation

The issues that are addressed in this category are in a sense dual to the problems in learning. Before one designs learning algorithms for a network architecture, more fundamental questions are: Given a desired mapping, is there a choice of parameters such that the resultant input-output relationship of the network is a representation of the mapping? Even if a network can represent a given mapping, how do the various computational costs increase with the problem size? In general, given any network architecture, it would be useful to determine the structure of the class of mappings that can be represented by the architecture, as well as the associated costs.

Example 1.2:   Consider again the learning architecture introduced in the preceding example. Given a set of pattern vectors P and a dichotomy of P into two subsets, X and Y, we want to determine whether the linear threshold element (LTE) can distinguish the two sets of vectors from each other (i.e., whether the patterns are linearly separable). The perceptron learning algorithm, however, will not terminate if the patterns are *not* linearly separable. For instance, the Parity function is a well-known example of a Boolean function that cannot be computed by an LTE. The training set for the Parity function consists of four vectors in two dimension: $\{(0,0),(0,1),(1,0),(1,1)\}$, and the dichotomy comprises the two sets $X = \{(0,0),(1,1)\}$ and $Y = \{(0,1),(1,0)\}$. It is easy to show that the patterns are not linearly separable. If this set of pattern vectors is used to train the perceptron, then the perceptron learning algorithm will not terminate.

There are other methods, including linear programming, that can determine
whether a set of patterns associated with a dichotomy is linearly separable,
and find the parameters of the LTE if appropriate. Further discussions on
the properties of pattern vectors that are not linearly separable appear in
Chapter 2.

The above-mentioned shortcoming is typical of many of the existing
learning algorithms. When the desired mapping cannot be represented by
the given architecture, either the learning algorithm will not terminate or it
will result in an erroneous solution.   □

Certainly, if a mapping cannot be represented by a given network architecture, then no algorithm can learn it. Therefore, it is useful to characterize
those mappings that cannot be represented by a given network architecture
before even attempting to learn any mapping. Such studies would indicate
the necessary modifications required in the architecture so that the resultant network can represent the desired mapping. One could also seek an
alternative approach and determine the optimal neural network architecture
necessary to represent a given mapping. Moreover, one might be able to indicate the minimal set of resources (as a function of the problem size), such
as the number of neural elements, the number of weights, and the precision
of the weights, so that a given set of mappings can be represented or learned
by the targeted network architecture. Based on such studies, one might be
able directly to design optimal networks for representing certain mappings
without going through any learning process.

In the 1960s, considerable research was focused on understanding the capabilities and the limitations of LTEs. For example, an exhaustive listing of
all Boolean functions with fewer than eight variables that can be computed
by an LTE was presented [85]. To show the limitations of an LTE as a classifier, Cover [31] and Winder [147] proved that the number of dichotomies
of n pattern vectors that could be realized by an LTE constitutes only a
vanishingly small fraction of all possible dichotomies as n increases. In these
early studies, most results were derived for a single LTE. The computational
capabilities of networks of LTEs and other multilayer models were not well
understood.

The objective of this book is to explore several representational issues
arising in discrete neural computation and to develop a theoretical framework for exploring the computational power and the limitations of discrete

neural networks. Examples of the specific issues to be addressed in this book are given in Section 1.4. Except for the results on learning issues for perceptrons presented in Chapter 2, we will not address any other issue in learning. Moreover, we will be primarily concerned with the *deterministic* aspects of discrete neural computation. Although our analysis employs probabilistic tools, the stochastic behavior of neural networks will not be addressed. For example, we will not present any results on the Boltzmann machine [2] and the statistical capacity of the Hopfield network. To clarify our points, we first review some of the basic concepts and the fundamental issues in neural computation.

1.2   Concepts in Discrete Neural Computation

1.2.1   Feedforward Model

One basic model of neural computation we shall examine in detail is the feedforward network.

Definition 1.1:   **(Feedforward Network)** A feedforward network can be modeled as a directed acyclic graph. The nodes are partitioned into three sets:

1. *input nodes* – with no in-coming edges,

2. *internal nodes* – with both in-coming and out-going edges,

3. *output nodes* – with no out-going edges.

Each directed edge $i \to j$ represents a connection in the network. Each input node is associated with an input to the circuit, and each internal or output node, also called a *gate*, computes a function of its inputs. (The inputs to the jth gate are the results of the functions computed at all gates i such that $i \to j$.)   □

   Since the results in this book will be mainly concerned with the computation of Boolean functions, the case when the feedforward network computes a Boolean function will receive special attention.

Definition 1.2:   **(Boolean Circuit)** A Boolean circuit is a feedforward network where all its inputs and outputs are binary.   □

From now on, we shall use the terms circuits and feedforward networks interchangeably.

Definition 1.3:   (**Circuit Size/Depth**) The *size* of a circuit is the number of gates it contains. The *depth* of a gate is the maximum number of edges along any directed path from the input nodes to that gate. The depth of the circuit is the maximum depth of all gates.   □

The notion of size as defined here is not the only notion accepted in the literature. Some authors define the size as the *edge complexity*, or the number of connections in the circuit. Most of the results in this book will be concerned with the *node complexity*, i.e., the number of gates in the circuit. Hence, we find it more convenient to adopt the above definition of circuit size. In Chapter 5, where we discuss both the node complexity and the edge complexity of a circuit, we shall explicitly distinguish these two terms to avoid ambiguity.

Definition 1.4:   (**Fan-in/Fan-out**) The number of connections leading into a node is the *fan-in* of that node. Similarly, the number of connections leading out of a node is the *fan-out* of that node. The fan-in and fan-out of a circuit are the maximum fan-in and fan-out, respectively, among all gates in the circuit.   □

1.2.2   Circuit as a Model of Parallel Computation

Interest in the study of the circuit model comes from the fact that it serves as a natural model for parallel computation. If we group all gates with the same depth together, then we can consider a circuit to be arranged in layers, where the first layer consists of all gates with depth 1, and the last layer consists of all gates with the maximum depth. Thus, the depth of a circuit is equal to the number of layers.

If each gate computes a Boolean function, then the circuit as a whole computes a multi-output Boolean function $\{0,1\}^{n_{in}} \rightarrow \{0,1\}^{n_{out}}$. The output computed by each gate will be propagated along all of its output edges, and all gates in the same layer compute concurrently. Under the simplifying assumption that all gates compute with equal delay, each layer in the Boolean circuit introduces a unit delay, and the circuit depth corresponds to the overall delay for the computation. Thus, the depth of a circuit can be interpreted as a measure of the time required for the parallel execution of the input-output mapping defined by the network. The size of the circuit can be interpreted as a measure of the required amount of hardware.

1.2.3 Types of Functional Elements

So far we have not specified any set of prescribed functions (also referred to as basis functions or activation functions) associated with the gates in a Boolean circuit. Clearly, the power of a Boolean circuit depends on the basis functions. In the traditional model for logic circuits, the basis consists of AND gates, OR gates, and the inverters (NOT gates). The output of an AND gate is 1 if all its inputs are 1, and is 0 otherwise. The output of an OR gate is 1 if at least one of its inputs is 1, and is 0 otherwise. The inverter (NOT gate) is a single input gate whose output is 1 if its input is 0, and vice versa. For convenience, we refer to this type of Boolean circuits as AND-OR circuits.

Definition 1.5: **(AND-OR Circuit)** An AND-OR circuit is a Boolean circuit in which every gate is an AND gate, OR gate, or an inverter.   □

We shall study in detail a feedforward model of neural networks where each gate is a linear threshold element. The Boolean function associated with each linear threshold element is naturally called the linear threshold function.

Definition 1.6:   **(Linear Threshold Function)** A *linear threshold function* $f(X)$ is a binary function such that

$$f(X) = \mathrm{sgn}(F(X)) = \begin{cases} 1 & \text{if } F(X) \geq 0, \\ 0 & \text{if } F(X) < 0, \end{cases}$$

where $F(X) = \sum_{i=1}^{n} w_i \cdot x_i - t.$ □

The real valued coefficients w_i's are commonly referred to as the *weights* of the threshold function, and t is referred to as the *threshold*. Without loss of generality, we can assume the threshold to be zero by having an extra input variable with weight t and with the value of that variable fixed to -1.

Definition 1.7:   (**Threshold Circuit**) A *threshold circuit* is a feedforward network in which every gate computes a linear threshold function.   □

Note that in our definition of threshold circuits, the inputs to the network do not need to be Boolean, though each element computes a binary function.

We shall also be interested in the computation of Boolean functions by networks of continuous-valued elements. To formalize this notion, we adopt the following definitions.

Definition 1.8:   (γ **Element/Network**) Let $\gamma : \mathbf{R} \to \mathbf{R}$. A γ element with weights $w_1, \ldots, w_m \in \mathbf{R}$ and threshold t computes the function $\gamma(\sum_{i=1}^{m} w_i x_i - t)$, where $(x_1, \ldots, x_m) \in \mathbf{R^m}$ is the input. A γ network is a feedforward network of γ elements.   □

Definition 1.9:   A γ network C is said to compute a Boolean function $f : \{0,1\}^n \to \{0,1\}$ with separation $\epsilon > 0$ if there exists $t_C \in \mathbf{R}$ such that for any input $X = (x_1, \ldots, x_m)$ to the network C, the output element of C outputs a value $C(X)$ with the following property: If $f(X) = 1$, then $C(X) \geq t_C + \epsilon$. If $f(X) = 0$, then $C(X) \leq t_C - \epsilon$.   □

A special case of γ network where each γ element computes a sigmoidal function is of particular interest in the study of neural computation; we state the corresponding definition separately.

Definition 1.10:   (**Sigmoidal Element/Network**) A *sigmoidal element* computes a function $f(X)$ of its input variables $X = (x_1, \ldots, x_n)$ such that

$$
\begin{aligned}
f(X) &= \sigma(F(X)) \\
&= \frac{1 - e^{-F(X)}}{1 + e^{-F(X)}} \, ,
\end{aligned}
$$

where

$$F(X) = \sum_{i=1}^{n} w_i \cdot x_i + w_0.$$

A *sigmoidal network* is a feedforward network of sigmoidal elements.   □

1.2.4 Hopfield Model

The Hopfield model differs from the feedforward model because feedback is incorporated in the connections between the nodes. Earlier we presented an informal description of the model; here we present a precise definition.

A Hopfield network can be viewed as a graph in which each node represents a neuron in the network. The *order* of the network is the number of neurons in the network. The connection strength between each pair of nodes (neurons) is represented by a *weight* associated with the edge connecting the corresponding pair of nodes. Formally, let H be a Hopfield network with n nodes denoted by $\{x_1, x_2, \ldots, x_n\}$. Let w_{ij} denote the weight associated with the edge (j, i) connecting node x_j to x_i, and t_i denote the threshold value attached to each node x_i. Let W be the $n \times n$ weight matrix whose $(i, j)^{th}$ entry is w_{ij}, and T be the n-dimensional threshold vector whose i^{th} entry is t_i.

The state of the ith node at time t is denoted by $x_i(t)$, $t = 0$, 1, 2, ... Each neuron (node) can assume one of two possible states, 1 or -1. The *state* of the network at time t is the vector $X(t) = [x_1(t) \ x_2(t) \ \ldots \ x_n(t)]^T$. In general, the state of the network at time $t + 1$ is a function of $\{W, T\}$ and the state $X(t)$ of the network at time t. The network is thus completely determined by the parameters $\{W, T\}$, the initial value $X(0)$ of the states, and the manner in which the nodes (neurons) are updated. If at time step t, node x_i is chosen to be updated, then at the next time step

$$x_i(t+1) = \text{sgn}(P_i(t)) = \begin{cases} 1 & \text{if } P_i(t) \geq 0, \\ -1 & \text{if } P_i(t) < 0, \end{cases} \qquad (1.1)$$

where we define

$$P_i(t) = \sum_{j=1}^{n} w_{ij} x_j(t) - t_i.$$

In other words, each node is updated by computing a linear threshold function of the current states of the other nodes. Note that we have used $\{1, -1\}$ notations instead of the usual $\{0, 1\}$ notations for the Boolean values. See Section 1.8 for a discussion on the use of such notations.

The next state of the network, $X(t+1)$, is updated from the current state $X(t)$ by evaluating Eqn. (1.1) for a prespecified subset of the nodes in the network. Obviously, the sequence of states of the network $X(1)$, $X(2)$, ...

depends on the subset of nodes that is selected to be updated at each time step.

Definition 1.11:   (Modes of Iteration)

- If at every iteration only a single node is updated (as in Eqn. (1.1)) while the states of the other nodes are left unchanged, then the network is said to be iterating in a *sequential* mode.

- If at every iteration all the nodes are chosen to be updated, then the network is said to be iterating in a *parallel* mode.

- In other cases, the network is said to be iterating in a *hybrid* mode. □

In matrix notation, each iteration of the parallel mode computes

$$X(t+1) = \text{sgn}(WX(t) - T),$$

where for vectors the sgn($\cdot$) function applies to every coordinate. Whereas there are many different modes of sequential iteration (depending on which node is selected to be updated at every iteration), it follows from the definition that there is only one mode of parallel iteration.

Definition 1.12:   (Stable State) A state vector $X \in \{1, -1\}^n$ is called a *stable* state of the network (with parameters $\{W, T\}$) if

$$X = \text{sgn}(WX - T). \qquad\qquad □$$

A stable state is also referred to as a *fixed point* in the state space.

It follows from Definition 1.12 that if at some time step t_0, the network has reached a stable state $X(t_0)$, then regardless of the mode of iteration $X(t) = X(t_0)$ for all $t > t_0$, i.e., the state of the network will remain unchanged after t_0.

Definition 1.13:   (Limit Cycle) The Hopfield network is said to enter a *limit cycle* of length m if after a finite number of iterations, the corresponding updating algorithm results in a sequence of states $X_1, \ldots, X_m, X_1, \ldots, X_m,$ $\ldots$ repeating indefinitely, where the states $\{X_1, \ldots, X_m\}$ are all distinct. □

Note that a stable state can be interpreted as the special case of a limit cycle whose length equals 1. However, the notion of a limit cycle is associated with the underlying updating algorithm, whereas the stable state is defined independently of the underlying mode of iteration in the network.

For certain choices of the network parameters and the mode of iteration, the Hopfield network goes through a succession of states until it reaches a stable state, when the outputs of the elements no longer change. The outputs of the network are then taken to be the steady-state outputs of the linear threshold elements. Such convergence issues for the Hopfield model and related applications are discussed in Chapter 11, and the complexity issues involved in the Hopfield model are introduced in Section 1.3.4.

1.3   Complexity Issues in Discrete Neural Computation

1.3.1   A Circuit Complexity Point of View

Given enough hardware resources (e.g., number of neural elements), one can always compute any input-output mapping with arbitrary precision using the feedforward models defined in the preceding section [33, 81]. For example, any Boolean function of n variables can be computed by a depth-2 threshold circuit or AND-OR circuit comprising at most $2^{n-1} + 1$ gates (see Chapter 2). However, such an implementation would require prohibitively large amounts of resources even for a small number of input variables. For a given function, the objective is therefore to minimize the amount of hardware and computation time. In our circuit model, this means it is desirable to design a circuit that has reasonable size and optimal depth. For example, if computing time is of primary concern, then one might want to design a circuit of smaller depth (thus a faster circuit) with the additional constraint that the size does not become prohibitively large. Similarly, if one is dealing with a known size constraint, then the objective would be to design a circuit with the minimum possible depth such that the available hardware resources are not exceeded.

In order to precisely address the above issues, one requires a rigorous framework where concepts such as 'reasonable size' and 'optimal depth' can be formally defined. Such a framework would enable one to differentiate designs based on their speed of computation and use of resources, and perhaps enable one precisely to answer questions such as: Given a function, what is the minimal amount of hardware resources required to compute it? Moreover, within the framework one could possibly design circuits that are provably more efficient than those one would otherwise design using ad hoc schemes. In this book, we shall adopt the framework of circuit complexity for addressing such issues.

Circuit complexity is concerned with the study of the resources required in circuits for computing Boolean functions. From a theoretical point of view, the circuit model is appealing due to its conceptual simplicity. Each type of resources corresponds to a complexity measure of a circuit. Several complexity measures such as the size, depth, and fan-in/fan-out, can be naturally defined for the circuit model, and well-defined optimization problems involving these measures can be outlined. A common objective in this framework is to minimize the size and the depth of a circuit that computes a desired function. We shall directly adapt this framework for the models of neural computation introduced earlier.

Before we study the complexity issues involved in discrete neural computation in more detail, we need to introduce some terminologies and notations to facilitate further discussions.

Asymptotic Notations

The notations $f(n) = O(g(n))$ or $g(n) = \Omega(f(n))$ mean that $f(n) \leq c \cdot g(n)$ for some constant $c > 0$, as $n \to \infty$. Also, $f(n) = o(g(n))$ if $\lim_{n \to \infty} (f(n)/g(n)) = 0$. Moreover, if $f(n) = O(g(n))$ and $f(n) = \Omega(g(n))$, then $f(n) = \Theta(g(n))$.

Family of Functions/Circuits

The complexity of circuits for Boolean functions is usually expressed in terms of the input size, which is the number of input variables of the circuit. We say that an n-variable function f_n has size complexity bounded above by $S(n)$ and depth complexity bounded above by $D(n)$, if for every n one can construct a circuit C_n with size $S(n)$ and depth $D(n)$ that computes f_n. Thus, one is dealing with a *circuit family* $\{C_n\}$ and a sequence of Boolean functions $\{f_n\}$ such that every function f_n is computed by a circuit C_n in the circuit family. For example, one can construct a family of circuits of size complexity $O(n \log n)$ and depth complexity $O(\log n)$ to add two n-bit numbers. To avoid cumbersome terminology, we often use the word 'circuit' to stand for the corresponding circuit family. We thus say that there is a circuit of size $O(n \log n)$ and depth $O(\log n)$ that computes the sum of two n-bit numbers.

With respect to some family of circuits $\{\hat{C}_n\}$, we say that an n-variable function $\hat{f}_n$ has size complexity $\Omega(\hat{S}(n))$ if there is a fixed constant $c > 0$ such that for sufficiently large n, any circuit in $\{\hat{C}_n\}$ computing f_n must have size at least $c \cdot \hat{S}(n)$. A function f_n has depth complexity $\Omega(\hat{D}(n))$ if there is a fixed constant $\alpha > 0$ such that for sufficiently large n, any circuit in $\{\hat{C}_n\}$ computing f_n has depth at least $\alpha \hat{D}(n)$.

Restrictions on the Circuit Model

The classical work of Shannon [119] demonstrates that all but a vanishingly small fraction of the possible n-variable Boolean functions require an exponential number (i.e., $\Omega(2^{cn})$ for some $c > 0$) of gates to be computed. Such lower bounds serve mainly to indicate that only a small fraction of the possible n-variable Boolean functions admits circuits of reasonable size, when n is large. For example, many commonly encountered functions, such as integer addition, multiplication, and division, do have efficient circuit implementations.

Although the circuit model for computation is elegant due to its conceptual simplicity, a great frustration of the theory is the paucity of results concerning *specific* functions. For example, no specific function is known for which the size of the corresponding circuit can be proved to be $\Omega(n \log n)$. In fact, the best known lower bound on the size of a Boolean circuit computing a specific function is still linear ($\Omega(n)$) [140]. On the other hand, Shannon's work indicates that there exist functions of n variables that require exponential-size circuits; we just do not know which specific ones.

A natural step to making progress on meaningful lower bound results is to limit the flexibility of the circuit model. The hope is that a more restricted model may yield nontrivial lower bound results, and more importantly, that the techniques and insights involved in proving these results may shed some light into the general model. A more practical motivation for studying circuits with limited flexibility is that any implementation of the circuit model will always have limited resources where one or more of the circuit parameters have to be restricted.

There are several parameters of a general Boolean circuit that can be constrained. First, the type of functions that each gate can compute in the circuit model is usually confined to a certain class of simple functions. For example, the AND-OR circuit model imposes the restriction that each gate can only be an AND gate, OR gate, or an inverter (NOT gate). On the other hand, the threshold circuit model allows each gate to compute a linear threshold function. Another important parameter of the Boolean circuit model that can be constrained is the circuit depth, leading to the bounded depth circuit model. Restrictions on several other circuit parameters will also be examined in more detail in the following sections.

1.3.2 Complexity Measures for Feedforward Model

Bounded vs. Unbounded Fan-in/Fan-out

An important parameter of a Boolean circuit as a model of parallel computation is the *fan-in* (i.e., the number of input connections) of each gate. Suppose that the fan-in of all gates in a Boolean circuit is bounded above by a constant k. It can be shown that if the circuit has size s, then it can depend on at most $s(k-1)+1$ inputs, and if the circuit has depth d, it can depend on at most k^d inputs. Hence, a circuit whose gates have fan-in bounded by a constant k computing any nondegenerate Boolean function of n variables (i.e., the function depends on all n variables) must have size $s \geq (n-1)/(k-1) = \Omega(n)$ and depth $d = \log n/\log k = \Omega(\log n)$. Therefore, if we want to design a circuit with depth less than $O(\log n)$ and size less than $O(n)$, we need to allow the fan-in to increase with the number of inputs to the circuit.

When the fan-in/fan-out of the gates in a Boolean circuit is not bounded above by a fixed constant, we say that the Boolean circuit has *unbounded fan-in/fan-out*. In such a case, the circuit model is also said to have unbounded fan-in/fan-out parallelism. Many computing devices can be modeled as circuits with unbounded fan-in/fan-out. For example, Programmable Logic Arrays (PLAs) can be considered as depth-2 unbounded fan-in/fan-out AND-OR circuits.

The theoretical study of unbounded fan-in/fan-out circuits may give us insights into devising faster algorithms for various computational problems than would be possible with bounded fan-in/fan-out parallelism. It is interesting to determine what can be saved with respect to hardware and time when there is no restriction on the fan-in/fan-out.

Our definition of threshold circuits does not impose any restriction on the fan-in and fan-out of each threshold gate in the circuit, i.e., each gate can have any number of inputs from previous layers and can feed its output to arbitrarily many gates in subsequent layers. Since neural networks are characterized by their massive parallelism, the unbounded fan-in/fan-out parallelism of the threshold circuit model seems to be a reasonable first assumption.

Polynomial vs. Exponential Size

The size (i.e., number of gates) in a circuit model corresponds to the amount of hardware needed to implement the physical circuit. It is obvious that the amount of hardware required also depends on the type of gates allowed in the circuit model. Intuitively, one expects the circuit to have better computing characteristics when a more powerful type of gates is permitted in the circuit. Given a certain type of gates and an n-variable Boolean function f_n, one would like to know the size complexity $S(n)$ of a circuit C_n that computes f_n. Often it is difficult to determine $S(n)$ precisely. We shall be satisfied with the knowledge of the asymptotic behavior of $S(n)$.

For example, a circuit C_n with size $\Theta(n^2)$ will require more gates to implement than a circuit $\tilde{C}_n$ with size $\Theta(n \log n)$ for sufficiently large n. In this example, we have deliberately ignored the constant factor involved in the complexity by using the asymptotic notations. However, such results are only meaningful when the effect of the constant factor is negligible compared to the asymptotic behavior. In the pathological case, if C_n has size $2n^2$ and $\tilde{C}_n$ has size $10^{10} n \log n$, then for all relevant input size n, $\tilde{C}_n$ will have a larger size than C_n.

Sometimes when we want to minimize other complexity measures such as the circuit depth, we might be satisfied with a circuit whose size does not prohibitively increase. In circuit complexity theory, the size of a circuit C_n of n input variables is considered to be reasonable if C_n has *polynomial size*, i.e., the size complexity $S(n)$ of C_n is bounded above by a fixed power in n, or more formally, there exists a constant $k > 0$ such that $S(n) = O(n^k)$. On the other hand, a circuit is of *exponential size* if its size is $\Omega(2^{n^\epsilon})$, for some fixed $\epsilon > 0$. In all the circuit models we consider in this book, polynomial number of gates also implies polynomial number of edges or connections. So polynomial-size circuits mean that both the node complexity and the edge complexity of the circuits are bounded by a polynomial.

There are a number of specific functions for which there are no known polynomial-size circuits. In general, the question of whether a function has a polynomial-size circuit depends on the type of gates allowed in the circuit model and the restrictions imposed on other complexity measures. For example, the n-variable Parity function can be computed by a bounded fan-in AND-OR circuit of $O(n)$-size and $O(\log n)$-depth. On the other hand, the Parity function requires exponential size in an unbounded fan-in depth-2 AND-OR circuit, but it can be computed by an unbounded fan-in depth-2 threshold circuit of $O(n)$-size. The interplay between the size and the depth in a circuit will be discussed next.

Bounded Depth vs. Unbounded Depth

As discussed in Section 1.2, the depth of a circuit corresponds to the time required for computing the outputs in parallel and hence determines its speed. Several branches of parallel computation are concerned with the design of circuits for which the depth is minimized and the size is constrained to be polynomially bounded. In other words, one of the issues in parallel computation is to determine how much hardware is required in a circuit to compute a specific function if the parallel time is restricted.

A circuit C_n is said to have *bounded depth* or *constant depth* if there is a fixed integer d such that for every n, the depth of C_n is bounded above by d; otherwise, a circuit is said to have unbounded depth. As we pointed out earlier, if the circuits have bounded fan-in gates, then the depth of the circuit must grow as $\Omega(\log n)$. Hence, for constant-depth circuits, it is interesting to consider only the case of unbounded fan-in gates.

We shall show in Chapter 9 that there are some simple functions not computable by constant-depth unbounded fan-in AND-OR circuits with polynomial size; however, these functions have efficient implementations in small-depth threshold circuits. In fact, significant progress has been made in understanding the power and the limitations of the class of constant-depth polynomial-size AND-OR circuits. We refer to this class of circuits as AC^0

and the functions computable by them as AC^0 functions. To understand how these known results on AND-OR circuits relate to the results on threshold circuits, we now give a more detailed discussion of the recent development in circuit complexity theory.

As mentioned earlier, a central issue in parallel computation can be stated as follows: Given that the number of gates in the Boolean circuit is bounded by a polynomial in the size of inputs, what is the minimum depth that is needed to compute certain functions? It is now known that for many basic functions, such as the Parity and the majority of n variables, or the multiplication of two n-bit numbers, any constant-depth unbounded fan-in AND-OR circuit computing these functions must be of exponential size. Another way of interpreting these results is that circuits of AND/OR gates computing these functions that use a polynomial amount of chip area must have unbounded delay (i.e., delay that increases with n). In fact, the lower bound results imply that the minimum possible delay for multipliers (with a polynomial number of AND/OR gates) is $\Omega(\log n / \log \log n)$. These results also give a theoretical justification for why it is impossible for circuit designers to implement small-depth Parity circuits or multipliers in a small chip area using AND/OR gates as the basic building blocks.

A natural extension is to study Boolean circuits that contain threshold gates. In fact, many of the functions provably not in AC^0 can be computed by small-depth threshold circuits. This is one of the motivations behind the study of threshold circuits in complexity theory. While substantial understanding of the limitations of the complexity class AC^0 has been acquired, efforts in obtaining nontrivial lower bound results on constant-depth threshold circuits have not been so successful.

A goal in this direction is to determine a specific function that is not computable in the class of constant-depth polynomial-size threshold circuits (TC^0 circuits) but computable in the class of $O(\log n)$-depth polynomial-size circuits with bounded fan-in gates (NC^1 circuits). It is well known that $TC^0 \subset NC^1$; moreover, it is a common conjecture that TC^0 is a proper subclass of NC^1 (i.e., $TC^0 \subsetneq NC^1$). The current lower bound techniques for showing the limitations of AC^0 circuits, however, do not seem to apply to the more powerful TC^0 circuits. As a first step toward proving this conjecture, the finer structure within the class of TC^0 has been studied. Let LT_d be the class of functions computable by depth-d polynomial-size threshold circuits. Clearly, LT_1 is simply the class of linear threshold functions and $LT_d \subseteq LT_{d+1}$ for every d. The union of the complexity classes LT_d for every constant d comprises TC^0. Many interesting functions have been shown to be computable in LT_d for $d \leq 3$. On the other hand, there is no specific function that is currently known to be not computable in LT_2.

Perhaps the reason why lower bound results on threshold circuits are difficult to derive is that threshold gates are computationally powerful. We shall explore in later chapters the power and the limitations of constant-depth threshold circuits.

Polynomial vs. Exponential Weights

In the definition of a linear threshold element (LTE), the weights are allowed to assume arbitrary real values. When the inputs are binary, the weights can always be chosen to assume only integer values, a result which will be proved in Chapter 2. In addition, we derive an upper bound of $2^{O(n \log n)}$ on the magnitudes of the integer weights for any linear threshold function of n Boolean variables.

The assumption of integer weights provides a basis to compare the complexities involved in the implementation of two different linear threshold elements. An LTE with exponentially large integer weights is likely to be more difficult to implement than an LTE with integer weights that are polynomially bounded.

In practice, we can only assume a certain degree of accuracy in the implementation of linear threshold elements with analog devices. To quantify this consideration of accuracy in theoretical terms, we consider a subclass of linear threshold elements, where each weight is an integer bounded by a polynomial in n (the input size), or equivalently, where only $O(\log n)$-bit accuracy is needed for each weight. In general, we use $\widehat{LT}_d$ to denote the class of functions that can be computed by depth-d polynomial-size threshold circuits where the weights are integers bounded by a polynomial in n. Informally, $\widehat{LT}_d$ circuits represent the class of depth-d polynomial-size threshold circuits that can be implemented with 'small' weights.

Note that an LTE with polynomially bounded integer weights is equivalent to an LTE with unit weights by duplicating the input variables at most polynomially many times. Thus, a polynomial-size threshold circuit with small weights can be simulated by another polynomial-size threshold circuit with unit weights and the same depth, where only the edge complexity is increased by a polynomial factor. Hence, the complexity class $\widehat{LT}_d$ is equivalent to the class of functions computable by depth-d polynomial-size circuits of threshold gates with unit weights.

1.3.3   Continuous-Valued Elements vs. Binary Elements

So far, we have only discussed the computation of Boolean functions by Boolean circuits, which are feedforward networks of Boolean gates. We shall also study feedforward networks of continuous-valued elements computing Boolean functions. Roughly speaking, a network of continuous-valued elements computes a Boolean function $f(X)$ if there is a fixed value θ such that for each input X, the output element of the network has value $> \theta$ when $f(X) = 1$, and $< \theta$ otherwise (for a more precise definition, see Section 1.2).

Our goal is to establish the computational power and the limitations of networks of continuous-valued elements such as sigmoidal networks, especially as compared to threshold circuits. For example, we would like to answer questions such as: How much added computational power does one gain by using sigmoidal elements or other continuous-valued elements to compute Boolean functions? Can the size of the network be reduced by using sigmoidal elements instead of threshold elements?

It has been shown that there is a Boolean function of n variables that can be computed by a depth-2 sigmoidal network with a fixed number of elements, but requires any threshold circuit (of even unbounded depth) to have size that increases at least logarithmic in n. In other words, there exist Boolean functions for which one can reduce the size of the network at least by a logarithmic factor by using continuous-valued elements instead of threshold elements with binary output values.

This result motivates the following question: Can one characterize a class of functions for which the corresponding threshold circuits have sizes at most a logarithmic factor larger than the sizes of the sigmoidal networks computing them? Because of the monotonicity of the sigmoidal functions, we do not expect sigmoidal elements to result in a substantial gain in computational power over the threshold elements. In fact, in Chapter 7, we show that the techniques of rational approximation for deriving lower bound results on threshold circuits can be extended to sigmoidal networks and other classes of feedforward neural networks. These extended techniques enable us to show that for symmetric Boolean functions of large strong degree (e.g., the Parity function), any depth-d network whose elements can be piecewise approximated by low-degree rational functions requires almost the same size as a depth-d threshold circuit computing the function.

1.3.4   Complexity Issues in Hopfield Model

Complexity issues in the Hopfield model are somewhat different from the feedforward model because of the special characteristics of its dynamics. Recall that the inputs to the Hopfield network are applied as the initial states of the network, and the outputs are taken as the steady-state values. Certain conditions need to be imposed on the network parameters and on the mode of iteration such that the network can always converge to stable states for meaningful interpretation of the network outputs.

A natural complexity measure is the time or the number of iterations it takes to perform computations in the Hopfield model. Since most computations in a Hopfield network are related to its convergence to a stable state, a natural question is: How long does it take for the network to converge to a stable state? The number of iterations for a Hopfield network to reach a stable state or a limit cycle is called the *transient period*; in general, it will depend on the mode of iteration and the initial state of the network. The transient period puts a limit on how fast a Hopfield network can solve a specific problem. We shall show in Chapter 11 that an upper bound on the transient period can be derived in terms of the weights of the network.

The size (i.e., the number of elements) also provides a measure of the complexity of the network and is determined by the specific applications. For example, the application of the Hopfield model for the Traveling-Salesman Problem (to be defined in Chapter 11) employs a network with size that grows quadratically with the number of cities in the problem.

In other applications such as associative memories, complexity measures will include parameters such as the number of stable states in the network. For example, a related issue is to determine the types and the number of stable states that can be programmed in the network. It turns out that the connectivity between the elements determines the types and the number of stable states or memories that can be stored in the network. Other issues such as spurious memories will also be discussed in Chapter 11.

1.4 Examples

In this section, we provide several examples to illustrate the problems and the related issues that will be addressed in this book. Since the theory and the general results are to be presented only in later chapters, the discussion in this section will be informal. Our aim is to give readers a rough idea of the complexity issues in neural computation.

1.4.1 The Parity Function

The n-variable Parity function $PAR_n(X)$, where $X = (x_1, \ldots, x_n) \in \{0,1\}^n$, is defined as

$$PAR_n(X) = \begin{cases} 1 & \text{if } \sum_{i=1}^{n} x_i \text{ is odd,} \\ 0 & \text{otherwise.} \end{cases} \tag{1.2}$$

We refer to the family of functions $\{PAR_n\}$ as Parity (see Section 1.8). Parity belongs to the class of functions whose values depend only on the sum of the input variables. Such functions are encountered quite frequently in logic circuit design and are called *symmetric* functions. Parity and symmetric functions serve as effective examples to illustrate many of the complexity issues to be addressed in this book.

It can be shown that any depth-2 AND-OR circuit computing Parity requires at least $(2^{n-1} + 1)$ gates. This implies that one would require exponential size if Parity is realized by any computing device that can be modeled by a depth-2 AND-OR circuit, such as a Programmable Logic Array (PLA). Even if the circuit depth is increased from 2 to any larger constant d, one can show that at least $2^{\Omega(n^{1/d})}$ AND/OR gates are still needed.

On the other hand, instead of using AND/OR gates, one can give a direct realization of Parity in a depth-2 threshold circuit comprising $(n+1)$ threshold gates. In this sense, a threshold gate is exponentially more powerful than an AND/OR gate. Figure 1.1 illustrates a depth-2 threshold circuit that computes the Parity function of four variables.

By generalizing the construction for Parity, one can realize any symmetric function of n variables in a depth-2 threshold circuit with at most $(n+1)$ gates. A more involved construction can reduce the size by a factor of two and yield a depth-2 threshold circuit for computing any symmetric function with at most $(\frac{n}{2} + 1)$ gates.

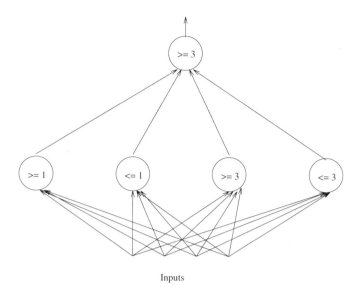

Inputs

Figure 1.1 Each node in the circuit represents a linear threshold gate, which computes the sum of the inputs (i.e., all weights equal 1) and compares it with the threshold value indicated within the node. One can verify that the sum of the outputs of the gates in the first layer equals either 2 or 3, depending on whether the input sum is even or odd.

Given the above results, one can explore several issues related to the efficiency of the circuits. For example, can the size of the depth-2 threshold circuit for Parity be reduced further without increasing the depth? One can show that any depth-2 threshold circuit computing Parity must have size $\Omega(n/\log^2 n)$. Thus, the preceding depth-2 constructions cannot be significantly improved upon and, hence, are almost optimal with respect to depth-2 realizations.

Can the size be reduced if we increase the depth of the threshold circuit from 2 to some larger constant d? The answer is yes. In fact, any symmetric function of n variables can be computed by a depth-3 threshold circuit with size $O(\sqrt{n})$. Fig. 3.1 in Chapter 3 illustrates a depth-3 threshold circuit that computes the Parity of 11 variables. This result indicates that for the class of symmetric functions, a small increase in the depth of threshold circuits can significantly decrease the size. Moreover, one can show that $\Omega(\sqrt{n/\log n})$ threshold gates are required to compute general symmetric functions even if there is no restriction on the circuit depth. This shows that the depth-3 construction for general symmetric functions is almost optimal with respect to threshold circuits of any depth.

For specific symmetric functions with periodic structures such as Parity, however, the size can be further reduced by increasing the depth of the threshold circuit. In particular, one can construct a depth-$(d+1)$ threshold circuit of size $O(dn^{1/d})$ for computing Parity. We shall discuss these depth/size trade-off issues in more detail in Chapter 5. Moreover, the size of such depth-$(d+1)$ threshold circuit for computing Parity is again almost optimal. More specifically, any depth-$(d+1)$ threshold circuit for computing Parity must have size $\Omega(dn^{1/d}/\log^2 n)$. The techniques used in deriving these lower bounds are discussed in Section 1.5 and Chapter 7.

So far we have introduced some of the complexity issues only in the context of optimizing the number of gates and the circuit depth. Similar optimization issues concerning the number of connections and fan-in/fan-out will be addressed in Chapter 5.

1.4.2   Multiplication and Other Arithmetic Functions

Given $(x_{n-1}, x_{n-2}, \ldots, x_0)$ and $(y_{n-1}, y_{n-2}, \ldots, y_0) \in \{0,1\}^n$ representing two n-bit integers $X = \sum_{i=0}^{n-1} 2^i x_i$ and $Y = \sum_{i=0}^{n-1} 2^i y_i$, the multiplication problem is to compute the $2n$-bit product of X and Y. As in the case of Parity, multiplication and other arithmetic functions such as powering, division, multiple product, and sorting cannot be computed by constant-depth polynomial-size AND-OR circuits. In Chapter 4, on the other hand, we derive small-depth polynomial-size threshold circuits for these functions. The fact that threshold circuits are more powerful as computational models than conventional AND-OR circuits will be a recurring theme in this book.

To address the fundamental issues of parallel computation, we focus on minimizing the circuit depth while allowing the circuit to have polynomial size. Many of these threshold circuits are depth optimal or nearly depth optimal. For example, it can be shown that multiplication of two n-bit integers cannot be computed by depth-2 polynomial-size threshold circuits with small weights (see the discussion on weights in the next section). In Chapter 4, we first derive a depth-4 polynomial-size threshold circuit for multiplication. This construction is based on the techniques developed for computing symmetric functions (as presented in Chapter 3) and some re-

sults from the theory of error-correcting codes. Similar techniques can be
applied for computing related arithmetic functions. For example, we also
show how to derive depth-3 polynomial-size threshold circuits for sorting n
n-bit integers.

Although multiplication cannot be computed by depth-2 polynomial-size
threshold circuits, it still leaves open the possibility of computing multiplication using depth-3 circuits. In fact, in Chapter 6, using a different set of
techniques, we derive threshold circuits of optimal depth (i.e., depth-3) for
multiplication and division.

1.4.3   The Comparison Function

Suppose we are given the binary representation of two (n-bit) non-negative
integers $X = (x_{n-1}, \ldots, x_0)$, $Y = (y_{n-1}, \ldots, y_0) \in \{0,1\}^n$, where x_0 and
y_0 are the least significant bits of X and Y, respectively. We would like to
compute if Y is greater than X. This seems to be an easy task, and one can
use a single linear threshold element to compute the answer as follows:

$$COMP_n(X,Y) = \text{sgn}(\sum_{i=0}^{n-1} 2^i(x_i - y_i)) = \begin{cases} 1 & \text{if } X \geq Y, \\ 0 & \text{if } X < Y. \end{cases} \qquad (1.3)$$

In other words, $COMP_n(X,Y)$ is a linear threshold function for every n.
We refer to the family of functions $\{COMP_n(X,Y)\}$ as Comparison.

Notice that the largest integer weight in Eqn. (1.3) grows exponentially
($\Omega(2^n)$) with the size of the input integers. If we want to implement such
a linear threshold element, the requirement of exponential integer weights
would imply the need for high ($O(n)$-bit) accuracy in the actual implementation.

Eqn. (1.3) for $COMP_n(X,Y)$ indicates only one way of choosing the
weights. Now the question is: Can Comparison be computed by a linear
threshold element with polynomially bounded integer weights? We show in
Chapter 6 that the answer is negative, i.e., any linear threshold element computing Comparison must have exponentially large integer weights. In fact, a
simple counting argument would show that most linear threshold functions
require exponential weights. On the other hand, as an application of harmonic analysis, we show in Chapter 4 that Comparison can be computed by
a depth-2 polynomial-size threshold circuit with only polynomially bounded
integer weights.

In general, a relevant question is: How limited is the computational power of a threshold circuit if each integer weight in the circuit is assumed to be 'small' (i.e., polynomially bounded)? We answer this question in Chapter 6 by showing a trade-off in the size and the depth between a threshold circuit with unrestricted weights and the other with small weights. In particular, we show that one can always simulate a depth-d polynomial-size threshold circuit with unrestricted weights by a depth-$(d+1)$ polynomial-size threshold circuit with small weights.

1.4.4 Symmetry Recognition/The Equality Function

In the *Epilogue* of [81], several complexity issues were illustrated via the example of *symmetry recognition*. Given a binary string of n-bits (n even), the problem of symmetry recognition is to determine whether the string is symmetric about the center of the string. For example, with $n = 6$, the string $(1, 1, 0, 0, 1, 1)$ is symmetric about its center, whereas the string $(1, 1, 0, 1, 1, 0)$ is not.

By relabeling the indexes of the binary string, symmetry recognition is equivalent to the following problem of the *Equality* (EQ) function:

$$EQ(x_1, \ldots, x_{\frac{n}{2}}, y_1, \ldots, y_{\frac{n}{2}}) = \begin{cases} 1 & \text{if } x_i = y_i \text{ for all } 1 \leq i \leq \frac{n}{2}, \\ 0 & \text{otherwise.} \end{cases}$$

To realize the Equality function, one only needs to check for each i if $x_i = y_i$, i.e., if $\overline{x_i \oplus y_i} = \text{sgn}\{x_i - y_i\} + \text{sgn}\{y_i - x_i\} - 1 = 1$. In other words,

$$EQ(x_1, \ldots, x_{\frac{n}{2}}, y_1, \ldots, y_{\frac{n}{2}}) = \bigwedge_{i=1}^{\frac{n}{2}} (\overline{x_i \oplus y_i}).$$

Thus, a straightforward depth-2 threshold circuit realization is to use n threshold gates in the first layer and a threshold gate in the second layer to realize the output AND ($\bigwedge$) function.

One can also apply the realization for $COMP_{\frac{n}{2}}(X, Y)$ to compute the Equality function (where $X = (x_1, \ldots, x_{\frac{n}{2}})$ and $Y = (y_1, \ldots, y_{\frac{n}{2}})$). Simply note that

$$EQ(x_1, \ldots, x_{\frac{n}{2}}, y_1, \ldots, y_{\frac{n}{2}}) = COMP_{\frac{n}{2}}(X, Y) \bigwedge COMP_{\frac{n}{2}}(Y, X).$$

This realization yields a depth-2 threshold circuit comprising three gates, with two threshold gates in the first layer computing the $COMP_{\frac{n}{2}}$ functions, and a threshold gate in the second layer computing the AND function.

The preceding two realizations of the Equality function have different circuit sizes and use different weights. The first realization uses n threshold gates each of fan-in 2 and unit weights in the first layer, while the second one uses two threshold gates each of fan-in n and weights of exponential magnitudes. These two realizations are described in [81] and the following two questions are raised: Must one use weights of exponential magnitudes if the circuit size is bounded? Is the first realization optimal in terms of size (i.e., if the magnitudes of the weights in each gate are nonexponential, can the circuit size be significantly reduced below $O(n)$)?

We answer the above questions in Chapter 10 by establishing rigorous lower bounds via communication complexity arguments. In particular, if the circuit size (number of gates) is bounded by a constant, then the magnitude of the weights must grow exponentially. On the other hand, if the weight magnitudes are polynomially bounded, then regardless of the circuit depth, the circuit size must grow almost linearly $\Omega(n/\log n)$. In the special case of a depth-2 realization, if the gates in the first layer have bounded fan-in, then the circuit size must grow linearly $\Omega(n)$. Therefore, the first realization for the Equality function has optimal size with respect to all depth-2 circuits with bounded fan-in gates in the first layer.

1.5   Analytical Techniques

Several analytical techniques have been developed to address complexity theoretic issues for the conventional circuit model comprising AND gates, OR gates, and inverters (NOT gates). These techniques do not seem to apply directly to the case of neural network models such as threshold circuits. However, a set of related techniques is emerging that yield meaningful results for threshold circuits and their variants. We next present brief discussions on some of these tools.

1.5.1   Spectral/Polynomial Representation of Boolean Functions

We shall make frequent use of the *spectral* or *polynomial representation* of Boolean functions and related concepts in our later discussions. A detailed discussion of the material is presented in Chapter 8. Here we introduce the basic concepts and present the spectral representations of some simple Boolean functions. For convenience, we adopt a $\{1, -1\}$ notation, where a (single-output) Boolean function $f(X)$ will be defined as $f : \{1, -1\}^n \rightarrow \{1, -1\}$. In the $\{1, -1\}$ notation, 0 corresponds to 1 and 1 corresponds to -1.

Every Boolean function f can be represented as a polynomial over the field of rational numbers as follows:

$$f(X) = \sum_{\alpha \in \{0,1\}^n} a_\alpha X^\alpha,$$

where $X^\alpha = x_1^{\alpha_1} x_2^{\alpha_2} \cdots x_n^{\alpha_n}$ and $\alpha = (\alpha_1, \alpha_2, \ldots, \alpha_n) \in \{0, 1\}^n$.

Such representation is unique and the coefficients of the polynomial, $\{a_\alpha | \alpha \in \{0, 1\}^n\}$, are called the *spectral coefficients* of f. Notice that since $x_i^2 = 1$, it suffices to consider only *multilinear* polynomials. The coefficients a_α's are computed as follows: Let P_{2^n} denote the (column) vector of the 2^n values of $f(X)$ in lexicographical order of X, and let A_{2^n} denote the vector comprising the spectral coefficients a_α's in lexicographical order of α. Then

$$A_{2^n} = \frac{1}{2^n} H_{2^n} P_{2^n}$$

where H_{2^n} denotes the *Sylvester-type Hadamard* matrix of order 2^n [76]. H_{2^n} can be defined recursively as follows:

$$H_2 = \begin{bmatrix} 1 & 1 \\ 1 & -1 \end{bmatrix},$$

$$H_{2^{n+1}} = \begin{bmatrix} H_{2^n} & H_{2^n} \\ H_{2^n} & -H_{2^n} \end{bmatrix}.$$

For example, consider the function $AND(x_1, x_2)$ of two variables. It is easy to check that this function has the following representation:

$$
\begin{aligned}
AND(x_1, x_2) &= \frac{1}{2}(1 + x_1 + x_2 - x_1 x_2) \\
&= \begin{cases} -1 & \text{if } x_1 = x_2 = -1, \\ 1 & \text{otherwise.} \end{cases}
\end{aligned}
\tag{1.4}
$$

Another example is the Majority function of 3 variables, which has the representation:

$$MAJ_3(x_1, x_2, x_3) = \frac{1}{2}(x_1 + x_2 + x_3 - x_1 x_2 x_3)$$

$$= \begin{cases} 1 & \text{if at least two } x_i\text{'s are 1,} \\ -1 & \text{otherwise.} \end{cases} \qquad (1.5)$$

In general, the spectral representation of the majority function of n variables involves monomial terms $x_1^{\alpha_1} x_2^{\alpha_2} \cdots x_n^{\alpha_n}$ with degree higher than one. Note also that the spectral representation of the n-variable Parity function is the product $x_1 x_2 \cdots x_n$.

Related to the spectral representation of Boolean functions is the issue of representing a Boolean function as the sign of a polynomial in the input variables. For example, the Majority function can be simply expressed as $\text{sgn}(\sum_{i=1}^{n} x_i)$. Thus, even though the spectral representation of the Majority function requires nonlinear monomials, only linear terms are sufficient to represent the function as the sign of a polynomial. For other Boolean functions, however, such a reduction in degree may not be possible. In Chapter 8, we show that if the n-variable Parity function is expressed as the sign of a polynomial, then the polynomial must have degree n. We say that the *strong degree* of the Parity function is n. The concept of the strong degree associated with a Boolean function will be used in Chapter 7 to derive key lower bound results.

We shall also use the concept of the L_1 *spectral norm* of a Boolean function f, which is defined to be

$$\|f\|_{\mathcal{F}} = \sum_{\alpha \in \{0,1\}^n} |a_\alpha|,$$

where the a_α's are the spectral coefficients of f as defined before. It will be shown that the L_1 spectral norm of a Boolean function can reveal its computational complexity. In particular, one can show that a function f with small $\|f\|_{\mathcal{F}}$ is always computable by a depth-2 threshold circuit of small size (see Chapter 4).

1.5.2   Rational Approximation

A major difficulty in analyzing neural networks lies in the nonlinearity of the neural elements in the network. We shall show that under certain conditions, the theory of rational approximation can be used as an effective tool in analyzing such nonlinear networks. There are two reasons why rational approximation theory is useful in the analysis of neural networks. First, many nonlinear functions (e.g., linear threshold functions and sigmoidal functions) can be closely approximated by rational functions of low degrees. Second, the class of rational functions is closed under functional compositions. In the following, we informally state some key results from the theory of rational approximation to illustrate these ideas.

Let the *degree* of a rational function $R(x) = P(x)/Q(x)$ be the maximum of the degrees of its numerator polynomial $P(x)$ and its denominator polynomial $Q(x)$. A central idea of the lower bound techniques comes from a result in the theory of rational approximation, which states the following: The function $\mathrm{sgn}(x)$ can be approximated with an error of $O(e^{-ck/\log(1/\epsilon)})$ by a rational function of degree k for $0 < \epsilon < |x| < 1$. This result allows us to approximate several layers of threshold gates by a rational function of low (i.e., logarithmic) degree when the size of the circuit is small. Then by deriving a lower bound on the degree of the rational function that approximates highly oscillating functions such as Parity, one can derive lower bounds on the size of the circuit.

It is natural to extend these techniques to sigmoidal networks by approximating each of the sigmoidal elements with rational functions. In fact, we can extend the techniques of rational approximation to general γ networks. For a general γ element, suppose there exists a good approximation of $\gamma(x)$ by piecewise rational functions of low degrees. One can then construct a single rational function that approximates $\gamma(x)$ by 'joining' the pieces together. With this approach, one can first obtain a piecewise low-degree rational approximation to the γ element. Then, one can derive a single rational function by joining the pieces together, without significantly increasing the degree. The details of such techniques are discussed in Chapter 7.

1.5.3   Geometric and Linear Algebraic Arguments

Geometric approaches are naturally suitable for analyzing threshold gates. An S-input threshold gate corresponds to a hyperplane in $\mathcal{R}^S$. This geometric interpretation, for example, can be used to count the number of Boolean functions computable by a single threshold gate, as well as to determine functions that cannot be computed by a single threshold gate. Such classical results are briefly discussed in Chapter 2.

Threshold circuits of depth 2 or more, however, do not carry a simple geometric interpretation in $\mathcal{R}^S$. The inputs to the gates in the second level are themselves threshold functions, hence the linear combination computed at the second level is a nonlinear function of the inputs. As described earlier, several analytical techniques have been invoked to understand such nonlinearities. We shall show that linear techniques are also applicable and lead to some insightful characterizations of threshold gates.

We shall outline a simple geometric relation between the output function of a threshold gate and its set of input functions. This relation applies to arbitrary sets of input functions and, along with some geometrical and linear algebraic tools, can be used to derive a variety of lower bound results. The key to this approach is to represent an n-variable Boolean function as a vector in $\mathcal{R}^{2^n}$. Simple geometric concepts such as linear subspaces, correlation, and projection can then be used to derive novel results on the realizability of Boolean functions using threshold gates.

The geometric framework will also be integrated with the theory of linear programming. It can be shown, for example, that the problem of determining the weights of a linear threshold element for computing a desired function can be solved using linear programming algorithms. We extend such studies and demonstrate that linear programming concepts, such as duality and Farkas' lemma, can be used to characterize functions that cannot be computed using a single linear threshold gate. Such tools are also used for analyzing the behavior of the well-known perceptron learning algorithm when the training set is linearly nonseparable.

1.5.4   Communication Complexity Arguments

The communication complexity problem can be described as follows. Let f be an n-variable Boolean function and let $\{X, Y\}$ be a partition of the input variables. A person P_X knows the input values of the variables in X and a person P_Y knows the input values of the variables in Y. The two communicate according to a predetermined protocol in order to find the value of f on their joint input. We are interested in determining a lower bound on the number of bits P_X and P_Y must transmit. A number of analytic tools for characterizing the communication complexity of various functions have been developed.

Communication complexity theory and related concepts have found applications in various disciplines, including fields such as VLSI complexity theory, linear decision trees, circuit complexity, distributed algorithms, and cryptography. We shall show that many commonly used concepts in communication complexity theory such as the *decomposition number* and the largest *monochromatic rectangle* can be used to derive linear $(\Omega(n))$ and almost linear $(\Omega(n/\log n))$ lower bounds on the size of unrestricted-depth threshold and related circuits. Several of these lower bounds match the best-known bounds in circuit complexity and often generalize other bounds derived using alternate approaches.

1.6   A Historical Perspective

Here we give a brief account of the major developments in neural network research since the pioneering work of McCulloch and Pitts [77]. Because of the vast literature, we do not attempt to give a comprehensive survey. Instead, we only describe the work of several key researchers who have directly or indirectly led to the formulation of the major ideas in discrete neural computation; whenever possible, we refer the readers to alternate sources that include a more comprehensive account of various aspects of neural networks.

It is commonly agreed that McCulloch and Pitts are the two pioneers in the modern era of neural network research. In their 1943 seminal paper [77], they formulated a mathematical model of brain mechanisms. Essentially, their models are finite state automata that embody logic statements in a form equivalent to computer programs. Because of the logic formulation of the model of McCulloch and Pitts, it comes as no surprise that their networks were designed to perform arithmetic operations such as negation, addition, and multiplication. In general, all finite arithmetic calculations can be carried out in their model.

Around 1960, there was a wave of research activities generated by Rosenblatt [109], who proposed the *perceptron* model. This model received considerable attention because of a simple learning algorithm invented by Rosenblatt. A similar model, called *Adaline*, that uses a different learning rule was proposed by Widrow and Hoff [144] around the same period. Their common feature is that the basic element in both models is a linear threshold element. These early models offered the promise of a basis for artificial intelligence.

The interest in a linear threshold element as an idealization of a neuron also led to considerable research efforts in the study of threshold logic. Cover [31], Muroga [85], and Winder [147] established several results related to the capacities of linear threshold elements in the 1960s. Chow [27], Kautz [64], Lewis and Coates [70], Muroga [86], and Nechiporuk [87] studied threshold logic in the context of conventional switching theory.

In 1969, Minsky and Papert presented in their book *Perceptrons* [81] the first comprehensive theoretical study that addresses the capabilities of perceptrons and related neural network architectures. The theory of perceptrons was discussed under the general framework of theory of computation; the objective was to gain more insights into the interconnected disciplines of parallel computation, pattern recognition, knowledge representation, and learning, by studying the mathematical structures of the perceptrons. Many of the issues in [81] were addressed from the perspective of computational complexity. Realizing that any function can be computed by a two-layer network comprising possibly an exponential number of elements, the book *Perceptrons* [81] studies the fundamental question of how the cost of a neural network scales with the problem size. Moreover, *Perceptrons* [81] hinted at the computational difficulty of training multilayer neural networks and questioned the existence of efficient learning algorithms for such networks. This view has been vindicated in part by recent complexity theoretic results indicating the intractability of training even a three-node network [16]. In contrast to the common perception, *Perceptrons* [81] does not focus on the limitations of neural networks, but rather provides a theoretical framework for exploring the classes of problems computable by neural networks, and for designing suitable network architectures. In this sense, the objectives of this book closely follow those of *Perceptrons* [81].

From the period of early '70s to early '80s, there were a number of researchers who proposed alternate paradigms for neural computation. To name a few, Kohonen [66] worked on the associative content-addressable memory, in which different input patterns are associated with each other if they are sufficiently similar. Grossberg [47] reformulated the general problem of learning in the framework of dynamical system theory.

The recent resurgence of interest in the study of neural networks is due in part to the paper of Hopfield in 1982 [58]. He proposed a model of neural computation that is based on the dynamics of mutually interacting neurons. The main result in Hopfield's influential paper indicates that there are emergent collective abilities at the network level that do not exist at the single neuron level. Hopfield's work was inspired in part by the early work of Hebb [52]. Moreover, Hopfield and Tank [59] demonstrated that the model can be used to solve combinatorial optimization problems. From a computational perspective, his model creates a novel paradigm of parallel computation. Other related neural network architectures such as the cellular neural networks have been proposed by Chua and Yang [28, 29], which have also found applications in signal and image processing. For discussions of results on neural networks from a physicist's perspective, see [54].

Another wave of interest in neural networks was created by the work of Rumelhart and his collaborators [116]. Their backpropagation algorithm extends the training rules of Rosenblatt and Widrow from a single neural element to multilayer feedforward networks. However, the elements used by Rumelhart et al. in their networks differ from the linear threshold elements used by Rosenblatt and Widrow. Instead of having elements with binary outputs, the networks of Rumelhart et al. use elements that have continuous-valued outputs. This reformulation allows the use of gradient descent techniques in learning. The backpropagation algorithm was developed independently in 1974 by Werbos [143] in his doctoral dissertation. In its original form as well as its variants, the backpropagation algorithm has proved to be effective in training multilayer neural networks, and it has been widely adopted in the scientific community.

The discrete models of neural computation have also received much attention recently in the computer science community. Results established by Furst, Saxe and Sipser [36], Chandra, Stockmeyer and Vishkin [26], Pippenger [104], Parberry and Schnitger [95, 96] demonstrate that threshold circuits are powerful models of computation. Hajnal, Maass, Pudlák, Szegedy and Turán [48] have derived the first exponential lower bound on the size of small depth threshold circuits. Related complexity results will be discussed and their sources will be indicated in the bibliographic notes of each chapter.

1.7   An Overview of the Book

1.7.1   Linear Threshold Element (LTE) and its Properties

The linear threshold element (LTE) model has been already introduced in Section 1.2.3, where we discussed different models for discrete neural computation. In Chapter 2, we first present some classical results on LTEs and then introduce a linear programming formulation for determining the weights of an LTE so that it computes a given function. We also present the well-known perceptron learning algorithm. Input patterns that can be classified by a single LTE, and hence can be learned by the perceptron learning algorithm, are referred to as *linearly separable*, and patterns that cannot be classified by a single LTE are referred to as *linearly nonseparable*.

Although the learning and convergence properties of perceptrons are well-studied topics when the input pattern vectors are linearly separable, the behavior of perceptrons when the input patterns are linearly nonseparable is not well understood. In Chapter 2, we develop a set of tools for analyzing the characteristics of linearly nonseparable input patterns and define the corresponding learning goals. Based on such analysis, we show that the perceptron can indeed learn some of the structures inherent in linearly nonseparable training sets.

In order to define learnable structures for linearly nonseparable patterns, we first develop a necessary and sufficient condition for linear nonseparability. Based on this characterization, we show that the set of input vectors can be classified into two classes: (1) *separable input vectors* and (2) *nonseparable input vectors*. We then define one of the learning goals for linearly nonseparable input patterns as determining the sets of separable and nonseparable vectors. Such learning would allow the system to determine the structure of the input patterns and identify those vectors that are responsible for nonseparability. We present convergence results, which show that if the perceptron algorithm is applied to linearly nonseparable input patterns, then it can learn the sets of separable and nonseparable vectors.

Chapter 2 also presents computational complexity results for the various learning problems that are defined in the context of linearly separable and nonseparable input patterns. For example, we show that one can determine the sets of separable and nonseparable vectors of a linearly nonseparable set in polynomial time by using linear programming algorithms. However, determining the maximum number of linearly separable vectors from a given linearly nonseparable set is shown to be computationally expensive. To indicate the inherent computational intractability of such a problem, we prove it to be *NP-Complete*.

We conclude Chapter 2 by addressing some classical questions on the capacity of LTEs. For example, we address the problem of determining the number of linearly separable binary functions defined over a set of n points in $\mathcal{R}^d$. Another question we address is: If one wants to compute each of the 2^n possible functions defined over a set of n points in $\mathcal{R}^d$ using a fixed feedforward network of LTEs, then how many connections (i.e., weights) are needed in the network? We show that the number of connections in the circuit has to be almost as large as the number of input vectors. We also address questions regarding the precision of the weights of an LTE. We shall show, for example, that if the inputs to an LTE are restricted to be bounded integers, then exponentially large integer weights are sufficient.

1.7.2 Computing Symmetric Functions

Chapter 3 presents a first step toward developing systematic methodologies for computing a number of functions using threshold circuits. In particular, Chapter 3 focuses on the computations of symmetric Boolean functions. The motivation comes from the fact that the computation of symmetric Boolean functions and their variants form an integral part of the methodologies required for computing arithmetic and related functions using threshold circuits. Thus, efficient computations of symmetric Boolean functions directly lead to realizations of several arithmetic functions. Moreover, we show that implementations of symmetric functions also yield almost optimal realizations of general Boolean functions.

We first present a fairly straightforward realization, which shows that any n-variable symmetric function can be computed by a depth-2 threshold circuit with at most $(n+1)$ threshold gates. We then introduce the 'telescopic' technique and show that the size can be reduced by a factor of 2 without increasing the depth, and present a depth-2 threshold circuit for computing symmetric Boolean functions with only $\frac{n}{2}+1$ gates. A lower bound result derived in Chapter 7 shows that any depth-2 threshold circuit computing the Parity function has size at least $\Omega(n/\log^2 n)$. This lower bound thus implies that increasing the depth of the network beyond 2 is essential in order to reduce the size significantly below $O(n)$.

We then show that a small increase in the depth of a threshold circuit computing symmetric Boolean functions can indeed lead to a significant decrease in the size of the circuits. In particular, we prove that any n-variable symmetric function can be computed by a depth-3 circuit with

$O(\sqrt{n})$ threshold gates. We also show that the size of a threshold circuit computing symmetric Boolean functions cannot be substantially reduced beyond $O(\sqrt{n})$, even if the circuits are allowed to be of unbounded depth. We establish a lower bound showing that any threshold circuit computing a general symmetric function has size $\Omega(\sqrt{n/\log n})$. These results are extended to the case of a more general class of functions, where the value of a function depends on a weighted sum of the inputs. Such functions are defined as generalized symmetric Boolean functions.

The results for symmetric and generalized symmetric functions are then applied to the case of general Boolean functions, and it is shown that any n-variable Boolean function can be computed by a depth-3 threshold circuit with $O(2^{n/2})$ gates. Moreover, this realization is shown to be almost optimal by proving that a threshold circuit computing any general Boolean function must have size $\Omega(2^{n/2}/\sqrt{n})$.

1.7.3 Depth-Efficient Arithmetic Circuits

Chapter 4 presents efficient realizations of common arithmetic functions by small-depth polynomial-size threshold circuits. The weights of the threshold circuits considered in this chapter are assumed to be integers bounded by a polynomial (polynomially bounded weights). A variety of techniques – including results from error-correcting codes, number theory, and spectral analysis – are utilized in the design of these depth-efficient arithmetic circuits. The emphasis of this chapter is on presenting small-depth circuits, where the circuit size is allowed to be polynomially large. Chapter 5 deals more with optimization issues, and the design of circuits with a restricted number of gates and connections are considered.

We first present depth-2 threshold circuits for computing the Comparison and Addition functions. The existence of depth-2 polynomial-size threshold circuits for these two functions is shown using harmonic analysis, and explicit constructions are presented using results from the theory of error-correcting codes. Depth-3 circuits are then developed for computing the sum of n n-bit integers using 'block-save' techniques and making judicious uses of efficient realizations of symmetric functions. Depth-4 circuits for computing the product of two n-bit integers are presented by reducing multiplication to an equivalent problem in which the sum of n $2n$-bit integers is computed. Other functions considered in Chapter 4 include powering, division, and sorting. For example, using results such as the Chinese Remainder Theorem, harmonic analysis, and approximation techniques, we show that both powering and division can be computed by depth-4 polynomial-size threshold circuits. Similarly, using results for the Comparison function, we show that the sorting of n n-bit integers can be computed by depth-3 polynomial-size threshold circuits.

We also address the issue of optimality of these circuits. For example, we show that the sorting of n n-bit numbers cannot be computed by depth-2 polynomial-size threshold circuits with polynomially bounded weights. Hence, the depth-3 circuits presented in Chapter 4 are optimal with respect to depth. We also show that multiplication and division of two n-bit integers cannot be computed by depth-2 polynomial-size threshold circuits with polynomially bounded weights. Thus, the depth-4 circuits presented in Chapter 4 are only unit depth away from being optimal. Chapter 6 uses a different set of techniques to show that multiplication and division of two n-bit integers can indeed be computed by depth-3 polynomial-size threshold circuits. The lower bounds on the depth of threshold circuits computing sorting and arithmetic functions are derived from a lower bound result on the depth of circuits computing the Inner Product Mod 2 function. In particular, we show in Chapter 8 that the Inner Product Mod 2 function cannot be computed by depth-2 polynomial-size threshold circuits with polynomially bounded weights.

1.7.4 Depth/Size Trade-offs and Optimization Issues

Chapter 5 deals with various optimization issues that arise in the design of threshold circuits. For example, we study the trade-offs between the depth and the size in threshold circuits. In a more general context, we ask the following question: Given a specific function, is it possible to reduce significantly the size of a threshold circuit computing it by increasing the circuit depth? In other words, can we reduce the required amount of hardware substantially by increasing the time for the parallel execution of the circuit? It is difficult to answer such questions in general, and we shall focus our study on the computations of specific symmetric Boolean functions and arithmetic functions studied in Chapters 3 and 4, respectively.

In the circuits derived in Chapters 3 and 4, we are primarily concerned with the number of gates as the measure of size. In Chapter 5, we also deal with the issue of minimizing the number of connections in the circuits, and present several efficient realizations as well as some trade-off results with respect to the depth. Furthermore, depth/size trade-off results under the additional constraint of restricted fan-in and fan-out will be presented.

The depth/size trade-off issues will be first explored in the context of periodic symmetric functions. The results in Chapter 3 show that for general symmetric functions, increasing the depth from 2 to 3 can lead to a reduction in size from $O(n)$ gates to $O(\sqrt{n})$. However, the lower bound results in the same chapter indicate that further significant reduction in the number of gates is not possible, even if unrestricted-depth threshold circuits are considered. We show in Chapter 5 that for periodic symmetric functions, such as the Parity and the Complete Quadratic functions, one can obtain depth/size trade-off results for every constant depth d. In particular, we show that for symmetric functions with periodic structures, one can construct threshold circuits of depth $(d+1)$ and size $O(dn^{1/d})$ for every constant $d \geq 2$. These trade-off results are useful in exploring the depth/size trade-offs for specific arithmetic functions such as Multiple Sum and Multiplication. Trade-off results are also presented for arithmetic functions when the number of edges or connections is optimized (instead of the number of gates) and circuits with restricted fan-in and fan-out gates are considered.

1.7.5   Computing with Small Weights

Although the definition of linear threshold functions allows the weights to be real numbers, it is shown in Chapter 2 that one can replace each of the real weights by integers of $O(n \log n)$ bits, where n is the number of input Boolean variables. The issues regarding exponentially large (integer) weights and polynomially bounded weights have been already introduced in the context of the Comparison function (see Sections 1.4.3 and 1.3.2, respectively). Chapter 6 addresses these issues in greater detail and explores the capabilities of threshold circuits in which the weights are polynomially bounded. In particular, it is shown that a depth-d polynomial-size threshold circuit with exponentially large weights can be simulated by another depth-$(d+1)$ polynomial-size threshold circuit with polynomially bounded weights. This shows that one can trade off exponentially large weights with polynomially bounded weights by increasing the depth of the circuit by at most one, while keeping the circuit size polynomially bounded.

We first show how a single threshold gate with exponentially large weights can be simulated by a depth-3 polynomial-size circuit with polynomially bounded weights. The ideas developed in Chapter 4 for designing depth-2 threshold circuits for computing the Comparison function will be used as basic tools for the simulation. We then show that any depth-d polynomial-size threshold circuit with unrestricted weights can be simulated by a depth-$(2d+1)$ polynomial-size threshold circuit with polynomially bounded weights. Using more advanced combinatorial arguments, we show how a single threshold gate with exponentially large weights can be simulated by a depth-2 threshold circuit with polynomially bounded weights. This leads to the desired result that any depth-d polynomial-size threshold circuit with arbitrary weights can be simulated by a depth-$(d+1)$ polynomial-size threshold circuit with polynomially bounded weights.

The above results are then applied to derive depth-optimal threshold circuits for computing both multiplication and division of two n-bit integers. In particular, we show that the sum of n n-bit integers can be computed by depth-2 polynomial-size threshold circuits and that both multiplication and division of two n-bit integers can be computed by depth-3 polynomial-size threshold circuits.

1.7.6 Rational Approximation and Optimal-Size Circuits

Chapter 7 addresses the issue of establishing lower bounds on the size of threshold circuits computing specific functions such as the Parity and the Complete Quadratic functions. The lower bound techniques developed in this chapter use results from the theory of rational approximation and harmonic analysis of Boolean functions. The resultant lower bounds show that many of the threshold circuits designed in Chapters 4 and 5 are almost optimal. Moreover, it is shown that the lower bound techniques can be also applied to γ networks that use continuous-valued elements. These extended results show that for highly oscillating functions, such as the Parity function, a substantial reduction in the number of gates is not possible, even if the linear threshold gates are replaced by seemingly more powerful continuous-valued elements such as the sigmoidal gates.

Approximating the output functions of threshold gates with rational functions of appropriate degrees, we prove that any depth-3 threshold circuit computing a general n-variable symmetric function must have $\Omega(\sqrt{n}/\log^2 n)$

gates. This lower bound almost matches the upper bound $O(\sqrt{n})$ presented in Chapter 4. We then generalize this result and show that any depth-$(d+1)$ threshold circuit computing Parity and Complete Quadratic (CQ) has size at least $\Omega(dn^{1/d}/\log^2 n)$. These lower bounds are almost tight, compared with the upper bound results of $O(dn^{1/d})$ presented in Chapter 5. We also derive almost tight lower bounds on the size of depth-d threshold circuits that approximate Parity. Furthermore, extending the techniques of rational approximation, we prove that in any constant-depth circuit, $n^{\Omega(1)}$ threshold gates are still needed to compute Parity, even if the circuit is augmented with an additional subexponential ($2^{n^{o(1)}}$) number of AND/OR gates.

The rational approximation techniques are then extended to the case of γ networks. A key lemma shows that if a continuous function can be piecewise approximated by low-degree rational functions over $k = \log^{O(1)} n$ consecutive intervals, then it can be approximated by a single low-degree rational function over the union of these intervals. Examples of continuous functions which can be piecewise approximated by low-degree rational functions include sigmoidal functions and radial basis functions. Using these extended results, we show that if each element of a depth-$(d + 1)$ neural network can be piecewise approximated by a low-degree rational function, and if the network computes the Parity function (or in general any function that has strong degree $\Omega(n)$), then it must have size $\Omega(dn^{1/d-\epsilon})$ for any fixed $\epsilon > 0$.

1.7.7   Spectral Analysis and Geometric Approach

Chapter 8 presents a variety of lower bound results using simple geometric concepts such as correlation, linear subspaces, and projection. The lower bound results are mostly expressed in terms of correlations of the output function of a threshold gate with the set of input functions. These basic results constitute some key lower bound results in threshold circuit complexity, and lead to an important separation result for constant-depth threshold circuits.

Viewing n-variable Boolean functions as vectors in $\mathcal{R}^{2^n}$, basic tools from linear algebra and linear programming are invoked to derive several results on the realizability of Boolean functions using threshold gates. Some of these results are: (1) a lower bound on the number of input functions required by a threshold gate computing a given function, (2) a lower bound on the error incurred when a Boolean function is approximated by a linear combination of a set of functions, (3) a limit on the effectiveness of a well-known technique (based on computing correlations among Boolean functions) for deriving lower bounds on the depth of threshold circuits computing Boolean functions, and (4) a construction showing that every n-variable Boolean function f is a threshold function of polynomially many input functions, none of which is significantly correlated with f.

A presentation of harmonic analysis and spectral/polynomial representation of Boolean functions is also provided in Chapter 8. The spectral representation is derived as a special case of the geometric framework, and results related to the spectral coefficients are derived as special cases of the more general results presented in the context of the geometric approach. Such an approach leads to simple proofs based on elementary linear algebraic arguments. Algebraic techniques for computing the spectrum of several Boolean functions are also outlined.

Recall that $\widehat{LT_d}$ denotes the class of Boolean functions computable by depth-d polynomial-size threshold circuits with polynomially bounded integer weights. The method of correlations presented in Chapter 8 leads to the following separation result: $\widehat{LT_2} \subsetneq \widehat{LT_3}$. This result is proved by showing that the Inner Product Mod 2 function cannot be computed by any depth-2 polynomial-size threshold circuit with polynomially bounded weights.

1.7.8 Limitations of AND-OR Circuits

In Chapter 9, we derive lower bound results to demonstrate the limitations of constant-depth polynomial-size AND-OR circuits (AC^0 circuits). In particular, we establish that any constant-depth AND-OR circuit with arbitrary fan-in that computes the n-variable Parity function must have size exponential in n. We adopt an algebraic approach that allows us to prove a more general result. More precisely, we show that any depth-d circuit with gates AND, OR, NOT, and MOD_p, where p is a prime, requires at least $2^{\Omega(n^{1/2d})}$ gates to compute the MOD_r function of n variables, where r is not a power of p. For simplicity, we give a proof for the special case when $p = 3$ and $r = 2$. This proof contains the essential ideas of the underlying techniques and can be generalized to the case of arbitrary prime p and r. We derive another result that also illustrates the computational limitations of AC^0 circuits. In particular, we show that any function that is computable by AC^0 circuits can also be computed by depth-3 threshold circuits of superpolynomial-size $(2^{\log^{O(1)} n})$.

1.7.9   Lower Bounds for Unbounded Depth Circuits

In Chapter 10, we use communication complexity concepts and techniques to derive linear and almost-linear lower bounds on the size of unrestricted-depth threshold and related circuits computing certain functions. The techniques utilize only basic features of the gates used and of the functions implemented, and hence apply to a large class of gates (including unbounded fan-in AND/OR, threshold, symmetric, and generalized symmetric) and to a large family of functions (including Equality, Comparison, and Inner Product Mod 2). Each of the bounds derived is shown to be tight and specific applications to threshold circuit complexity are indicated.

The communication complexity concepts used in Chapter 10 include the decomposition number and the largest monochromatic rectangle of a function. These are simple attributes that have proven useful in analyzing the communication complexity of various functions. For example, we consider circuits comprising polynomially rectangular gates. These gates, which include symmetric, generalized symmetric, and polynomially-bounded-weight threshold gates, compute functions with small decomposition numbers. We show that functions computed by small circuits of polynomially rectangular gates have small decomposition numbers. It follows that functions with high decomposition numbers require large circuits of polynomially rectangular gates. We use effective techniques that have been developed to derive lower bounds on the decomposition numbers and prove almost-linear lower bounds on the circuit complexity of several functions. We then strengthen the results for triangular gates. These gates, which include all threshold gates, compute functions with large monochromatic rectangles. We show that any function computed by a small circuit of triangular gates must contain a large monochromatic rectangle. Therefore, functions with only small monochromatic rectangles require large circuits of triangular gates.

1.7.10   The Hopfield Model

The Hopfield model is discussed in Chapter 11, where we derive basic properties of the model and present some applications. Unlike the threshold circuit model whose computation is combinational, the Hopfield model can be considered as a sequential machine in which the computation is performed at discrete time steps.

The concept of an energy function for a Hopfield network is introduced in Chapter 11, and it is shown that if the model iterates in a sequential mode under certain restrictions imposed on its parameters, then it always converges to a stable state. When the network iterates in the parallel mode, we show that it converges to a limit cycle of length at most 2. The proof for the parallel iteration mode is presented by reducing the parallel mode of iteration to a sequential mode of iteration. We then derive an upper bound on the transient period of the Hopfield network as a function of the network's parameters. In particular, we show that when the thresholds and the connection weights among the nodes are bounded in magnitude by some fixed constant, then a Hopfield network (in a cyclic mode) converges to a stable state in at most $O(n^3)$ steps, where n is the number of nodes in the network. However, when the magnitudes of the weights are not restricted, then the transient period could be exponential in the number of nodes.

The convergence properties of the Hopfield model suggest the application of the model as an associative memory, where the stables states of the network correspond to the stored information in the memories. We relate the properties of the stored patterns to the connectivity of the network and demonstrate that the number of undesirable spurious memories produced by a popular method called the outer-product rule can be exponential, thus demonstrating the limitations of this method.

In the last section of the chapter, we demonstrate the applications of the Hopfield network to combinatorial optimization. Two important problems in combinatorial optimization, namely the Min-Cut Problem and the Traveling-Salesman Problem, are targeted for implementation on the networks. The key idea in applying the Hopfield networks to these problems is to map the cost functions of the optimization problems to the energy functions. The stable states of the network that are local minima of the energy functions will then correspond to the local minima of the cost functions. It is shown that the minimization of a quadratic form is equivalent to finding a minimum cut in an undirected graph. From this correspondence, it follows that searching for the global minimum of the energy function in a Hopfield network is equivalent to finding a minimum cut. We prove a lemma stating that the minimization of a quadratic form subject to the constraints of some linear equalities is equivalent to the minimization of another unconstrained quadratic form. We then make use of this lemma to demonstrate how the Traveling-Salesman Problem can be mapped to the Hopfield network.

1.8   Notes on Terminology

Most of the definitions and general concepts related to the study of discrete neural computation have been introduced in Sections 1.2 and 1.3, where the neural models and related complexity measures are discussed. More topical concepts are introduced in individual chapters, and the glossary contains definitions of most of these technical concepts introduced throughout the book. We present here discussions on some additional issues that pertain to all the chapters.

An n-variable Boolean function f is usually defined as $f : \{0,1\}^n \to \{0,1\}$. In this book, we find it occasionally more convenient to let 0 correspond to 1, and 1 correspond to -1. Thus, in the $\{1,-1\}$ notation, a Boolean function is defined as $f : \{1,-1\}^n \to \{1,-1\}$. We have already adopted the $\{1,-1\}$ notation in Section 1.5.1 while introducing the spectral representation of Boolean functions. If x_i represents a variable in the $\{0,1\}$ notation, and $\tilde{x}_i$ in the $\{1,-1\}$ notation, then they could be related by the following two equivalent transformations:

$$\tilde{x}_i = (-1)^{x_i},$$

$$\tilde{x}_i = 1 - 2x_i.$$

For example, the Parity of $(\widetilde{x_1}, \widetilde{x_2}, \ldots, \widetilde{x_n})$ in the $\{1,-1\}$ notation is given by $\prod_{i=1}^{n} \tilde{x}_i = (-1)^{\sum_{i=1}^{n} x_i}$.

Recall that in the $\{0,1\}$ notation, $\mathrm{sgn}(y)$ has been defined to equal 1 if $y \geq 0$, and 0 if $y < 0$. In the $\{1,-1\}$ notation, we adopt the commonly used definition of the function $\mathrm{sgn}(y)$, where it is 1 if $y \geq 0$, and -1 otherwise. A linear threshold function f, which is given by $\mathrm{sgn}(\sum_{i=1}^{n} w_i x_i + w_0)$ over the $\{0,1\}$ notation, can be represented as

$$f = -\mathrm{sgn}(\sum_{i=1}^{n} \frac{w_i}{2}(1 - \tilde{x}_i) + w_0)$$

$$= \ \mathrm{sgn}(\sum_{i=1}^{n} \frac{w_i}{2} \widetilde{x}_i - (w_0 + \sum_{i=1}^{n} \frac{w_i}{2}))$$

in the $\{1, -1\}$ notation, where without loss of generality we assume $\sum_{i=1}^{n} w_i x_i + w_0 \neq 0$ (see Chapter 2). Hence, threshold circuits designed using the $\{0, 1\}$ notation can be transformed into threshold circuits in the $\{1, -1\}$ notation (and vice versa) without any increase in size or depth.

As mentioned in Section 1.3.2, we are concerned with the computational complexity of a family of functions f_n parameterized by the input size n. We use capitalized words to denote such families of functions. For example, we say that there is a depth-3 threshold circuit of size $O(n^2)$ computing Addition. It means that we have a family of threshold circuits $\{C_n\}$, such that for every n, C_n computes the sum of 2 n-bit numbers, where the size and the depth of each C_n are $O(n^2)$ and 3, respectively.

In later chapters, we make frequent use of some terminologies regarding upper bounds on the size of a circuit. We say that a circuit is of subexponential size if its size is $2^{s(n)}$ for some function $s(n)$ such that $s(n) = o(n^\epsilon)$ for all $\epsilon > 0$, and for convenience, we simply express the circuit size as $2^{n^{o(1)}}$. A circuit is of superpolynomial size if its size is $2^{O(\log^k n)}$ for some fixed k, and in this case, we also express the circuit size as $2^{\log^{O(1)} n}$. Note that if a circuit is of superpolynomial size, then it is of subexponential size, but the converse does not always hold.

Chapter 2

Linear Threshold Element

2.1 Introduction

This chapter introduces the Linear Threshold Element (LTE) model and study its properties. The topics covered include well-known concepts such as the perceptron learning algorithm, capacity of perceptrons, and limitations of the computing power of single LTEs. In addition, we present several results on the analysis of linearly nonseparable training sets and the behavior of the perceptron learning algorithm for such inputs.

Section 2.2 introduces the basic model of an LTE and some of its properties. For example, we show that any Boolean function can be computed by a depth-2 feedforward network of LTEs. Examples of simple functions that cannot be computed by a single LTE are also presented.

Sections 2.3 to 2.6 explore issues involved in learning. The problem of learning in a perceptron can be stated as follows: Given a set of m input vectors $\{X_1, \ldots, X_m\}$ in $\mathcal{R}^d$ (this set of input vectors is often referred to as the training set), determine a hyperplane such that each vector X_i lies on a pre-assigned side of the hyperplane. If such a hyperplane exists for the given training set, then the set of input vectors is referred to as *linearly separable*, and the hyperplane is referred to as a *separating hyperplane*. On the other hand, if no such hyperplane exists, then the training set is referred to as *linearly nonseparable*.

In Section 2.3, we introduce the perceptron learning algorithm and show that it determines a separating hyperplane if the training set is linearly separable. In Section 2.4, we study properties of linearly nonseparable training sets. Using a linear programming framework and associated tools, we develop necessary and sufficient conditions for linear nonseparability. Based on such characterizations, structures within linearly nonseparable training sets are identified. In particular, it is shown that the vectors in a given training set can be naturally decomposed into two subsets: (1) the set of nonseparable vectors, and (2) the set of separable vectors. Motivations for such a decomposition and the properties of these two subsets of vectors are established.

In Section 2.5, learning problems for linearly nonseparable training sets are defined and related computational complexity issues are explored. For example, one of the learning problems introduced in Section 2.5 involves determining the sets of separable and nonseparable vectors for a given training set. Efficient polynomial-time algorithms for solving this problem are established in Section 2.5. A related problem, where one wants to learn a linearly separable subset (of the given linearly nonseparable training set) of maximum cardinality, is shown to be NP-complete. Results presented in Sections 2.4 and 2.5 are independent of any learning algorithms and relate to the properties of training sets.

Section 2.6 analyzes the behavior of the perceptron learning algorithm when the training sets are linearly nonseparable. The results show that if the well-known perceptron learning algorithm is applied to linearly nonseparable input patterns, then it can 'learn' the sets of separable and nonseparable vectors in the input patterns. Other related results that augment the power of the perceptron learning algorithm are also discussed.

Section 2.7 addresses some fundamental questions regarding the computational power of LTEs. For example, one of the questions addressed is: How many binary functions defined over a set of n points in $\mathcal{R}^d$ can an LTE compute? Given a set of n points or vectors in $\mathcal{R}^d$, there are 2^n possible binary functions that can be defined over it. In Section 2.7, it is shown that: (1) if $n \leq d$, then all the binary functions can be computed using a single LTE, provided the points are in general position, (2) if $n = 2d$, then at most half of the 2^n possible functions can be computed by a single LTE, and (3) if $n >> d$, then at most $O(n^d)$ of the 2^n possible functions can be computed using a single LTE. So the fraction of the total number of binary functions defined over a set of n points in $\mathcal{R}^d$ rapidly decreases from 1 to almost 0 around the value $n = 2d$. Because of this phenomenon, we define the *capac-*

ity of an LTE to be $2d$, where d is the dimension of the input vectors to the LTE. In Section 2.7.2, we also establish an asymptotically tight lower bound on the number of Boolean functions computable by a single LTE.

Section 2.7.4 addresses a related question: If one wants to compute each of the 2^n possible functions using a fixed feedforward network of LTEs, then how many connections (i.e., weights) are needed in the network? Using the capacity results for a single LTE, a lower bound is derived showing that the required number of weights is $\geq n/(1 + \log_2 n)$. The chapter concludes by showing that if the inputs to a threshold gate are bounded integers, then only a finite precision in the weights is sufficient.

2.2 Linear Threshold Element and its Basic Properties

Recall from Section 1.2.3 that a linear threshold element with d inputs is parameterized by a set of d weights, $w_1, \ldots, w_d$, and a threshold t. Given a set of inputs $X_i = [x_{i1}\ x_{i2}\ \cdots x_{id}]^T$, where $x_{ij} \in \mathcal{R}$, the output, y_i, of the element is either 1 or 0, and is determined as follows:

$$y_i = \begin{cases} 1 & \text{if } (\sum_{k=1}^{d} w_k x_{ik} - t) \geq 0, \\ 0 & \text{if } (\sum_{k=1}^{d} w_k x_{ik} - t) < 0. \end{cases} \tag{2.1}$$

Equivalently,

$$y_i = \text{sgn}(\sum_{k=1}^{d} w_k x_{ik} - t) = \text{sgn}(\mathbf{w}^T X_i - t),$$

where $\mathbf{w} = [w_1 \cdots w_d]^T$ is referred to as the weight vector.

Lemma 2.1: Given a set of input vectors $X_1, \ldots, X_m$, $X_i \in \mathcal{R}^d$, and a d-input LTE with weight vector $\mathbf{w} \in \mathcal{R}^d$, and threshold t, there exists a threshold t' such that $(\sum_{k=1}^{d} x_{ik} w_k - t') = (\mathbf{w}^T X_i - t') \neq 0$ and $\text{sgn}(\mathbf{w}^T X_i - t) = \text{sgn}(\mathbf{w}^T X_i - t')$ for all $i = 1, \ldots, m$.

Proof: Without loss of generality, assume that the input vectors are partitioned into two sets $\{X_1, \ldots, X_k\}$ and $\{X_{k+1}, \ldots, X_m\}$, such that every vector in the first set has output $y_i = 1$ and every vector in the second set has output $y_i = 0$. Let,

$$\delta \;=\; \min\{|(\mathbf{w}^T X_i - t)| \;:\; (k+1) \leq i \leq m\}.$$

Since $y_i = 0$ for $i = k+1, \ldots, m$, it follows from Eqn. (2.1) that $\delta > 0$. Let $t' = t - \dfrac{\delta}{2}$, then one can easily verify that $(\mathbf{w}^T X_i - t') \neq 0$ for $i = 1, \ldots, m$, and that the output of the LTE (with the modified threshold, t') remains unchanged for every input vector X_i.   □

Lemma 2.1 suggests that without loss of generality, one can always assume that $\mathbf{w}^T X_i - t \neq 0$ for all input vectors to an LTE, and that the output y_i of an LTE can be defined as follows:

$$
y_i \;=\;
\begin{cases}
1 & \text{if } \left(\displaystyle\sum_{k=1}^{d} w_k x_{ik} - t\right) > 0, \\[2ex]
0 & \text{if } \left(\displaystyle\sum_{k=1}^{d} w_k x_{ik} - t\right) < 0.
\end{cases}
$$

For a given weight vector $\mathbf{w}$ and a threshold t, an LTE defines a binary function, which maps $\mathcal{R}^d$ to the set $\{0, 1\}$. Different choices of $\mathbf{w}$ and t can result in the LTE computing different functions. Functions that can be computed by an LTE will be referred to as *linearly separable functions* or *linear threshold functions*.

Example 2.1:   *Computing a 2-input NAND function*
The 2-input NAND function, often denoted as $\overline{x_1 \wedge x_2}$, equals 0 if both the inputs are 1, and equals 1 if at least one of the inputs is 0. Thus, the NAND function can be computed using a 2-input LTE by choosing $w_1 = w_2 = -1$, and $t = -3/2$, i.e., $\overline{x_1 \wedge x_2} = \text{sgn}(-x_1 - x_2 + 3/2)$.   □

Example 2.2:   *Computing n-input AND and OR functions*
The n-input AND function, also represented as $x_1 \wedge x_2 \wedge \cdots \wedge x_n$, equals 1 if and only if every input variable equals 1, and equals 0 otherwise. It can be computed by choosing the weights as $w_1 = w_2 = \cdots = w_n = 1$ and $t = n - 1/2$, i.e., $x_1 \wedge \cdots \wedge x_n = \text{sgn}(\displaystyle\sum_{i=1}^{n} x_i - (n - 1/2))$. Since any

complemented variable $\overline{x_i}$ can be expressed as $\overline{x_i} = (1-x_i)$, an AND function in which some of the variables are complemented can be also realized by a single LTE. For example,

$$AND(x_1, \overline{x_2}, x_3) = \text{sgn}(x_1 + (1 - x_2) + x_3 - 5/2) = \text{sgn}(x_1 - x_2 + x_3 - 3/2).$$

The n-input OR function, also denoted as $x_1 \vee x_2 \vee \cdots \vee x_n$, equals 1 if any of the inputs is 1, and equals 0 if every input variable is 0. Thus, $x_1 \vee \cdots \vee x_n = \text{sgn}(\sum_{i=1}^{n} x_i - 1/2)$. One can again show that any OR function in which some of the input variables are complemented can be also realized by a single LTE.   □

The preceding two examples (2.1 and 2.2) naturally lead to the issue of determining functions that cannot be computed by a single LTE; such functions are referred to as linearly nonseparable functions. In order to understand the computing power of networks of LTEs, one must first understand the limitations and the computing power of a single LTE. The following example illustrates that a simple two-input Boolean function cannot be computed by a single LTE. In Section 2.4, we study the characteristics of training sets and functions, which cannot be computed by a single LTE. Moreover, in Section 2.7, we establish upper bound and lower bound results on the number of functions computable by a single LTE.

Example 2.3:   The Parity or the Exclusive-OR (XOR) function of two input variables x_1 and x_2 (also represented as $x_1 \oplus x_2$) is defined as follows:

x_1	x_2	$x_1 \oplus x_2$
0	0	0
0	1	1
1	0	1
1	1	0

If the above XOR function is to be computed by an LTE, then the input vectors and the corresponding desired outputs are as follows:

$$X_1 = \begin{bmatrix} 0 \\ 0 \end{bmatrix}, \ y_1 = 0; \quad X_2 = \begin{bmatrix} 1 \\ 1 \end{bmatrix}, \ y_2 = 0;$$

$$X_3 = \begin{bmatrix} 0 \\ 1 \end{bmatrix}, \ y_3 = 1; \quad X_4 = \begin{bmatrix} 1 \\ 0 \end{bmatrix}, \ y_4 = 1 \ .$$

Suppose there exists an LTE with weights w_1, w_2, and threshold t such that it computes the XOR function. Then, the constraints imposed on the weights and the threshold by the input vectors X_3, X_4, and X_1 are $w_1 > t$, $w_2 > t$, and $t > 0$, respectively. Thus, $w_1 + w_2 > t$, which implies that for the input vector X_2, the output is 1. This leads to a contradiction, since by the definition of the XOR function, the output for the input vector X_2 is 0. Hence, there exists no set of weights and threshold values that can realize the XOR function. In other words, XOR is not linearly separable.   □

While Example 2.3 suggests that a single LTE cannot compute some simple functions, the following theorem shows that a simple depth-2 network of LTEs can compute any given Boolean function.

Theorem 2.1:   Every n-variable Boolean function

$$f(x_1, \ldots, x_n) : \ \{0, 1\}^n \to \{0, 1\}$$

can be computed by a depth-2 threshold circuit with at most $(2^{n-1} + 1)$ linear threshold elements.

Proof: Every Boolean function $f(x_1, \ldots, x_n)$ can be written in a canonical Sum Of Products (SOP) form as follows:

$$f(x_1, \ x_2, \ldots, x_n) \ = \ P_1 \lor P_2 \lor \cdots \lor P_k,$$

where each product-term P_i is a distinct n-variable AND function, where a subset of the variables are complemented, and $k \leq 2^n$. In the canonical SOP representation of f, each product-term P_i represents a distinct input assignment for which the value of f is 1. Alternatively, f can be written in a canonical Product Of Sums (POS) form as follows:

$$f(x_1, \ x_2, \ldots, x_n) \ = \ S_1 \land S_2 \land \cdots \land S_l,$$

in which each sum-term S_i is a distinct n-variable OR function, where a subset of the variables are complemented, and $l \leq 2^n$. In the canonical POS representation of f, each sum-term S_i represents a distinct input assignment for which the value of f is 0.

Since there are 2^n distinct input assignments, it follows that $k + l = 2^n$, and either $k \leq 2^{n-1}$ or $l \leq 2^{n-1}$. If $k \leq 2^{n-1}$, then f can be computed by a depth-2 threshold circuit with at most $(2^{n-1} + 1)$ gates as follows: Each product-term P_i in the SOP expression is computed by a single LTE (as illustrated in Example 2.2), and then f is computed by using an output LTE in the second layer that computes an OR function of the outputs of the LTEs in the first layer. If $l \leq 2^{n-1}$, then one can compute f in a depth-2 threshold circuit using at most $(2^{n-1} + 1)$ threshold gates by appropriately implementing its POS representation. $\square$

The preceding depth-2 construction, however, requires exponentially many gates in the number of input variables, and one of the objectives of this book is to develop tools so that certain classes of functions can be efficiently computed using only polynomial-size threshold circuits.

2.3 Perceptron Learning Algorithm

In this section, we introduce the perceptron learning algorithm, which attempts to solve the following problem: Given a binary function defined over a set of m vectors, how does one determine the weights and the threshold value of an LTE that computes the given function?

Definition 2.1: A binary function defined over a set of vectors in $\mathcal{R}^d$ and specified by the input-output pairs $\{(X_1, y_1), (X_2, y_2), \ldots, (X_m, y_m)\}$, where $X_i \in \mathcal{R}^d$ and $y_i \in \{0, 1\}$, is said to be a *linearly separable* function if there exists a vector $\mathbf{w} \in \mathcal{R}^d$ and a threshold $t \in \mathcal{R}$ such that $y_i = \text{sgn}(X_i^T \mathbf{w} - t)$ for all $i = 1, \ldots, m$. A binary function is said to be *linearly nonseparable* if it is not linearly separable. $\square$

Before exploring the problem of determining a weight vector for a linearly separable function, let us make the following remarks, which will facilitate the presentation of the results in the rest of this chapter.

Remark 2.1: Without loss of generality, the threshold value t can be assumed to equal zero. This can be achieved by increasing the dimension of every input vector by augmenting it with an entry that equals -1, and by increasing the dimension of the weight vector by augmenting it with t. Then the output of the LTE can be written as

$$y_i = \text{sgn}(\sum_{k=1}^{d+1} w_k X_{ik})$$
$$= \text{sgn}(\mathbf{w}^{\mathbf{T}} X_i),$$

where $\mathbf{w}^{\mathbf{T}} = [w_1 \cdots w_d\ t]$ and $X_i^T = [X_{i1} \cdots X_{id}\ -1]$.   □

If the given vectors are assumed to be already augmented, then the problem of learning in a perceptron can be defined as follows.

Problem 2.1: Given a set of vectors $\{X_1, X_2, \ldots, X_m\}$, $X_i \in \mathcal{R}^d$, and a set of desired output values $\{y_1, y_2, \ldots y_m\}$, $y_i \in \{0,1\}$, determine a weight vector $\mathbf{w} \in \mathcal{R}^d$ (if there exists one) such that $\text{sgn}(\mathbf{w}^{\mathbf{T}} X_i) = y_i$ for $i = 1, \ldots, m$.   □

Remark 2.2: Let every input vector, X_i, that is assigned to $y_i = 0$ be replaced by $-X_i$. Moreover, without loss of generality, let us assume that $\mathbf{w}^{\mathbf{T}} X_i \neq 0$ for $i = 1, \ldots, m$ (see Lemma 2.1), then Problem 2.1 can be equivalently stated as Problem 2.2 (stated below).   □

Problem 2.2: Given a set of vectors $\{X_1, X_2, \ldots, X_m\}$, $X_i \in \mathcal{R}^d$, determine a weight vector $\mathbf{w} \in \mathcal{R}^d$ such that $\mathbf{w}^{\mathbf{T}} X_i > 0$ for $i = 1, \ldots, m$.   □

Remark 2.3: In Problem 2.2, the learning problem has been reduced to determining a hyperplane in $\mathcal{R}^d$ that is constrained to go through the origin such that all the input vectors lie on one side of it; see Fig. 2.1. As explained in Remarks 2.1 and 2.2, this is a *general* situation if the vectors are appropriately preprocessed. In the rest of this chapter, we assume that the given input vectors X_i have been already modified to satisfy the properties mentioned in Remarks 2.1 and 2.2. Example 2.4 illustrates a case where such preprocessing of vectors is carried out.   □

Assuming that the vectors are already preprocessed, linearly separable and nonseparable sets can be defined as follows.

Definition 2.2: A set of vectors $\{X_1, X_2, \ldots, X_m\}$, $X_i \in \mathcal{R}^d$, is a *linearly separable* set if there exists a vector $\mathbf{w} \in \mathcal{R}^d$ such that $X_i^T \mathbf{w} > 0$ for $i = 1, \ldots, m$. A set of vectors $\{X_1, X_2, \ldots, X_m\}$, $X_i \in \mathcal{R}^d$, is a *linearly nonseparable* set if it is not linearly separable.   □

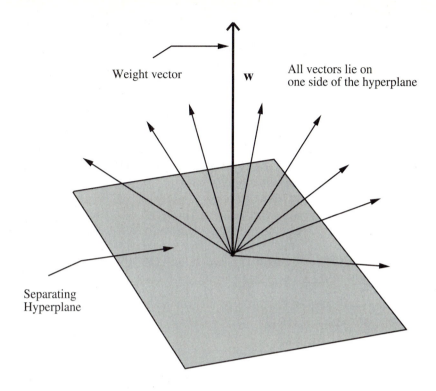

Weight vector   **w**   All vectors lie on
                                              one side of the hyperplane

Separating
Hyperplane

Figure 2.1 For a linearly separable set of vectors, a separating hyperplane can be determined such that all the vectors lie on one side of it. The normal to the hyperplane is determined by the weight vector **w**. Moreover, if the vectors are modified according to Remarks 2.1, 2.2, and 2.3, then without loss of generality, the hyperplane can be assumed to pass through the origin.

For a linearly separable set, the hyperplane determined by the weight vector **w** is referred to as the separating hyperplane. We also say that **w** separates the vectors in a linearly separable set.

Given a training set, $\{X_1, X_2, \ldots, X_m\}$, $X_i \in \mathcal{R}^d$, Rosenblatt [109] (see also [81, 31]) proposed an algorithm that learns a weight vector **w** for a given set of linearly separable input vectors. The *perceptron learning algorithm* can be simply stated as follows:

START : Let $\mathbf{w}_0 = \mathbf{0}$.

TEST : Select an arbitrary training vector X_i,

If $X_i^T \mathbf{w}_l \leq 0$, then go to ADD; or else go to TEST.

ADD : $\mathbf{w}_{l+1} = \mathbf{w}_l + X_i; \, l = l + 1$.
   Go to TEST.

The selection of X_i in the step [TEST] must guarantee that every X_i appears infinitely often in the selected sequence. For example, a valid sequencing is to select the vectors in the following cyclic order: $X_1, X_2, \ldots, X_m, X_1, X_2, \ldots$.

The algorithm is said to *converge* if for some l, $\mathbf{w}_l^T X_i > 0$ for every vector X_i in the training set. Theorem 2.2 shows that if the set of input vectors is linearly separable, then the above learning algorithm converges in a finite number of steps.

Theorem 2.2:   If the set of input vectors $\{X_1, X_2, \ldots, X_m\}$ is linearly separable, then the perceptron learning algorithm determines a weight vector $\mathbf{w}$ in a finite number of steps, such that $\mathbf{w}^T X_i > 0$ for $i = 1, \ldots, m$.

Proof: Since the set of input vectors is linearly separable, there exist a weight vector $\mathbf{w}^*$, with $||\mathbf{w}^*|| = 1$, and a constant $\delta > 0$ such that for $i = 1, \ldots, m$,

$$\mathbf{w}^{*T} X_i > \delta > 0,$$

Let us also assume that $||X_j|| \le L$, for $i = 1, \ldots, m$, i.e., the length of all the input vectors are bounded above by L. Let k denote the number of times the vector $\mathbf{w}$ in the perceptron learning algorithm has been updated, and let $\mathbf{w}_k$ denote the value of the weight vector after the kth update. In other words, k is the number of times step [ADD] is executed. Our objective is to show that k is a bounded number.

Note that if an input vector X_i is used for the kth update in the algorithm, then $\mathbf{w}_k$ can be recursively written as:

$$\mathbf{w}_k = \mathbf{w}_{k-1} + X_i,$$

where $X_i^T \mathbf{w}_{k-1} \le 0$. First, we establish an upper bound on $||\mathbf{w}_k||$:

$$
\begin{aligned}
||\mathbf{w}_k||^2 &= ||(\mathbf{w}_{k-1} + X_i)||^2 \\
&= (\mathbf{w}_{k-1} + X_i)^T(\mathbf{w}_{k-1} + X_i) \\
&= \mathbf{w}_{k-1}^T \mathbf{w}_{k-1} + X_i^T X_i + 2\mathbf{w}_{k-1}^T X_i \\
&\le ||\mathbf{w}_{k-1}||^2 + L^2 \quad (\text{because } X_i^T \mathbf{w}_{k-1} \le 0 \text{ and } ||X_i||^2 \le L^2) \\
&\le kL^2 \quad (\text{since } \mathbf{w}_0 = 0).
\end{aligned}
$$

Thus,

$$||\mathbf{w}_k|| \leq L\sqrt{k}. \tag{2.2}$$

Next, we prove a lower bound on $||\mathbf{w}_k||$:

$$
\begin{aligned}
\mathbf{w}^{*T}\mathbf{w}_k &= \mathbf{w}^{*T}\mathbf{w}_{k-1} + \mathbf{w}^{*T}X_i \\
&\geq \mathbf{w}^{*T}\mathbf{w}_{k-1} + \delta \\
&\geq k\delta \ .
\end{aligned}
$$

Applying the Cauchy-Schwarz inequality, we have

$$||\mathbf{w}^*|| \cdot ||\mathbf{w}_k|| \geq \mathbf{w}^{*T}\mathbf{w}_k \geq k\delta.$$

Since $||\mathbf{w}^*|| = 1$,

$$||\mathbf{w}_k|| \geq k\delta. \tag{2.3}$$

Combining Eqns. (2.2) and (2.3), we get $L\sqrt{k} \geq k\delta$, or equivalently,

$$k \leq \frac{L^2}{\delta^2} \ . \qquad\qquad \square$$

In the perceptron learning algorithm, the initial weight vector $\mathbf{w}_0$ can be chosen to be an arbitrary non-zero vector. For a non-zero $\mathbf{w}_0$, the convergence of the perceptron learning algorithm can be established by introducing simple modifications in the above proof. For linearly nonseparable training sets, the perceptron learning algorithm cannot converge. Based on an analysis of linearly nonseparable training sets, as presented in Sections 2.4 and 2.5, results on the behavior of the algorithm for linearly nonseparable training sets are presented in Section 2.6.

2.4   Analysis of Linearly Nonseparable Sets of Input Vectors

In this section, we study the case where the set of input vectors, $S = \{X_1, \ldots, X_m\}$, $X_i \in \mathcal{R}^d$, is not linearly separable. We first determine necessary and sufficient conditions for a set to be linearly nonseparable, and then identify structures within such sets. These results are independent of any learning algorithms and relate to inherent properties of linearly nonseparable sets.

2.4.1   Necessary and Sufficient Conditions for Linear Non-separability

Let us first develop a linear programming formulation for the learning problems discussed in this chapter. Recall that in Problem 2.2, the goal is to determine a weight vector $\mathbf{w}$, such that $\mathbf{w}^T X_i > 0$, for $i = 1, \ldots, m$. If the weight vector $\mathbf{w}$ is properly scaled, then Problem 2.2 reduces to determining a vector $\mathbf{w}$ (if such a vector exists) such that the following matrix inequality is satisfied:

$$\mathbf{w}^T [X_1 \; X_2 \; \cdots X_m] \; \geq \; [1 \; 1 \; \cdots \; 1]. \tag{2.4}$$

We refer to Eqn. (2.4) as a linear programming (LP) formulation of Problem 2.2. Clearly, Eqn. (2.4) represents a problem whereby a feasible solution to a set of linear inequalities has to be determined. Hence, it can be solved by using linear programming algorithms in polynomial time (in m and d) [94]. Thus, from a computational perspective, the problem of learning in perceptrons (Problem 2.2) can be solved efficiently by using any of the polynomial-time algorithms for linear programming [94].

We should note here that from the point of view of computational requirements, the perceptron learning algorithm is a much simpler algorithm than any linear programming algorithm and has the essence of learning. That is, it uses simple operations, is iterative, and makes only 'local' decisions. There are, however, significant advantages to using an LP formulation. As we shall show in the rest of this chapter, the analysis of linearly nonseparable training sets as well as the study of the behavior of the perceptron learning algorithm are greatly facilitated by using it.

The following lemma presents a sufficient condition for linear nonseparability.

Lemma 2.2:   If there exists a positive linear combination (PLC) of the given set of input vectors $\{X_1, \ldots, X_m\}$ (where, $X_i \in \mathcal{R}^d$) that equals $\mathbf{0}$, then the given set of vectors is linearly nonseparable.

Proof: If a PLC of the vectors equals $\mathbf{0}$, then there exist $q_i \geq 0$, such that $q_j > 0$ for some j, $1 \leq j \leq m$, and

$$\sum_{i=1}^{m} q_i X_i = 0.$$

We prove the lemma by contradiction. Suppose that the given set of vectors is linearly separable, then there exists a weight vector $\mathbf{w}$, such that $\mathbf{w}^T X_i > 0$ for $i = 1, \ldots, m$. Since $q_i \geq 0$ and there exists at least one $q_j > 0$ (for some j, $1 \leq j \leq m$), we have

$$\mathbf{w}^T \sum_{i=1}^{m} q_i X_i = \sum_{i=1}^{m} q_i(\mathbf{w}^T X_i) > 0.$$

However, this is a contradiction since $\sum_{i=1}^{m} q_i X_i = 0$. Hence, there exists no weight vector that separates all the input vectors. □

Example 2.4: Lemma 2.2 can be applied to show that the 2-input Exclusive-OR (XOR) function cannot be computed by an LTE. We show in Example 2.3 that in order to compute the XOR function, the input vectors and the corresponding desired output for an LTE should be as follows:

$$X_1 = \begin{bmatrix} 0 \\ 0 \end{bmatrix}, y_1 = 0; \quad X_2 = \begin{bmatrix} 1 \\ 1 \end{bmatrix}, y_2 = 0;$$

$$X_3 = \begin{bmatrix} 0 \\ 1 \end{bmatrix}, y_3 = 1; \quad X_4 = \begin{bmatrix} 1 \\ 0 \end{bmatrix}, y_4 = 1.$$

Let us preprocess the vectors so that they are in the form stated in Remarks 2.1 and 2.2. That is, if we eliminate the threshold (by augmenting the input vectors with -1) and negate the input vectors for which the desired output is 0, then the problem of computing the two-input XOR function by a single LTE is equivalent to determining a separating hyperplane for the following vectors:

$$X_1 = \begin{bmatrix} 0 \\ 0 \\ 1 \end{bmatrix}; \quad X_2 = \begin{bmatrix} -1 \\ -1 \\ 1 \end{bmatrix}; \quad X_3 = \begin{bmatrix} 0 \\ 1 \\ -1 \end{bmatrix}; \quad X_4 = \begin{bmatrix} 1 \\ 0 \\ -1 \end{bmatrix}.$$

However $\sum_{i=1}^{4} X_i = 0$, and Lemma 2.2 implies that there exists no separating hyperplane for $\{X_1, X_2, X_3, X_4\}$. □

We next show that the sufficient condition for linear nonseparability stated in Lemma 2.2 is also necessary.

Theorem 2.3:   A set of vectors $\{X_1, \ldots, X_m\}$, $X_i \in \mathcal{R}^d$, is linearly non-separable if and only if there exists a positive linear combination of the vectors that equals $\mathbf{0}$, i.e., there exists $q_i \geq 0$, $1 \leq i \leq m$, such that

$$\sum_{i=1}^{m} q_i X_i = 0,$$

and $q_j > 0$ for some j, $1 \leq j \leq m$.

Proof: Let D be a $(d \times m)$ matrix whose columns are the input vectors $X_1, X_2, \ldots, X_m$, i.e., $D = [X_1 \ X_2 \ \cdots X_m]$. From Eqn. (2.4), we know that the given set of vectors is linearly separable if and only if the following LP has a feasible solution:

$$\begin{array}{c} \text{Minimize} \quad \mathbf{0}^T \mathbf{w} \quad \text{such that} \\ \mathbf{w}^T D \ \geq \ [1\,1\,1\cdots 1]. \end{array} \tag{2.5}$$

The feasibility of an LP can be often determined by studying the dual LP problem of the original LP formulation. The dual LP for Eqn. (2.5) can be stated as follows (see [94]):

$$\begin{array}{c} \text{Maximize} \quad [1\,1\cdots 1]\mathbf{q} \quad \text{such that for } i = 1, \ldots, m \\ D\mathbf{q} = 0; \quad q_i \geq 0. \end{array} \tag{2.6}$$

The quantity $[1\,1\cdots 1]\mathbf{q}$ is referred to as the *cost* or *objective* function of the linear program. Note that the column vector, $D\mathbf{q}$, $q_i \geq 0$, represents a non-negative linear combination of the columns of D. Since the columns of D are the input vectors X_i, $D\mathbf{q}$ ($q_i \geq 0$) represents non-negative linear combination of the vectors X_i.

It follows from the duality theorem of linear programming (or Farkas' Lemma) that an original or primal LP (e.g., Eqn. (2.5)) has a feasible solution if and only if its dual LP (e.g., Eqn. (2.6)) has a *bounded* objective function.

In Eqn. (2.6), the objective function is finite (in fact $= 0$) if and only if the only solution to the equation $D\mathbf{q} = 0$, $q_i \geq 0$, is $\mathbf{q} = 0$. If there is a solution $\mathbf{q} \neq 0$, then $[1\,1 \ \cdots 1]\mathbf{q} > 0$. Moreover, for any $\alpha > 0$, $\mathbf{q}' = \alpha\mathbf{q}$ is also a feasible solution for the LP in Eqn. (2.6). Hence, the objective function $[1\,1 \ \cdots 1]\mathbf{q}'$ $(> \alpha)$ can be made unbounded by choosing α arbitrarily large.

A proof for Lemma 2.2 (showing the sufficiency part) follows immediately. If there is a positive linear combination (PLC) of the vectors X_i that equals

0, then there is a non-zero **q** satisfying the constraints $D\mathbf{q} = 0$, $\quad q_i \geq 0$. Hence, Eqn. (2.6) has an unbounded objective function. Applying the duality theorem we obtain that Eqn. (2.5) is infeasible, which implies that the set of vectors is linearly nonseparable.

Next, consider the case where the set of vectors is linearly nonseparable, and Eqn. (2.5) has no feasible solution. It follows then from Farkas' Lemma that Eqn. (2.6) has an unbounded objective function. However, from the preceding discussions, we know that Eqn. (2.6) has an unbounded objective function if and only if there is a solution $\mathbf{q} \neq 0$ that satisfies $D\mathbf{q} = 0$, $\quad q_i \geq 0$. Thus, there exists a PLC of the vectors X_i that equals zero.   □

2.4.2   Structures Within Linearly Nonseparable Training Sets

In this section, we study possible structures within a set of linearly nonseparable vectors. Closer observation would reveal that if a set of input vectors is linearly nonseparable, then it is not necessary that all the vectors in the set are 'responsible' for nonseparability. In the rest of this section, we shall establish a formal framework for determining when a vector is 'responsible' for linear nonseparability.

Example 2.5:   Consider the following set of three vectors:

$$X_1 = \begin{bmatrix} 1 \\ 1 \end{bmatrix}; \ \ X_2 = \begin{bmatrix} 1 \\ -1 \end{bmatrix}; \ \ X_3 = \begin{bmatrix} -1 \\ 1 \end{bmatrix}.$$

The above set is linearly nonseparable, because $X_2 + X_3 = 0$. However, if we want to solve the following equation

$$aX_1 + bX_2 + cX_3 = 0; \quad a, b, c \geq 0,$$

then one can verify that it is necessary that a is 0, i.e., there is no positive linear combination (PLC) of X_1, X_2, and X_3 that equals **0** and in which X_1 'participates.' In other words, only X_2 and X_3 are responsible for linear nonseparability of the above set of vectors.   □

This observation and the results proved later in Theorem 2.4 motivate the following classification of the input vectors.

Definition 2.3:   Given a set $S = \{X_1, \ldots, X_m\}$, $X_j \in \mathcal{R}^d$, a vector $X_i \in S$ is defined to be *separable* if it never participates in a positive linear combination that equals **0**. That is,

$$\sum_{j=1}^{m} X_j q_j = 0; \quad q_j \geq 0 \quad \Longrightarrow \quad q_i = 0 .$$

A vector $X_i \in S$ is defined to be *nonseparable* if it participates in a positive linear combination that equals **0**, i.e., for some $q_i > 0$, we have

$$\sum_{j=1}^{m} X_j q_j = 0,$$

where $q_j \geq 0$ for $j = 1, \ldots, m$.   □

Example 2.6:   In Example 2.5, the set of separable vectors comprises only X_1, and the set of nonseparable vectors consists of X_2 and X_3.
In Example 2.4, however, the set of separable vectors is empty and every vector is nonseparable.   □

Let us denote the separable vectors as $X_1, \ldots, X_k$, and the nonseparable vectors as $X_{k+1}, \ldots, X_m$. If $k = 0$, then the set of separable vectors is empty. Similarly, if $k = m$, then the set of nonseparable vectors is empty (in other words, the given set of vectors is linearly separable).

In the following discussions, we shall establish a number of properties of the sets of separable and nonseparable vectors of a given set. The following lemma follows directly from Definition 2.3.

Lemma 2.3:

1. If the set of nonseparable vectors is nonempty, then it must consist of at least two vectors.

2. There exists a single PLC of the nonseparable vectors that equals **0**, and in which *all* the nonseparable vectors participate. That is, there exist $q_i > 0$, for $i = k+1, \ldots, m$, such that

$$\sum_{i=k+1}^{m} q_i X_i = \mathbf{0}.$$
                                      □

Remark 2.4:   Whether a given vector X_i in S is separable or nonseparable is determined by the other vectors in the set. For example, if some vectors are deleted from S, then in the reduced set, say S', vectors that are nonseparable in S might become separable in S'.   □

Example 2.7:   Consider the set of vectors $S = \{X_1, X_2, X_3, X_4\}$ introduced in Example 2.4. As mentioned in Example 2.6, the set of separable vectors in S is empty, and every vector is nonseparable. However, if, for example, X_1 is deleted from the set, then one can verify that in the reduced set $S' = \{X_2, X_3, X_4\}$, all the vectors are separable. In fact, if any of the vectors in the set $\{X_1, X_2, X_3, X_4\}$ is deleted from the set, then the rest of the vectors form a linearly separable set.

Similarly, in Example 2.5, if X_2 is deleted from the set, then in the reduced set $S' = \{X_1, X_3\}$, both the vectors are separable.   □

Lemma 2.4 shows that the set of separable vectors can indeed be arbitrarily large in a linearly nonseparable training set.

Lemma 2.4:   For every $n \geq 2$ there exists a linearly nonseparable set of vectors of size $2^{n-2} + 2$, such that there are 2^{n-2} separable vectors and only two nonseparable vectors.   □

The proof is constructive and is left as an exercise (Exercise 2.4). The following example illustrates the construction for $n = 4$.

Example 2.8:   Consider the following set of vectors in $\mathcal{R}^4$:

$$X_1 = [1\ 1\ 1\ 1]^T; \quad X_2 = [1\ 1\ -1\ -1]^T$$

$$X_3 = [1\ 1\ -1\ 1]^T; \quad X_4 = [1\ 1\ 1\ -1]^T$$

$$X_5 = [1\ -1\ -1\ -1]^T; \quad X_6 = [-1\ 1\ 1\ 1]^T.$$

One can show (see Exercise 2.4) that the set of separable vectors is $\{X_1, X_2, X_3, X_4\}$ and the set of nonseparable vectors is $\{X_5, X_6\}$.   □

Remark 2.5:   Even if all the vectors in a given training set are nonseparable, the analysis presented in this section might be useful. For example, in Section 2.5.1, an algorithm to determine large linearly separable subsets of a given linearly nonseparable set is described. The algorithm reduces the size of the nonseparable set by successively deleting nonseparable vectors. As the size decreases, the subsets are going to have *nonempty* sets of separable vectors. Section 2.5.1 shows how the sets of separable vectors can be used to reduce the search for large linearly separable subsets.   □

The following theorem proves two key properties of the separable and non-separable sets of vectors, which further clarify the motivation for their definitions.

Theorem 2.4:   Consider a set of vectors $\{X_1, \ldots, X_m\}$, $X_i \in \mathcal{R}^d$, where $\{X_1, \ldots, X_k\}$ is the set of separable vectors and $\{X_{k+1}, \ldots, X_m\}$ is the set of nonseparable vectors.

1. If $\mathbf{w}^T X_i > 0$ for some i, $k + 1 \leq i \leq m$ (i.e., X_i is a nonseparable vector), then there exists another nonseparable vector, X_l, $k + 1 \leq l \leq m$, such that $\mathbf{w}^T X_l < 0$.

2. Let $k \geq 1$ (i.e., there is at least one separable vector), then there exists a weight vector $\mathbf{w}_0$ such that:
   (a) $\mathbf{w}_0^T X_i \geq 1$, for $i = 1, \ldots, k$, and
   (b) $\mathbf{w}_0^T X_j = 0$, for $j = k + 1, \ldots, m$.

Proof: In order to show the first part, let X_i $(k + 1 \leq i \leq m)$ be such that $\mathbf{w}^T X_i = \gamma > 0$. It follows from the definition of nonseparable vectors (see Lemma 2.3) that there exists $q_j > 0$, $(k + 1) \leq j \leq m$, such that $\sum_{j=k+1}^{m} q_j X_j = 0$. Hence,

$$\mathbf{w}^T \sum_{j=k+1}^{m} q_j X_j = q_i \mathbf{w}^T X_i + \sum_{j=k+1; j \neq i}^{m} \mathbf{w}^T q_j X_j = 0.$$

In other words,

$$\sum_{j=k+1; j \neq i}^{m} q_j \mathbf{w}^T X_j = -\gamma < 0.$$

Since $q_j > 0$, it implies that there must exist an $l \neq i$ $(k + 1 \leq l \leq m)$ such that $\mathbf{w}^T X_l < 0$.

In order to prove the second part, we first show that the following LP always has a feasible solution.

$$\begin{aligned} \text{Minimize} \quad & \mathbf{0}^T \mathbf{w} \quad \text{such that} \\ & \mathbf{w}^T [X_1 \ X_2 \cdots X_k \ X_{k+1} \cdots X_m] \geq [1 \ 1 \cdots 1 \ 0 \cdots 0], \end{aligned} \tag{2.7}$$

where the vector $[1 \ 1 \cdots 1 \ 0 \cdots 0]$ has the first k entries (i.e., the entries corresponding to separable vectors) as 1 and the last $m - k$ entries (i.e., the entries corresponding to nonseparable vectors) as 0. As discussed in the proof of Theorem 2.3, one can show that the above LP admits a feasible solution by showing that its dual has a bounded objective function. The dual of Eqn. (2.7) is as follows:

$$\text{Maximize} \quad [1\ 1 \cdots 1\ 0 \cdots 0]\mathbf{q} \quad \text{such that}$$

$$[X_1\ X_2 \cdots X_k\ X_{k+1} \cdots X_m] \begin{bmatrix} q_1 \\ \vdots \\ q_k \\ q_{k+1} \\ \vdots \\ q_m \end{bmatrix} = \mathbf{0}; \quad q_i \geq 0. \tag{2.8}$$

Since $X_1, \ldots, X_k$ are separable vectors, it follows from Definition 2.3 that any feasible solution vector $\mathbf{q}$ for Eqn. (2.8) has the first k entries equal to 0, i.e., $q_i = 0$, for $i = 1, \ldots, k$.

The objective function of Eqn. (2.8) is $\sum_{i=1}^{k} q_i$. Since $q_i = 0$ for $i = 1, \ldots, k$, it follows that the objective function of Eqn. (2.8) equals 0 and hence, is bounded. Thus, Farkas' Lemma implies that Eqn. (2.7) admits a feasible solution. In other words, there always exists a $\mathbf{w}_0$ such that (1) $\mathbf{w}_0^T X_i \geq 1 > 0$, for $i = 1, \ldots, k$, and (2) $\mathbf{w}_0^T X_j \geq 0$, for $j = k+1, \ldots, m$.

We next show that any feasible solution, $\mathbf{w}_0$, of Eqn. (2.7) must satisfy $\mathbf{w}_0^T X_j = 0$, for $j = k+1, \ldots, m$. Let

$$\mathbf{w}_0^T[X_1\ X_2 \cdots X_k\ X_{k+1} \cdots X_m] = [\delta_1 \ \cdots \delta_k\ \delta_{k+1} \cdots \delta_m],$$

where $\delta_i \geq 1$ for $i = 1, \ldots, k$, and $\delta_i \geq 0$ for $i = k+1, \ldots, m$. We have to show that $\delta_i = 0$, $\forall\ k+1 \leq i \leq m$. Since X_i, $(k+1) \leq i \leq m$ are nonseparable vectors, there exists a non-negative vector $\mathbf{q}$ such that

$$[X_1\ X_2 \cdots X_k\ X_{k+1} \cdots X_m]\mathbf{q} = 0,$$

where $q_i > 0$ for $i = k+1, \ldots, m$ and $q_i = 0$ for $i = 1, \ldots, k$. Hence,

$$\mathbf{w}^T[X_1\ X_2 \cdots X_k\ X_{k+1} \cdots X_m]\mathbf{q} = \sum_{i=k+1}^{m} \delta_i q_i = 0.$$

For $i = k+1, \ldots, m$, since $q_i > 0$, it must be the case that $\delta_i = 0$. Hence, $\mathbf{w}_0$ must satisfy $\mathbf{w}_0^T X_j = 0$, for $j = k+1, \ldots, m$. $\qquad\square$

It follows from the definition of a linearly nonseparable set that if one attempts to determine a separating hyperplane for the given set of vectors, then it must be the case that some of the vectors do not lie on the designated side of it. If a vector lies on the designated side of the hyperplane, then we will consider the given vector to have been 'learned.' In this context, the results in Theorem 2.4 can be interpreted as follows:

1. The first part of Theorem 2.4 shows the following: If a *nonseparable vector* lies on the designated side of a hyperplane, then there must exist another vector which must lie on the wrong side. Thus, if one wants to learn a nonseparable vector, then one must commit an error in learning another vector.

2. On the other hand, the second part of Theorem 2.4 shows that one can separate all the separable vectors without committing errors. In fact, the result of the second part of Theorem 2.4 is rather surprising: There exists a hyperplane such that all the separable vectors lie on one side of it and all the nonseparable vectors lie on it. This result can also be interpreted as follows: There is a vector $\mathbf{w_0}$ that separates all the separable vectors and remains ambiguous with respect to the nonseparable ones. This is illustrated in Fig. 2.2.

2.5 Learning Problems for Linearly Nonseparable Training Sets and Their Computational Complexity

Given the analysis in Section 2.4, we define two learning objectives for a linearly nonseparable set of input vectors. The first learning problem can be stated as follows:

Problem 2.3:   Given a set of m vectors in $\mathcal{R}^d$ $\{X_1, X_2 \cdots X_m\}$, determine the set of separable vectors and the set of nonseparable vectors. $\quad\square$

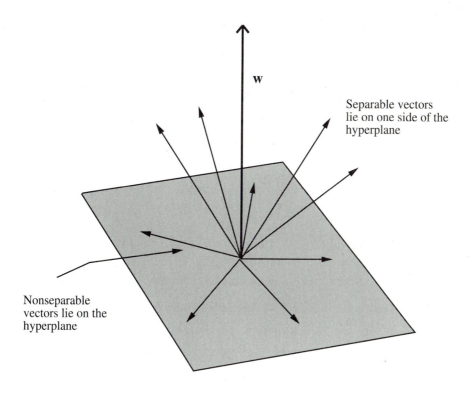

Figure 2.2 A hyperplane that can be determined for a linearly nonseparable training set, with at least one separable vector. The separable vectors can be assigned to one side of the hyperplane and all the nonseparable vectors can be made to lie on the hyperplane. If any of the nonseparable vectors is assigned to the designated side of a trial hyperplane, then there must exist other vectors that will lie on the wrong side of it, thereby leading to errors. We have assumed, without loss of generality, that the hyperplane passes through the origin; see Remarks 2.1, 2.2, and 2.3 for justification.

A solution to Problem 2.3 will give information about the input vectors that are responsible for linear nonseparability and the input vectors that are not. We show that Problem 2.3 can be solved by executing at most m linear programming problems and hence can be solved in polynomial time.

Problem 2.3 can be also motivated in the context of a learning system, which we shall refer to as noncommittal. Given a learning problem where all the inputs cannot be simultaneously learned, a learning system is said to be noncommittal if it does not commit errors. In other words, if an input cannot be learned without forcing errors on other inputs, then the noncommittal learning system should remain ambiguous with respect to it, and thereby not commit errors on other inputs.

The results of Theorem 2.4 imply that if a linearly nonseparable set is to be learned by a noncommittal learning system, then the best the system can do is to learn the separable vectors and remain ambiguous with regard to all the nonseparable ones. This is because if the system decides to learn any of the nonseparable vectors, then it must commit errors on some other vectors (see Theorem 2.4), thereby committing itself to make a decision about which of the nonseparable vectors to learn and which ones not to learn. Hence, Problem 2.3 can be considered a natural problem in the context of a noncommittal system.

A second learning objective can be stated as follows:

Problem 2.4:   Given a set of m vectors in $\mathcal{R}^d$ $\{X_1, X_2, \ldots, X_m\}$, determine a linearly separable subset of *maximum* cardinality.   □

Example 2.9:   In order to distinguish between Problem 2.3 and Problem 2.4, consider the set of vectors $S = \{X_1, X_2, X_3, X_4\}$ defined in Example 2.4. As stated in Example 2.6, the set of separable vectors in S is empty and the set of nonseparable vectors comprises all the four vectors. However, as discussed in Example 2.7, if any vector is deleted from the given set, then the reduced set becomes linearly separable. Hence, an example of a linearly separable subset of maximum cardinality is the set $S' = \{X_2, X_3, X_4\}$.   □

Solving Problem 2.4 would allow one to choose the 'best' set of input vectors as far as linear separability is considered. It turns out that Problem 2.4 is a computationally hard problem and we show it to be NP-complete. Proving a problem NP-complete shows that solving the problem is as hard as solving many infamous hard problems such as the Traveling-Salesman Problem. Although no proof is known showing that polynomial-time algorithms do not exist for NP-complete problems, the general conjecture is that it is highly unlikely that such algorithms would exist [37].

Theorem 2.5:   Problem 2.3 can be solved by solving at most m linear programming problems. Hence, there is a polynomial-time algorithm for solving Problem 2.3.

Proof: To determine whether a given vector is separable or nonseparable, we have to determine whether it participates in a PLC of the given set of vectors that equals $\mathbf{0}$. Given a vector X_i, there are several ways of formulating the above query in terms of an LP problem, and following is one particular formulation. Define a unit vector $e_i = [0 \cdots 0\ 1\ \cdots 0]$, i.e., its ith entry is 1 and the rest of the entries are 0. Now consider the following LP:

$$\text{Maximize}\quad e_i^T \mathbf{q}\quad \text{such that}$$
$$[X_1 \cdots X_m]\,\mathbf{q} = \mathbf{0};\quad q_i \geq 0. \tag{2.9}$$

X_i is a separable vector if and only if the objective function of the above LP is bounded $(= 0)$. The reasonings can be summarized as follows: (1) the objective function is q_i, and (2) it will be unbounded if and only if there is a non-negative vector $\mathbf{q}$ satisfying $[X_1 \cdots X_m]\mathbf{q} = \mathbf{0}$ and $q_i > 0$ (which implies that X_i is nonseparable).

One can thus determine the sets of separable and nonseparable vectors by solving at most m LP problems.   □

We next show that Problem 2.4 is NP-complete. In order to prove the NP-completeness result, we reduce the following NP-complete problem to Problem 2.4.

Problem 2.5:   *Feedback Arc Set Problem* [94]
Given a directed graph $G = (V,\ E)$ (where V is the set of nodes and E is the set of directed edges), determine the minimum number of edges that needs to be removed from E so that the resultant subgraph is *acyclic*.   □

Theorem 2.6:   Problem 2.4 is NP-complete.

Proof: First, let us observe that Problem 2.4 is in the class NP. This is because if a trial solution is given for the corresponding decision problem, then using a polynomial-time algorithm for linear programming [94], one can verify whether the given subset is linearly separable or not. The next step in showing NP-completeness is to show that a known NP-complete problem (the Feedback Arc Set Problem for our case) can be reduced in polynomial time to Problem 2.4.

Before we present the reduction, let us review some basic concepts in the scheduling of directed graphs. Given a directed graph $G = (V, E)$, a scheduling is an indexing of the vertices of V such that if there is a directed edge $v_i \leftarrow v_j$ in E, then the schedule assigned to v_i, say S_{v_i}, should be greater than the schedule assigned to v_j, i.e., $S_{v_i} - S_{v_j} \geq 1$. Let $S^T = [S_{v_1}\ S_{v_2} \cdots S_{v_{|V|}}]$ be the scheduling vector (where S_{v_i} represents the schedule for node v_i). If one writes the constraints that the schedules must satisfy for each edge in E, then one obtains the following matrix inequality:

$$S^T \mathbf{C} \geq [1\ 1 \cdots 1], \qquad (2.10)$$

where $\mathbf{C}$ is referred to as the *connection matrix* and has the following properties: (1) it is of dimension $|V| \times |E|$, i.e., it has one row for every node and one column for every edge in G, and (2) the entries of the ith column, defined by the ith edge $e_i = (v_k,\ v_l) \in E$, are defined as follows: it has a -1 entry at the kth row (i.e., the row corresponding to node v_k, where e_i originates), it has a $+1$ entry at the lth row (i.e., the row corresponding to node v_l, where e_i terminates), and the rest of the entries are set to 0.

It is easy to show that there is a valid schedule for a given graph G, if and only if G is acyclic. In other words, there is a solution to Eqn. (2.10) if and only if the underlying graph is acyclic. Hence, the problem of determining the minimum number of edges to be deleted so that the resultant graph is acyclic is equivalent to the problem of determining the minimum number of edges to be deleted so that the resultant subgraph admits a valid schedule.

Given the above discussions, a polynomial-time reduction from the Feedback Arc Set problem to Problem 2.4 follows rather directly. Given a directed graph $G(V, E)$, first determine its connection matrix $\mathbf{C}$; this can be done in linear time in $|V|$ and $|E|$. Assign the columns of $\mathbf{C}$ as the input vectors to Problem 2.4, i.e., set the vector X_i as the ith column of $\mathbf{C}$.

Now, determining whether there is a weight vector $\mathbf{w}$ that separates $X_1, \ldots, X_m$ is equivalent to determining whether G is acyclic. This is because if such a $\mathbf{w}$ exists, then it must satisfy (see Eqn. (2.4))

$$\mathbf{w}^T [X_1 \; X_2 \; \cdots \; X_m] \;\geq\; [1 \; 1 \; \cdots \; 1].$$

Since X_i's are the columns of the connection matrix C, $S = \mathbf{w}$ would satisfy Eqn. (2.10) and, hence, can be considered as a schedule for G. On the other hand, a schedule exists for a graph G if and only if it is acyclic.

Therefore, if the graph G is cyclic, then the set of vectors $X_1, \ldots, X_m$ is linearly nonseparable. However, if a subset of $\{X_1, \ldots, X_m\}$ is linearly separable, then the corresponding subgraph (defined by deleting those edges in G that do not appear in the subset) will be acyclic. Hence, the problem of determining a linearly separable subset of $\{X_1, \ldots, X_m\}$ of maximum cardinality corresponds directly to the problem of deleting the minimum number of edges in G so that the resultant subgraph of G is acyclic.   □

Remark 2.6:   Geometrically, Theorem 2.6 implies that given a set of vectors, the problem of determining a hyperplane passing through the origin that assigns a maximum number of the input vectors to one side of it is NP-complete. Exercise 2.7 shows that the corresponding problem in which the hyperplane is not constrained to pass through the origin is also NP-complete.
□

2.5.1 A Heuristic Algorithm for Solving Problem 2.4

As shown in Theorem 2.6, Problem 2.4 is inherently harder than Problem 2.3. However, based on the properties of separable and nonseparable vectors, one can give a heuristic algorithm for solving Problem 2.4, as described next.

Input: A set of vectors: $S = \{X_1, X_2, \ldots, X_m\}$.

Output: A linearly separable set of vectors: $V = \{X_{i_1}, X_{i_2}, \ldots, X_{i_k}\} \subseteq S$.

In the following algorithm, let ϕ denote the empty set.

Algorithm:

Let $S = \{X_1, X_2, \ldots, X_m\}$, and $V = \phi$

While $S \neq \phi$ **Do**

Begin

Decompose S into the set of separable vectors, say S_1, and the set of non-separable vector, say S_2;

Set $V = V \bigcup S_1$;

If $S_2 \neq \phi$, then randomly pick a vector $X_k \in S_2$;

Set $S = S_2 - \{X_k\}$

End

At each step of the above algorithm, one can use a polynomial-time algorithm (or any other learning algorithm that can solve Problem 2.3) to determine the sets S_1 (the set of separable vectors) and S_2 (the set of non-separable vectors). Since the vectors in S_1 are not responsible for linear nonseparability, one can append it directly to the desired output set V; it is left as an exercise (Exercise 2.6) to show that the output set V of the algorithm is linearly separable. On the other hand, since the vectors in S_2 are responsible for linear nonseparability, one needs to delete at least one vector to make the rest of the vectors linearly separable. The analysis in Section 2.4 thus enables us to make an 'intelligent choice' of vectors that need to be considered for deletion. In particular, only the vectors in S_2 need to be considered for deletion, and all the vectors in S_1 can be directly appended to the output set. In the above algorithm, a randomly chosen vector (X_k) is deleted from S_2 and the reduced set $(S_2 - \{X_k\})$ is again checked for linear nonseparability.

By varying the choice of vectors being deleted at every step in the above algorithm, one can obtain approximate solutions to Problem 2.4. In fact, one can show (Exercise 2.6) that in the above algorithm, there always exists a choice of vectors that can be deleted (i.e., X_k's) so that the resultant output set V is a linearly separable subset of $\{X_1, X_2, \ldots, X_m\}$ of maximum cardinality.

The above algorithm can be further improved in several ways. For example, once a linearly separable set V is obtained, one can try to increase its size by checking whether any of the deleted vectors can be added to V without making it linearly nonseparable. Since the size of V is determined by the choice of vectors being deleted at each step (as implied in Exercise 2.6), one may be able to increase its size by adding some of the vectors that were deleted during the execution of the algorithm.

Remark 2.7:   The heuristic algorithm in this section shows how the analysis in Section 2.4 might be useful even if all the vectors in a given set

$S = \{X_1, X_2, \ldots, X_m\}$ are nonseparable. In such a case, during the first pass of the algorithm, S_1 will be empty; however, as vectors are deleted from S, the set of separable vectors (S_1) in subsequent passes will become nonempty. $\qquad\qquad\qquad\qquad\qquad\qquad\qquad\qquad\qquad\qquad\qquad\qquad\qquad$ $\square$

2.6 Analysis of the Perceptron Learning Algorithm for Linearly Nonseparable Sets

The results in this section show that the perceptron learning algorithm can indeed be used to learn the set of separable vectors and identify the set of nonseparable vectors in a finite number of steps. Hence, when viewed in the context of Problem 2.3, the perceptron learning algorithm can be as effective in learning linearly nonseparable patterns as it is in learning separable ones. In Section 2.6.1, a dual problem based on the null-space of the input training set is introduced, and it is shown that the power of the perceptron learning algorithm can be enhanced if one simultaneously runs an independent learning algorithm on the dual problem. Section 2.6.2 indicates how the perceptron learning algorithm can be used to determine large linearly separable subsets of any given nonseparable training set.

The perceptron learning algorithm (as stated in Section 2.3) does not converge if the input vectors are linearly nonseparable. The following theorem, however, states that even if the algorithm iterates indefinitely, the length of the weight vector $\mathbf{w}_l$ will remain bounded.

Theorem 2.7: If the perceptron learning algorithm is applied to a linearly nonseparable set of vectors $S = \{X_1, \ldots, X_m\}$, then the length of the weight vector $\mathbf{w}_l$ remains bounded, i.e., there exists a constant N_S such that $||\mathbf{w}_l|| \leq N_S$ for all $l \geq 0$. $\qquad\qquad\qquad\qquad\qquad\qquad\qquad$ $\square$

The above theorem is referred to as the *perceptron cycling theorem*, and a proof for it can be found in [81].

In the perceptron algorithm, the step ([ADD]) where a vector X_i is added to the current value of $\mathbf{w}_l$ to generate the next value, $\mathbf{w}_{l+1}$, is referred to as an *update step*. The algorithm is said to have *converged* with respect to a vector X_i after the kth step if $X_i^T \mathbf{w}_l > 0$ for all $l > k$. That is, after a finite number of updates $(= k)$, X_i will not be used to update the weight vector $\mathbf{w}_l$. In such a case, we also say that *the algorithm has learned* the vector X_i.

Let $\mathbf{w}_l$ be the value of the weight vector after the lth update. The total number of updates, l, can be written as $l = l_1 + l_2$, where l_1 is the number of updates using only the separable vectors and l_2 is the number of updates using only the nonseparable vectors. The following theorem proves that l_1 is finite, which implies that after a finite number of steps, the separable vectors are never used for updating the weight vector $\mathbf{w}_l$. Equivalently, we can say that the perceptron learning algorithm learns all the separable vectors after a finite number of updates.

Theorem 2.8:   Given a set of vectors $S = \{X_1, \ldots, X_m\}$ $(X_i \in \mathcal{R}^d)$, let l_1 be the total number of updates of the weight vector $\mathbf{w}_l$ using only the separable vectors (of S) in the perceptron learning algorithm. Then, l_1 is finite.

Proof: Without loss of generality, assume that the first k vectors, $X_1, \ldots, X_k$, are the separable vectors, and the rest, $X_{k+1}, \ldots, X_m$, are the nonseparable vectors. We can always write the weight vector $\mathbf{w}_l$ (after l updates) as

$$\mathbf{w}_l = \sum_{i=1}^{k} \alpha_i X_i + \sum_{i=k+1}^{m} \alpha_i X_i,$$

where integers $\alpha_i \geq 0$ represent the number of times the vector X_i has been used in updating the weight vector. Let $\sum_{i=1}^{k} \alpha_i = l_1$ and $\sum_{i=k+1}^{m} \alpha_i = l_2$. Thus, l_1 is the total number of updates using only separable vectors and l_2 is the total number of updates using only nonseparable vectors.

We know from Theorem 2.4 that there exists a weight vector $\mathbf{w}_0$ such that $\mathbf{w}_0^T X_i \geq 1$ for $i = 1, \ldots, k$, and $\mathbf{w}_0^T X_j = 0$ for $j = k+1, \ldots, m$. Hence,

$$\mathbf{w}_0^T \mathbf{w}_l = \sum_{i=1}^{k} \alpha_i \mathbf{w}_0^T X_i + \sum_{j=k+1}^{m} \alpha_j \mathbf{w}_0^T X_j = \sum_{i=1}^{k} \alpha_i \mathbf{w}_0^T X_i \geq \sum_{i=1}^{k} \alpha_i = l_1.$$

If $||\mathbf{w}_0|| \leq L$, then applying Cauchy-Schwarz inequality, we have

$$l_1 \leq ||\mathbf{w}_0|| \, ||\mathbf{w}_l|| \leq L ||\mathbf{w}_l||.$$

However, Theorem 2.7 implies that $||\mathbf{w}_l|| \leq N_S$. Hence, we obtain

$$l_1 \leq L N_S .$$

Thus, the perceptron learning algorithm learns the separable vectors in a finite number of updates.   □

As shown in Theorem 2.9, for certain choices of the sequencing of the input vectors, the perceptron learning algorithm never converges with respect to the nonseparable inputs. Hence, the nonseparable inputs can be distinguished by the fact that for every nonseparable input X_i, the inner-product $\mathbf{w}_l^T X_i$ will always become negative during some updating after a finite number of steps. Thus, the perceptron learning algorithm can be used to learn the structure of linearly nonseparable training sets in the following manner:

> Apply the perceptron learning algorithm to the given set of vectors and record the vectors that are being used for updating the weight vector. As the algorithm keeps iterating, separate the vectors into two sets: (1) a set of vectors that are no longer being used for updating the weight vector (call it the set B), and (2) a set of vectors that are being recurringly used to update the weight vector (call it the set C).

Theorem 2.8 shows that the set B will contain all the separable vectors after a finite number of updates. Moreover, the property of a nonseparable vector (as discussed next) implies that as the algorithm keeps iterating, it must end up in the set C. Thus, there exists a finite number of steps such that if the perceptron learning algorithm is stopped, then sets B and C will respectively correspond to the sets of separable and nonseparable vectors.

Definition 2.4:   Given a set of vectors $S = \{X_1, X_2, \cdots, X_m\}$, $X_j \in \mathcal{R}^d$, a subset $S_i \subseteq S$ is a *minimal linearly nonseparable set* for any vector $X_i \in S$ if: (1) $X_i \in S_i$, (2) S_i is linearly nonseparable, and (3) the set $S_i - \{X_i\}$ is linearly separable.   □

Example 2.10:   In Example 2.4, $S = \{X_1, X_2, X_3, X_4\}$ is a linearly nonseparable set, and every vector X_i is nonseparable. One can easily verify that if X_1 is removed from the set, then the reduced set $S - \{X_1\}$ is a linearly separable set. Hence, the minimal linearly nonseparable set for X_1 is the whole set S. One can similarly verify that for any vector $X_i \in S$, the minimal linearly nonseparable set is the whole set S.   □

Lemma 2.5:   If the perceptron learning algorithm is applied to a set S_i, which is a minimal linearly nonseparable set for a given vector X_i, then the algorithm will never converge with respect to X_i, i.e., X_i is used to update the weight vector in the algorithm infinitely often.

Proof: By definition, $S_i - \{X_i\}$ is a linearly separable set. Hence, if X_i is not used in updating the weight vector, then the perceptron learning algorithm works on a linearly separable set. From Theorem 2.2 [81], we know that the algorithm is going to converge in a finite number of steps over the reduced set $S_i - \{X_i\}$. Thus, after a finite number of steps, the vector X_i will again be used to update the weight vector in the algorithm. $\square$

We next prove the existence of a minimal linearly nonseparable set for every nonseparable vector X_i in S.

Lemma 2.6:   Given a linearly nonseparable set $S = \{X_1, \cdots, X_m\}$, and a nonseparable vector $X_i \in S$, there always exists a subset $S_i \subseteq S$, which is a minimal linearly nonseparable set for X_i.

Proof: Since X_i is a nonseparable vector, there exist vectors $S' = \{X_{i_1}, \cdots, X_{i_l}\}$ and positive numbers $q_{i_j} > 0$ such that $\sum_{j=1}^{l} q_{i_j} X_{i_j} + X_i = 0$. If the set S' is linearly separable, then the set $S_i = S' \cup \{X_i\}$ is a minimal linearly nonseparable set for X_i.

If S' is linearly nonseparable, then we first show that one always eliminates at least one vector from the set S'. That is, one can always choose a nonempty set of vectors $U_i \subset S'$ such that

$$\sum_{X_{i_j} \in S'} q_{i_j} X_{i_j} = \sum_{X_{i_j} \in S' - U_i} q'_{i_j} X_{i_j},$$

and $q'_{i_j} > 0$. Thus,

$$\sum_{X_{i_j} \in S' - U_i} q'_{i_j} X_{i_j} + X_i = 0,$$

where $q'_{i_j} > 0$. If the set $S'' = S' - U_i$ is linearly separable, then clearly the set $S_i = S'' \cup \{X_i\}$ is a minimal linearly nonseparable set for X_i. If this condition is not met, then one can continue this elimination process on the set $S' - U_i$ until the reduced set becomes linearly separable.

Since S' is assumed to be linearly nonseparable, it must have a PLC of a subset of its vectors that equals $\mathbf{0}$. That is, there exist $p_{i_j} \geq 0$ (with at least one $p_{i_j} > 0$) such that

$$\sum_{j=1}^{l} p_{i_j} X_{i_j} = 0.$$

Considering only those $p_{i_j} > 0$, choose a vector X_{i_k} such that

$$\frac{q_{i_k}}{p_{i_k}} = \min\{\frac{q_{i_j}}{p_{i_j}} | \ 1 \leq j \leq l, \text{ and } p_{i_j} > 0\}. \tag{2.11}$$

Now using the relationship $X_{i_k} = - \sum_{j=1;j\neq k}^{l} \frac{p_{i_j}}{p_{i_k}} X_{i_j}$, we get that

$$\sum_{j=1}^{l} q_{i_j} X_{i_j} = \sum_{j=1;j\neq k}^{l} q_{i_j}(1 - \frac{q_{i_k}}{p_{i_k}}\frac{p_{i_j}}{q_{i_j}}) X_{i_j}.$$

If $q_{i_j}' = q_{i_j}(1 - \frac{q_{i_k}}{p_{i_k}}\frac{p_{i_j}}{q_{i_j}})$, then it follows from Eqn. (2.11) that $q_{i_j}' \geq 0$. Let $U_i = \{X_{i_k}\} \cup \{X_{i_l} | \ q_{i_j}' = 0\}$ (i.e., the set of vectors for which $q_{i_j}' = 0$) then

$$\sum_{j=1}^{l} q_{i_j} X_{i_j} = \sum_{X_{i_j} \in S'-U_i} q_{i_j}' X_{i_j},$$

where $q_{i_j}' > 0$. Hence, for Lemma 2.6, we have shown that if S' is linearly nonseparable, then one can always eliminate a nonempty subset of vectors, U_i, such that

$$\sum_{X_{i_j} \in S'-U_i} q_{i_j}' X_{i_j} + X_i = 0,$$

where $q_{i_j}' > 0$. □

Theorem 2.9:   Given a nonseparable set S, there exists a sequencing of the input vectors such that the perceptron learning algorithm will not converge with respect to any of the nonseparable vectors in S.

Proof: If the perceptron algorithm is run such that it restricts itself for a finite number of steps to a minimal linearly nonseparable set for every nonseparable vector, then it follows from Lemma 2.5 and the Perceptron Cycling Theorem that every nonseparable vector will be used infinitely often for updates. One way of making sure that the algorithm runs on at least one minimal linearly nonseparable set for every nonseparable vector is to restrict the algorithm to every subset of the set of input vectors for a finite number of steps. □

Example 2.11: Consider the set of input vectors studied in Example 2.5, where the input vectors are given as

$$X_1 = \begin{bmatrix} 1 \\ 1 \end{bmatrix}; \quad X_2 = \begin{bmatrix} 1 \\ -1 \end{bmatrix}; \quad X_3 = \begin{bmatrix} -1 \\ 1 \end{bmatrix}.$$

Without loss of generality, let us assume that the perceptron learning algorithm chooses the vectors in the cyclic order: $X_1, X_2, X_3, X_1, \ldots$. One can verify that the weight vector converges with respect to X_1 in one step. That is, $\mathbf{w}_l^T X_1 > 0$ for all $l \geq 1$. As the algorithm keeps iterating indefinitely, only vectors X_2 and X_3 are used alternately for updating the weight vector. In fact, the weight vector $\mathbf{w}_l$ would start repeating its value after three iterations. One can conclude then from Theorem 2.8 that X_1 is a separable vector, and X_2, X_3 are nonseparable vectors. Thus, the perceptron algorithm correctly learns the structure of the training set in only a few number of iterations.

One can verify that a similar behavior can be observed for the set of vectors introduced in Example 2.8. In fact, if the vectors are chosen in a cyclic order, then after the first pass (i.e., after all the vectors have been checked once), the perceptron algorithm will learn the set of separable vectors (X_1, X_2, X_3, X_4), and from then on, only X_5 and X_6 will be used alternately for updating the weight vector indicating that these are the nonseparable vectors. □

Example 2.12: If the perceptron learning algorithm is applied to the four vectors in Example 2.4, then the algorithm will not converge with respect to any of the vectors. In fact, if the vectors are input in a cyclic order (i.e., $X_1, X_2, X_3, X_4, X_1, \ldots$) during the execution of the algorithm, then the weight vector $\mathbf{w}_l$ will become $\mathbf{0}$ every four iterations, thereby indicating that every vector will be used for updates infinitely often. Thus, Theorem 2.8 implies that none of the vectors is a separable vector. Hence, within a few iterations, the perceptron learning algorithm would reach the correct conclusion that all the input vectors are linearly nonseparable.   □

2.6.1   A Dual Learning Problem

We present a dual problem that can enhance the perceptron learning algorithm in classifying linearly nonseparable training sets. The formulation and properties of the dual problem is independent of any particular learning algorithm and, hence, the related concepts and results can perhaps be used for enhancing the performance of learning algorithms other than the perceptron learning algorithm.

Let $D = [X_1\ X_2 \cdots X_k\ X_{k+1} \cdots X_m]$ be the matrix formed by the set of input vectors, and without loss of generality assume that D has full row-rank $(= d)$. Then we can denote the right null-space of D as follows:

$$\mathbf{N}_D^T = [Y_1\ \cdots\ Y_k\ Y_{k+1}\ \cdots\ Y_m], \tag{2.12}$$

where $Y_i \in \mathcal{R}^{m-d}$, and $D\mathbf{N}_D = \mathbf{0}$. In other words, the columns of $\mathbf{N}_D$ form a basis for the right null-space of D. Note that the null-space of a matrix D can be computed relatively easily using well-known techniques such as the Gaussian elimination algorithm and other iterative algorithms.

Recall that since $\{X_1, \ldots, X_k\}$ is the set of separable vectors and $\{X_{k+1}, \ldots, X_m\}$ is the set of nonseparable vectors, there exists a vector $\mathbf{q}_0 \geq 0$ such that $D\mathbf{q}_0 = 0$, $q_{0i} = 0$, $1 \leq i \leq k$, and $q_{0i} \geq 1$, $k + 1 \leq i \leq m$. Also, since $\mathbf{q}_0$ is in the right null-space of D, there exists a vector $\alpha \in \mathcal{R}^{m-d}$ such that $\mathbf{N}_D\alpha = \mathbf{q}_0$; equivalently,

$$\alpha^T \mathbf{N}_D^T = \alpha^T\ [Y_1 \cdots Y_k\ Y_{k+1} \cdots Y_m]\ = \mathbf{q}_0^T. \tag{2.13}$$

We can prove the following properties about the null-space vectors Y_i.

Theorem 2.10:   For the set $A = \{Y_1, Y_2, \ldots, Y_k, Y_{k+1}, \ldots Y_m\}$, $Y_i \in \mathcal{R}^{m-d}$, as defined in Eqn. (2.12), $\{Y_1, Y_2, \ldots, Y_k\}$ is the set of nonseparable vectors and $\{Y_{k+1}, Y_{k+2}, \ldots, Y_m\}$ is the set of separable vectors.

Proof: We first show that the set $A' = \{Y_{k+1}, Y_{k+2}, \ldots, Y_m\}$ is a subset of the set of separable vectors of A. Then, we argue that the set of separable vectors cannot be any larger than A'. In order to show that the vectors $Y_{k+1}, Y_{k+2}, \ldots, Y_m$ are separable, it suffices to show that the following LP has a bounded objective function (in fact, $= 0$) (see the definition of separable vectors and the LP formalism introduced in Section 2.4):

$$\text{Maximize} \quad [0 \ \cdots \ 0 \ 1 \ 1 \cdots 1]\mathbf{p} \quad \text{such that}$$

$$[Y_1 \ Y_2 \cdots Y_k \ Y_{k+1} \cdots Y_m] \begin{bmatrix} p_1 \\ \vdots \\ p_k \\ p_{k+1} \\ \vdots \\ p_m \end{bmatrix} = \mathbf{0}; \quad p_i \geq 0, \tag{2.14}$$

where the first k entries in $[0 \ \cdots \ 0 \ 1 \ 1 \cdots 1]^T$ equal 0 and rest equal 1. From the duality theorem, we know that the LP in Eqn. (2.14) will have a bounded objective function if and only if its dual has a feasible solution:

$$\text{Minimize} \quad \mathbf{0}^T \mathbf{b} \quad \text{such that}$$
$$\mathbf{b}^T[Y_1 \ Y_2 \cdots Y_k \ Y_{k+1} \cdots Y_m] \geq [0 \ \cdots 0 \ 1 \ 1 \cdots 1]. \tag{2.15}$$

However, a feasible solution to the LP in Eqn. 2.15 is already given in Eqn. (2.13), i.e., set $\mathbf{b} = \alpha$. Then,

$$\mathbf{b}^T[Y_1 \ Y_2 \cdots Y_k \ Y_{k+1} \cdots Y_m] = \mathbf{q}_0 \geq [0 \ \cdots 0 \ 1 \ 1 \cdots 1].$$

The proof of the theorem can be completed by showing that any other vector not in A' is nonseparable. One way of proving this would be by contradiction. For example, if we assume that Y_i is also separable for some i, $1 \leq i \leq k$, then following arguments very similar to the ones previously used, one can show that the corresponding vector X_i is a nonseparable vector, which contradicts the assumption that X_i is a separable vector. $\square$

Theorem 2.10 shows that if a vector X_i is separable in any given set $\{X_1, \ldots, X_m\}$, then the corresponding vector, Y_i (as defined in Eqn. 2.12), is nonseparable in the set $\{Y_1, \ldots, Y_m\}$, and vice-versa.

Given the set $\{X_1, \ldots, X_m\}$, let us define the *dual learning problem* as the learning problem for the null-space vectors $\{Y_1, \ldots, Y_m\}$ (as defined in Eqn. (2.12)).

It follows from Theorem 2.10 that the structure of the set $\{X_1, \ldots, X_m\}$ can be directly obtained by learning the structure of the set $\{Y_1, \ldots, Y_m\}$.

That is, if $\{Y_{i_1}, \ldots Y_{i_k}\}$ is the set of nonseparable vectors in $\{Y_1, \ldots, Y_m\}$, then the set of separable vectors in $\{X_1, \ldots, X_m\}$ is the corresponding set $\{X_{i_1}, \ldots X_{i_k}\}$. Moreover, the set of nonseparable vectors comprises the rest of the vectors.

A special case where the dual learning problem will obviously be useful is when all the vectors in $\{X_1, \ldots, X_m\}$ are nonseparable. In such a case, the perceptron learning algorithm applied to $\{X_1, \ldots, X_m\}$ will not converge with respect to any of the vectors. However, the set $\{Y_1, \ldots, Y_m\}$ for the dual problem is linearly separable (as a consequence of Theorem 2.10), and the dual learning algorithm will converge for all the vectors. Hence, the structure of the learning problem can be learned without any errors by the outcome of the dual problem; see Example 2.14.

Remark 2.8:   In general, enhanced performance can be achieved by simultaneously applying the perceptron learning algorithm on the original vectors ($\{X_1, \ldots, X_m\}$) and on the null-space vectors ($\{Y_1, \ldots, Y_m\}$). One can keep iterating the two algorithms until their predictions match. The set of separable vectors predicted by the algorithm operating on the original vectors should match the set of nonseparable vectors predicted by the dual learning problem. That is, if X_j is predicted as a separable vector in the original learning problem, then the corresponding vector Y_j should be predicted as a nonseparable vector in the dual learning problem. Similarly, the set of separable vectors predicted by the dual learning problem must match the set of nonseparable vectors predicted by the learning algorithm applied to the original vectors.   □

Example 2.13:   Consider the set of vectors, $S = \{X_1, X_2, X_3\}$, in Example 2.5. The null-space matrix $\mathbf{N}_D$ is of dimension 1 $(= m - d)$ and is given by

$$N_D = \begin{bmatrix} 0 \\ 1 \\ 1 \end{bmatrix}.$$

Hence, $Y_1 = [0]$, $Y_2 = [1]$, $Y_3 = [1]$. Since $Y_1 = [0]$, $\mathbf{w}^T Y_1 = 0$ for every choice of a weight vector. Thus, one can conclude from inspection that Y_1 is a nonseparable vector. In fact, if the perceptron learning algorithm is run on the Y_i's, then one can easily observe that the weight vector $\mathbf{w}_l = [1]$ for all $l \geq 2$ (assuming that $\mathbf{w}_0 = 0$). Hence, within at most two iterations, one can determine that the learning algorithm has converged with respect to Y_2, Y_3 and will never converge with respect to Y_0. Theorem 2.8 would then imply that $\{Y_2, Y_3\}$ is the set of separable vectors, and Y_1 is a nonseparable vector. Hence, it follows from Theorem 2.10 that X_2, X_3 are the nonseparable vectors, and X_1 is the separable vector in S, which is proved in Example 2.5.

This example thus shows that applying the perceptron learning algorithm on the dual problem can sometimes lead to efficient means for determining the sets of separable and nonseparable vectors. $\square$

Example 2.14:   Consider the set of vectors $S = \{X_1, X_2, X_3, X_4\}$ in Example 2.4. The dimension of the null-space is again 1, and the corresponding matrix is given by

$$N_D = \begin{bmatrix} 1 \\ 1 \\ 1 \\ 1 \end{bmatrix}.$$

Hence, $Y_1 = Y_2 = Y_3 = Y_4 = [1]$, and the perceptron learning algorithm would converge in a single iteration, e.g., $\mathbf{w} = [1]$ is a solution. Hence, $\{Y_1, Y_2, Y_3, Y_4\}$ forms a linearly separable set. Theorem 2.10 then implies that X_1, X_2, X_3, X_4 are all nonseparable vectors.

This example again illustrates that applying the perceptron learning algorithm on the null-space vectors can lead to efficient means of identifying the structure of a linearly nonseparable set. $\square$

2.6.2   Determining Linearly Separable Subsets

Since the perceptron learning algorithm can learn the sets of separable and nonseparable vectors, it can be applied to obtain approximate solution to Problem 2.4. In other words, one can use the perceptron algorithm to learn large linearly separable subsets of any given nonseparable training set by following the heuristic algorithm outlined in Section 2.5.

Example 2.15:   Consider the set of vectors in Example 2.4. If the algorithm in Section 2.5.1 is applied to the set, then in the first pass, $S_1 = \phi$ and $S_2 = \{X_1, X_2, X_3, X_4\}$. Let X_1 be the vector deleted from S_2. In the next pass of the algorithm, one can verify that one will have $S_1 = \{X_2, X_3, X_4\}$

and $S_2 = \phi$. Thus, the output of the algorithm will be $V = \{X_2, X_3, X_4\}$, which is indeed a linearly separable subset of maximum cardinality.

One can also apply the algorithm to the linearly nonseparable set in Example 2.8. In the first pass, one will have $S_1 = \{X_1, X_2, X_3, X_4\} = V$ (set of separable vectors), and $S_2 = \{X_5, X_6\}$ (set of nonseparable vectors). Let X_5 be the vector deleted from S_2. Then in the second pass one will have $S_1 = \{X_6\}$ and $S_2 = \phi$. Thus, the output linearly separable set becomes $V = \{X_1, X_2, X_3, X_4, X_6\}$, which is again a linearly separable subset of maximum cardinality. $\qquad\Box$

2.7 Capacity of Linear Threshold Elements

In this section, we continue our exploration of the computational power of LTEs. Consider n vectors, $X_1, X_2, \ldots, X_n$, in $\mathcal{R}^d$. The number of binary functions that can be defined on this set of points is 2^n. The first question we address is: How many of these 2^n binary functions can be computed by a single LTE? Based on the answer to this query, we address a related question: How many weights must a feedforward network of LTEs have so that it can compute *any* of these functions? Given a set of input vectors, a network that can compute (by varying the weights) all of the 2^n possible binary functions is called a *universal network*.

2.7.1 Number of Binary Functions Implementable by an LTE

We begin by establishing an upper bound on the number of linearly separable functions over n vectors in $\mathcal{R}^d$. Let $B(n, d)$ be the total number of binary functions over the set of given points that can be computed by a single LTE. In the following theorem, without loss of generality, we consider the case where threshold $t = 0$; see Section 2.3 for justification.

Theorem 2.11:   The number of homogeneously linearly separable functions over n vectors in $\mathcal{R}^d$ is

$$B(n, d) \leq C(n, d) = 2 \sum_{i=1}^{d-1} \binom{n-1}{i},$$

where $C(n, d)$ is the maximum number of regions that $\mathcal{R}^d$ can be partitioned into by n hyperplanes (passing through the origin).

Proof: Consider the space of weight vectors $\mathbf{w}$ in $\mathcal{R}^d$. For every vector X_i, plot a hyperplane $\mathbf{w}^T X_i = 0$. Each hyperplane divides the space into two regions – on one side $\mathbf{w}^T X_i < 0$ and on the other side $\mathbf{w}^T X_i > 0$. The n hyperplanes (one for each vector X_i) passing through the origin partition $\mathcal{R}^d$ into a number of regions. The $\mathbf{w}$ vectors in any one of these regions yield the same linearly separable function over the n vectors. Moreover, $\mathbf{w}$ vectors in different regions yield different linearly separable functions.

Given any n vectors, $X_1, X_2, \ldots, X_n$ in $\mathcal{R}^d$, let $C(n, d)$ be the maximum number of regions that $\mathcal{R}^d$ can be divided into by n hyperplanes, $\mathbf{w}^T X_i = 0$, for $i = 1, \ldots, n$. It follows that $C(n, d)$ serves as an upper bound for $B(n, d)$. We next determine $C(n, d)$ by induction.

The maximum number of regions that $\mathcal{R}^d$ can be partitioned into by any set of $n - 1$ vectors is $C(n - 1, d)$. We next add an nth hyperplane, and the objective is to add this n^{th} hyperplane so that the number of new regions is maximized.

The nth hyperplane intersects the other $n-1$ hyperplanes in no more than $n - 1$ *hyperlines*. These hyperlines all pass through the origin and partition the nth hyperplane to at most $C(n - 1, d - 1)$ regions. Each one of these regions of the hyperplane can partition at most one of the regions already defined by the first $(n - 1)$ hyperplanes in $\mathcal{R}^d$ into two regions. Therefore,

$$C(n, d) \; = \; C(n - 1, d) \; + \; C(n - 1, d - 1).$$

The initial condition is that for $n = 2$ (i.e., with only two vectors) $C(2, d) = 4$. One can verify that the following expression for $C(n, d)$ satisfies the above recursion.

$$C(n, d) \; = \; 2 \sum_{i=0}^{d-1} \binom{n - 1}{i}. \qquad \square$$

Definition 2.5:   A set of n points in $\mathcal{R}^d$ is said to be in *general position* if every subset of d or fewer points forms a linearly independent set. $\square$

The next corollary states some of the special cases of Theorem 2.11.

Corollary 2.1:

1. If the n vectors in Theorem 2.11 are in general position, then the upper
   bound established in Theorem 2.11 is tight, i.e., $B(n, d) = C(n, d)$.

2. Let $d = n$ (i.e., there are n n-dimensional vectors), then

$$C(n, d) = 2 \sum_{i=0}^{n-1} \binom{n-1}{i} = 2^n.$$

3. If the n points are chosen to be in general position, then the probability
   that a randomly chosen function is linearly separable is

$$P(n, d) = \frac{C(n, d)}{2^n}.$$

   Hence, $P(n, d) = 1$ if $n \leq d$, $P(n, d) = 1/2$ if $n = 2d$, and $P(n, d) < 1/2$
   if $n > 2d$.

4. If the input vectors are restricted to be binary, then the number of
   input vectors for a d-input LTE is given as $n = 2^d$. Thus, the total
   number of d-variable Boolean linear threshold functions is bounded
   above by

$$
\begin{aligned}
B(n, d+1) \leq C(n, d+1) &= 2 \sum_{i=0}^{d} \binom{2^d - 1}{i} \\
&\leq 2(d+1) \binom{2^d - 1}{d} \\
&\leq 2^{d^2}.
\end{aligned}
$$
□

Remark 2.9: Corollary 2.1 shows that if n points are chosen in general
position in $\mathcal{R}^n$, then all binary-valued functions over these points are linearly
separable. It also shows that $P(2d, d) = 1/2$; hence, if the number of input
vectors is $\leq 2d$, then with probability $\geq 1/2$, an LTE can compute any
randomly chosen binary function over the set of inputs. Moreover, it can be
shown that $P(n, d)$ decreases sharply for values of $n > 2d$. Because of this
phenomenon, we define the *capacity* of a d-input LTE to be $2d$.   □

2.7.2   A Lower Bound on the Number of Linearly Separable Functions

Corollary 2.1 shows that the maximum number of linearly separable Boolean functions of n variables is 2^{n^2}. In this section, we derive an effective lower bound by proving that the number of linearly separable functions of n variables is at least $2^{\frac{n(n-1)}{2}+16}$. In Section 2.7.3, using deeper mathematical tools, we strengthen the lower bound to $2^{n^2-O(n)}$. Since the total number of Boolean functions of n variables is 2^{2^n}, the fraction of Boolean functions that is linearly separable is therefore very small. In other words, most Boolean functions are linearly nonseparable.

In this section, we assume the following definition of a linear threshold function $f(x_1,\ldots,x_n) = \mathrm{sgn}(\sum\limits_{i=1}^{n} w_i x_i - t)$, which is equivalent to the one stated in Eqn. (2.1):

$$f = \begin{cases} 1 & \text{if } \sum\limits_{k=1}^{n} w_k x_k \ge t, \\ 0 & \text{if } \sum\limits_{k=1}^{n} w_k x_k \le t-1. \end{cases} \qquad (2.16)$$

Definition 2.6:   A Boolean function f is *positive* if it can be expressed in a Sum of Product (SOP) form (i.e., in a disjunctive form) without using any negated variable as a literal.

The *dual* of a Boolean function $f(x_1,\ldots,x_n)$ is defined as

$$f^d(x_1,\ldots,x_n) = \overline{f}(\overline{x_1},\ldots,\overline{x_n}).$$

If $f(x_1,\ldots,x_n) = f^d(x_1,\ldots,x_n)$, then f is called *self-dual*.   □

It is easy to see that a threshold function with positive weights is a positive Boolean function. The following lemma establishes a key property of self-dual linear threshold functions.

Lemma 2.7:   A linear threshold function $f(x_1,\ldots,x_n) = \mathrm{sgn}(\sum\limits_{i=1}^{n} w_i x_i - t)$ is self-dual if

$$2t - 1 = \sum\limits_{i=1}^{n} w_i. \qquad (2.17)$$

Moreover, if a linear threshold function $f(x_1, \ldots, x_n) = (\sum_{i=1}^{n} w_i x_i - t)$ is self-dual, then one can always choose a threshold t' such that $f(x_1, \ldots, x_n) = \text{sgn}(\sum_{i=1}^{n} w_i x_i - t')$ and Eqn. (2.17) is satisfied by w_i's and t'.

Proof: If $f(x_1, \ldots, x_n) = \text{sgn}(\sum_{i=1}^{n} w_i x_i - t)$, then it follows from Eqn. (2.16) that $\overline{f}(x_1, \ldots, x_n) = \text{sgn}(-\sum_{i=1}^{n} w_i x_i - (1-t))$. Since $\overline{x_i} = 1 - x_i$, it follows from the definition of the dual of a function that

$$f^d(x_1, \ldots, x_n) = \overline{f}(\overline{x_1}, \ldots, \overline{x_n}) = \text{sgn}(\sum_{i=1}^{n} w_i x_i - (1 - t + \sum_{i=1}^{n} w_i)). \quad (2.18)$$

If $2t - 1 = \sum_{i=1}^{n} w_i$, then $t = (1 - t + \sum_{i=1}^{n} w_i))$, and $f^d(x_1, \ldots, x_n) = f(x_1, \ldots, x_n)$.

If $f(x_1, \ldots, x_n)$ is self-dual, then one can show that $f(x_1, \ldots, x_n) = \text{sgn}(\sum_{i=1}^{n} w_i x_i - t')$, where $t' = (1 + \sum_{i=1}^{n} w_i)/2$, and hence, Eqn. (2.17) is satisfied. The details are left as an exercise (Exercise 2.10). □

In the rest of this section, we assume without loss of generality that the threshold value t of any linearly separable self-dual function satisfies Eqn. (2.17).

We shall also need the results of the following lemma, showing that without loss of generality one can assume every partial sum of the weights of a linear threshold function to have a different value.

Lemma 2.8:   Given a threshold function $f(x_1, \ldots, x_n) = \text{sgn}(\sum_{i=1}^{n} w_i x_i - t)$, there exists a choice of the weights such that every partial sum is different, i.e., $\sum_{i \in S_1} w_i \neq \sum_{j \in S_2} w_j$, where $S_1, S_2 \subseteq \{1, \ldots, n\}$, and $S_1 \neq S_2$.

Proof: Suppose that there exist different partial sums that have the same value. Consider a pair $\sum_{i \in S_1} w_i = \sum_{j \in S_2} w_j$, where $S_1, S_2 \subseteq \{1, \ldots, n\}$ and $S_1 \neq S_2$. Let $\delta > 0$ be the minimum difference among different values of partial sums, as well as between any partial sum $(< t)$ and the threshold t. Choose a k such that $k \in S_1$ but $k \notin S_2$, and for some $0 < \epsilon < \delta$, assign $w_k = w_k + \epsilon$. With the modified weights, $\sum_{i \in S_1} w_i \neq \sum_{j \in S_2} w_j$; moreover, the choice of ϵ guarantees that different values of partial sums remain different and the partial sums, which are less than t, also remain less than t. Thus, the modified weights compute the same function f. One can now repeat this procedure for all pairs of partial sums that have the same value.   □

Thus, Lemmas 2.8 and 2.7 together imply that given a positive self-dual function, we can assume without loss of generality that the weights w_i's and the threshold t are such that every partial sum of the weights is different and Eqn. (2.17) is satisfied.

We shall prove a lower bound on the number of linearly separable Boolean functions, by first proving a lower bound on the number of positive, self-dual n-variable Boolean functions that are linearly separable.

Definition 2.7:   An n-variable Boolean function f is *nondegenerate* if it depends on all the n variables, i.e., for each input variable, there is an input assignment for the function such that changing the variable changes the value of f.   □

Lemma 2.9:   Let $f_i(x_1, \ldots, x_{n-1}) = \text{sgn}(\sum_{i=1}^{n-1} w_i x_i - t)$ be a nondegenerate positive $(w_i > 0)$ self-dual threshold function of $(n-1)$ variables, such that every partial sum of $w_1, \ldots, w_{n-2}$ is different and Eqn. (2.17) is satisfied. Then, for any non-zero C that satisfies $(C + w_{n-1}) > 0$ and for some partial sum $\sum_{i \in S} w_i$ of $w_1, \ldots, w_{n-2}$ (i.e, $S \subseteq \{1, \ldots, (n-2)\}$),

$$t = (C + w_{n-1}) + \sum_{i \in S} w_i, \qquad (2.19)$$

the function

$$f_{iC}(x_1, \ldots, x_n) = \text{sgn}(\sum_{i=1}^{n-2} w_i x_i + |C| x_{n-1} + (C + w_{n-1}) x_n - (t + (C + |C|)/2))$$

is a nondegenerate positive self-dual threshold function of n variables, and is different for each distinct C. Moreover, two distinct $(n-1)$-variable functions, f_i and f_j, will lead to different n-variable functions f_{iC}'s and f_{jC}'s.

Proof: Since Eqn. (2.17) is satisfied by f_i, we have

$$2t - 1 = \sum_{i=1}^{n-1} w_i. \tag{2.20}$$

Each partial sum of $w_1, \ldots, w_{n-2}$ has a different value, and therefore Eqn. (2.19) implies that there is a different C for each partial sum. Lemma 2.10 states a lower bound on how many of these 2^{n-2} C's also satisfy the condition: $(C + w_{n-1}) > 0$.

Let us first assume that $C > 0$. Then $(C + |C|)/2 = C$ and the threshold value for f_{iC} equals $(t + C)$. One can verify that f_{iC} is self-dual by showing that its weights and threshold satisfy Eqn. (2.17):

$$2(t + C) - 1 = 2t - 1 + 2C = \sum_{i=1}^{n-2} w_i + C + (C + w_{n-1}).$$

Moreover, from Eqn. (2.19) we have

$$t + C = C + (C + w_{n-1}) + \sum_{i \in S} w_i. \tag{2.21}$$

Since C (for $C > 0$) and $(C + w_{n-1})$ are the weights for the variables x_{n-1} and x_n, respectively, Eqn. (2.21) implies that $f_{iC} = 1$ when $x_n = x_{n+1} = 1$ and $x_i = 1$ for all $i \in S$ (irrespective of the other inputs).

We next show that each distinct C will generate a different f_{iC}. Let f_{iC_1} and f_{iC_2} be two functions generated respectively by C_1, which is determined by the partial sum $\sum_{i \in S_1} w_i$, and by C_2, which is determined by the partial sum $\sum_{i \in S_2} w_i$. Since each partial sum is different, assume without loss of generality that $\sum_{i \in S_1} w_i < \sum_{i \in S_2} w_i$. Consider the input assignment where $x_n = x_{n+1} = 1$, $x_i = 1$ for all $i \in S_1$, and $x_i = 0$ for all $i \in \overline{S_1}$, where $\overline{S_1} = \{1, \ldots, (n-2)\} - S_1$ (i.e., the remaining inputs are set to 0). Let us call this input assignment X. Eqn. (2.21) implies that $f_{iC_1}(X) = 1$. However,

$$
\begin{aligned}
f_{iC_2}(X) &= \operatorname{sgn}\left((C_2 + (C_2 + w_{n-1}) + \sum_{i \in S_1} w_i) - (t + C_2)\right) \\
&= \operatorname{sgn}\left(\sum_{i \in S_1} w_i - \sum_{i \in S_2} w_i\right) \quad \text{(from Eqn. (2.21))} \\
&= 0.
\end{aligned}
$$

Hence, f_{iC_1} and f_{iC_2} are different functions. Now, Eqn. (2.21) shows that f_{iC} depends on both x_n and x_{n-1}. Moreover, $f_{iC}(x_1,\ldots,\overline{x_n},x_n) = f_i(x_1,\ldots,x_{n-2},x_n)$, and f_i is dependent on all the variables $x_1,\ldots,x_{n-2}$. Thus, f_{iC} is a nondegenerate function of n variables.

Next, let us consider the case where $(C + w_{n-1}) > 0$ but $C < 0$. Then $(C+|C|)/2 = 0$ and the threshold for f_{iC} equals t. One can again show that f_{iC} is self-dual by showing that Eqn. (2.17) is satisfied by its weights and threshold:

$$2t - 1 = \sum_{i=1}^{n-1} w_i = \sum_{i=1}^{n-2} w_i + |C| + (C + w_{n-1}).$$

Now from Eqn. (2.19), we have

$$t = (C + w_{n-1}) + \sum_{i \in S} w_i, \qquad (2.22)$$

which implies that $f_{iC} = 1$ when $x_n = 1$ and $x_i = 1$ for all $i \in S$.

One can now use arguments similar to those used for the case $C > 0$ to show that f_{iC} is different for different values of C. Consider two different C's: C_1, which is determined by the partial sum $\sum_{i \in S_1} w_i$, and C_2, which is determined by the partial sum $\sum_{i \in S_2} w_i$. If both $C_1 < 0$ and $C_2 < 0$, then assume without of loss of generality that $\sum_{i \in S_1} w_i < \sum_{i \in S_2} w_i$. Now consider the assignment where $x_n = 1$, $x_i = 1$ for all $i \in S_1$, and $x_i = 0$ for all $i \in \overline{S_1}$; call this the input assignment Y. Eqn. (2.22) implies that $f_{iC_1}(Y) = 1$. However, one can verify that $f_{iC_2}(Y) = 0$; thus, $f_{iC_1} \neq f_{iC_2}$.

Next consider the case where $C_1 > 0$ and $C_2 < 0$. It follows from Eqn. (2.19) that $\sum_{i \in S_1} w_i < \sum_{i \in S_2} w_i$. Then, consider the input assignment X, where $x_n = x_{n+1} = 1$, $x_i = 1$ for all $i \in S_1$, and $x_i = 0$ for all $i \in \overline{S_1}$. As mentioned for the $C > 0$ case, $f_{iC_1}(X) = 1$. However, the following arguments show that $f_{iC_2}(X) = 0$. Since $C_2 < 0$, we have

$$
\begin{aligned}
f_{iC_2}(X) &= \operatorname{sgn}((|C_2| + (C_2 + w_{n-1}) + \sum_{i \in S_1} w_i) - t) \\
&= \operatorname{sgn}(w_{n-1} + \sum_{i \in S_1} w_i - (C_1 + w_{n-1} + \sum_{i \in S_1} w_i)) \\
&\qquad \text{(from Eqn. (2.22))} \\
&= \operatorname{sgn}(-C_1) = 0.
\end{aligned}
$$

Next we show that f_{iC} is a nondegenerate function of n variables. Eqn. (2.22) shows that f_{iC} is dependent on x_n. Also, $f_{iC}(x_1, \ldots, x_{n-1}, x_{n-1})$ $= f_i(x_1, \ldots, x_{n-1})$, hence, f_{iC} is dependent on the rest of the variables as well.

The above arguments show that for any C, f_{iC} can be reduced to f_i by appropriately assigning values to x_{n-1} and x_n. Hence, if one starts with two distinct $(n-1)$-variable Boolean functions f_i and f_j, then the corresponding functions f_{iC}'s and f_{jC}'s will also be distinct. $\qquad\square$

In the above construction, different functions of n-variables are generated from the same $(n-1)$-variable function by choosing distinct C's, and therefore, partial sums $\sum_{i \in S} w_i$, $S \subseteq \{1, \ldots, (n-2)\}$, with different values. The following lemma gives a lower bound on the number of different functions that can be generated by the procedure outlined in Lemma 2.9.

Lemma 2.10: Let $R(n)$ be the number of nondegenerate positive self-dual linearly separable Boolean functions of n variables. Then $R(n) > 2^{n-3} R(n-1)$, and for $n \geq 9$,

$$R(n) > 2^{\frac{n(n-5)}{2}+17}.$$

Proof: Lemma 2.9 shows that given an $(n-1)$-variable positive self-dual threshold function f_i, one can construct a distinct n-variable function for each different C that satisfies the conditions stated in the lemma. It follows from Eqn. (2.19) that for each distinct partial sum of $w_1, \ldots, w_{n-2}$, there is a distinct C. Since there are 2^{n-2} different partial sums of $w_1, \ldots, w_{n-2}$, one can possibly have 2^{n-2} different C's; however, each C in Lemma 2.9 has to also satisfy the condition that $(C + w_{n-1}) > 0$. We next show that there are at least 2^{n-3} C's that satisfy the conditions in Lemma 2.9.

It follows from Eqns. (2.19) and (2.20) that given a partial sum $\sum_{i \in S} w_i$, where $S \subseteq \{1, \ldots, n-2\}$, C satisfies the following:

$$2C = 2t - 2w_{n-1} - 2\sum_{i \in S} w_i$$

$$= \sum_{i=1}^{n-1} w_i + 1 - 2w_{n-1} - 2\sum_{i \in S} w_i$$

$$= \left(\sum_{i \in \overline{S}} w_i - \sum_{i \in S} w_i\right) - w_{n-1} + 1,$$

where $\overline{S} = \{1, \ldots, (n-2)\} - S$, is the complement of the set S. Hence,

$$(C + w_{n-1}) = \frac{w_{n-1} + 1}{2} + \frac{1}{2}(\sum_{i \in \overline{S}} w_i - \sum_{i \in S} w_i). \qquad (2.23)$$

Clearly, for half of all the partial sums, $(\sum_{i \in \overline{S}} w_i - \sum_{i \in S} w_i) > 0$. Hence, Eqn. (2.23) implies that for at least half of the partial sums of $w_1, \ldots, w_{n-2}$, $(C + w_{n-1}) > 0$. Thus, there are at least 2^{n-3} different C's that satisfy the conditions in Lemma 2.9. Therefore, $R(n) > 2^{n-3}R(n-1)$.

It has been shown by enumeration [86] that $R(9) > 2^{35}$. Thus, for $n \geq 9$,

$$
\begin{aligned}
R(n) &> 2^{n-3} \cdot 2^{n-4} \cdot \ldots \cdot 2^7 \cdot R(9) \\
&> 2^{n-3} \cdot 2^{n-4} \cdot \ldots \cdot 2^7 \cdot 2^{35} \\
&= 2^{\frac{n(n-5)}{2}+17}.
\end{aligned}
$$
$\qquad\square$

Theorem 2.12:   Let $N(n)$ be the number of nondegenerate threshold functions of n variables. Then, for $n \geq 8$,

$$N(n) > 2^{\frac{n(n-1)}{2}+16}.$$

Proof: For Theorem 2.12, a lower bound on $N(n)$ can be obtained by counting the number of nondegenerate, nonself-dual threshold functions of n variables, which can be derived from positive nondegenerate, self-dual functions of $(n+1)$ variables. Given a self-dual $(n+1)$-variable function, one can generate 2^{n+1} different n-variable nonself-dual functions by assigning one of the $(n+1)$ variables to 1 or 0, and then complementing the remaining variables. Hence, $N(n) > 2^{n+1} \cdot 2^{\frac{(n+1)(n-4)}{2}+17} = 2^{\frac{n(n-1)}{2}+16}$.
$\qquad\square$

2.7.3   Tighter Lower Bound on the Number of Linearly Separable Functions

We have shown earlier that the number of linear threshold functions defined over $\{1, -1\}^n$ is bounded above and below by 2^{n^2} and $2^{0.5n^2 - O(n)}$, respectively. While these results require nontrivial arguments, the techniques in their derivation are elementary. In order to narrow the gap between these upper and lower bounds, deeper mathematical tools are needed. Using techniques from the theory of geometric lattices, we show here that the upper bound derived in Section 2.7.1 is almost tight, as stated in the following theorem.

Theorem 2.13:   Let N_n be the number of linear threshold functions defined over $\{1, -1\}^n$. Then, as $n \to \infty$,

$$\frac{\log_2 N_n}{n^2} \to 1 \; . \qquad \Box$$

The proof of the above theorem uses results from the theory of geometric lattices. Geometric lattice is a partially ordered set (poset) with a special structure. Let us first introduce the terminologies and concepts from the theory of posets.

Definition 2.8:   A *partially ordered set* (or *poset*) is a set S on which an order relation $\preceq$ is defined, satisfying

1. $x \preceq x$ for all $x \in S$.

2. If $x \preceq y$ and $y \preceq x$, then $x = y$,

3. If $x \preceq y$ and $y \preceq z$, then $x \preceq z$. $\qquad \Box$

Examples of posets include the collection of subsets of a given set under set inclusion, and the set of divisors of an integer under divisibility.

It is not necessary that all pairs of elements in a poset are related by the partial order. We say that x and y in a poset are *comparable* if $x \preceq y$ or $y \preceq x$. An *interval* $[x, y]$, for x and y in a poset, is the set of all elements z such that $x \preceq z \preceq y$. We sometimes use open or half-open intervals such as $[x, y)$, where one of the endpoints is to be omitted. If $x \preceq y$ and the interval $[x, y]$ contains exactly two elements, we say that y *covers* x. If $y \preceq x$ implies $y = x$, then x is a *minimal* element. Similarly, x is a *maximal* element if $x \preceq y$ implies $x = y$. If a poset has a unique minimal element or a unique maximal element, the minimal element is denoted by O and the maximal element is denoted by I.

If $x_1 \preceq x_2 \preceq \cdots \preceq x_k$ and the x_i's are all distinct, then we say that $\{x_1, \ldots, x_k\}$ forms a *chain*. A *saturated chain* is a chain $\{x_1, \ldots, x_k\}$ such that x_{i+1} covers x_i for each $i < k$. Some posets with O and I have the property that for $x \preceq y$ and $x \neq y$, all saturated chains from x to y have the same number of elements. Under these circumstances, we can define the *rank* $r(x)$ of an element x to be 1 less than the cardinality of any saturated chain from O to x. For example, in the poset of subsets of a given set S, the empty set ϕ represents the minimal element O, the entire set S represents the maximal element I, and the rank of each subset equals its cardinality. A *maximal chain* in a ranked poset is a chain containing one element of each rank. Clearly, a maximal chain must be saturated, but not vice versa. The *height* of a ranked poset is the cardinality of any maximal chain. In Section 3.5, we shall use the notion of an *antichain*, which is a subset $\mathcal{A}$ of a poset S such that no two elements in $\mathcal{A}$ are comparable.

A *lattice* is a set where two commutative and associative binary operations *max* (also called *join*) denoted by $\vee$ and *min* (also called *meet*) denoted by $\wedge$ are defined such that for any two elements a and b in the lattice, we have $a \wedge (a \vee b) = a$ and $a \vee (a \wedge b) = a$. In every lattice, the relation $\preceq$ defined by

$$a \preceq b \text{ if and only if } a \wedge b = a$$

imposes a partial order on the lattice. A finite lattice L is a *geometric lattice* if there exists a rank function r defined on L satisfying $r(a \wedge b) + r(a \vee b) \leq r(a) + r(b)$.

A fundamental result in the theory of (finite) poset is the Möbius inversion formula. Before we derive this formula, we need to introduce some more terminologies. Given a finite poset S, one can define the *incidence algebra* of S as follows: Consider the set of all real-valued functions of two variables $f(x, y)$ defined for all x and y in S, and with the property that $f(x, y) = 0$ if $x \not\preceq y$. The sum of two such functions f and g, as well as multiplication by scalars, are defined as usual. The product $h = fg$ is defined as

$$h(x, y) = \sum_{z \in [x, y]} f(x, z) g(z, y) \ .$$

Since S is finite, the sum above is well defined. It is easy to see that the incidence algebra is associative over the real field and has an identity element $\delta(x, y)$, which is 1 if and only if $x = y$. The *zeta function* $\zeta(x, y)$ of the incidence algebra is defined as $\zeta(x, y) = 1$ if $x \preceq y$ and $\zeta(x, y) = 0$ otherwise.

The *Möbius function* $\mu(x, y)$ of the incidence algebra is defined by induction over the number of elements in the interval $[x, y]$, as follows: First, set $\mu(x, x) = 1$ for all x in S. Suppose now that $\mu(x, z)$ has been defined for all z in the open interval $[x, y)$. Then set

$$\mu(x, y) = - \sum_{z \in [x, y)} \mu(x, z) \ .$$

Note that the Möbius function has integer values.

Lemma 2.11:   The Möbius function $\mu(x, y)$ and the zeta function $\zeta(x, y)$ are inverses of each other in the incidence algebra.

Proof: Exercise 2.12.   □

Theorem 2.14:   (**Möbius inversion formula**) Let $f(x)$ be a real-valued function defined for all x in a finite poset S with an element p such that $f(x) = 0$ unless $p \preceq x$. Suppose

$$g(x) = \sum_{y \preceq x} f(y) \ .$$

Then,

$$f(x) = \sum_{y \preceq x} g(y)\mu(y, x) \ .$$

Proof: Substituting the right side of the first equation into the right side of the second one, we get

$$\sum_{y \preceq x} g(y)\mu(y, x) \ = \ \sum_{y \preceq x}\sum_{z \preceq y} f(z)\mu(y, x)$$

$$= \ \sum_{y \preceq x}\sum_{z} f(z)\zeta(z, y)\mu(y, x) \ .$$

Interchanging the order of summation, we have

$$\sum_{z} f(z) \sum_{y \preceq x} \zeta(z, y)\mu(y, x) = \sum_{z} f(z)\delta(z, x) = f(x) \ .$$ □

Suppose P (with partial order $\preceq_P$) and Q (with partial order $\preceq_Q$) are two posets. A function $f : P \to Q$ is *monotonic* if $f(x) \preceq_Q f(y)$ whenever $x \preceq_P y$.

Theorem 2.15:   Let P and Q be two posets with unique minimal elements, and $f : P \to Q$ be a monotonic function. Assume that the inverse image of every interval $[O, a]$ in Q is an interval $[O, x]$ in P, and that the inverse image of the minimal element O in Q contains at least two elements. Then,

$$\sum_{x : f(x) = a} \mu(O, x) = 0$$

for every a in Q.

Proof: The proof for Theorem 2.15 is by induction on the set Q. Since the inverse image of O ($= [O,O]$) in Q is an interval $[O,p]$ in P with $p \neq O$, we have

$$\sum_{x:f(x)=O} \mu(O,x) = \sum_{x\in[O,p]} \mu(O,x) = 0 \,.$$

Now suppose the statement is true for all $b \neq a$ and $b \preceq_Q a$ in Q. Then,

$$\sum_{b\preceq a, b\neq a} \sum_{x:f(x)=b} \mu(O,x) = 0 \,,$$

and

$$\sum_{x:f(x)=a} \mu(O,x) = \sum_{b\preceq a} \sum_{x:f(x)=b} \mu(O,x) \,.$$

The last sum equals the sum over some interval $[O,r]$ in P, which is the inverse image of the interval $[O,a]$, i.e.,

$$\sum_{b\preceq a} \sum_{x:f(x)=b} \mu(O,x) = \sum_{x\in[O,r]} \mu(O,x) = O \,,$$

since $r \neq O$.   □

Corollary 2.2:

(a) Let $a \neq O$ in a finite lattice P. Then, for any b in P,

$$\sum_{x\vee a=b} \mu(O,x) = 0 \,.$$

(b) Let $a \neq I$ in a finite lattice P. Then, for any b in P,

$$\sum_{x\wedge a=b} \mu(x,I) = 0 \,.$$

Proof: For part (a), let $f(x) = x \vee a$ for all x in P. Then, f is a monotonic function from P onto $f(P)$. It is easy to verify that f also satisfies the other two conditions of Theorem 2.15. Part (b) follows by inverting the order. □

Theorem 2.16:   Let μ be the Möbius function of a finite geometric lattice L and $r(x)$ be the rank function of L. Then:

(a) $\mu(x,y) \neq 0$ for any x, y in L, provided $x \preceq y$.

(b) If y covers z, then $\mu(x,y)$ and $\mu(x,z)$ have opposite signs. Hence, for $x \preceq y$, the sign of $\mu(x,y)$ is $(-1)^{r(y)-r(x)}$.

Proof: To prove the theorem, it suffices to assume that $x = O$ and $y = I$, since any interval of a geometric lattice is also a geometric lattice. We proceed by induction on the height of the lattice.

The theorem is true for lattices of height 2, where $\mu(O,I) = -1$. Assume it is true for all lattices of height $(n-1)$, and let L be a geometric lattice of height n. By Corollary 2.2, with $b = I$ and a being an element that covers O, we have

$$\mu(O,I) = - \sum_{x \vee a = I,\ x \neq I} \mu(O,x) . \tag{2.24}$$

Now from the subadditivity of the rank function in a geometric lattice,

$$r(x \wedge a) + r(x \vee a) \le r(x) + r(a) ,$$

it follows that if $x \vee a = I$, then $n \le r(x) + r(a)$, and $r(x) \ge (n-1)$. Hence, if $x \neq I$, x must be covered by I. By the inductive hypothesis, we conclude that all the $\mu(O,x)$ in the sum on the right side of Equation (2.24) have the same sign, and none of them is zero. Therefore, $\mu(O,I)$ is not zero, and its sign is the opposite of that of $\mu(O,x)$ for any x that is covered by I.   □

Now let us relate the above results on geometric lattices to the number of linear threshold functions over all n-dimensional binary vectors. Recall the proof of Theorem 2.11, where we derive an upper bound by considering the number of regions in the space of weight vectors partitioned by the set of hyperplanes whose direction vectors are given by the input vectors to the linear threshold functions. The underlying idea is that these regions in the weight space form a one-to-one correspondence with the linear threshold functions defined over the input vectors. This allows us to reduce the problem of counting the number of linear threshold functions to the computation of the number of regions in the weight space. We prove the upper bound stated in Theorem 2.11 using a simple induction. Here we derive a lower bound by relating the regions to a geometric lattice generated by the intersection of the hyperplanes and applying the properties of the Möbius function.

Given a set of m direction vectors $X_i \in \mathcal{R}^n$, $i = 1, \ldots, m$, let us consider a set of m hyperplanes in the n-dimensional space of weight vectors $\mathbf{w} = [w_1 \cdots w_n]^T$, given by

$$\mathbf{w}^T X_i = \sum_{j=1}^n x_{ij} w_j = 0 ,$$

for $i = 1, \ldots, m$. Let $\mathcal{L}$ be the set of linear subspaces obtained as the nonempty intersections of these hyperplanes together with the entire weight space.

Lemma 2.12: $\mathcal{L}$ forms a geometrical lattice under set inclusion.

Proof: $\mathcal{L}$ clearly forms a poset under $\subseteq$, with the unique minimal element O being the origin and the maximal element I being the entire weight space. A lattice structure can be imposed upon $\mathcal{L}$ by defining the $\wedge$ and $\vee$ operations as follows. For s and t in $\mathcal{L}$,

$$s \wedge t = s \cap t ,$$

$$s \vee t = \bigcap \{h \in \mathcal{L} : s \cup t \subseteq h\} .$$

Thus, each element of the lattice $\mathcal{L}$ is generated by the intersection of some subset of the hyperplanes, i.e., is an orthogonal complement to the subspace spanned by the direction vectors of these hyperplanes. The rank $r(s)$ of each element s in $\mathcal{L}$ is defined as the dimension of s. Any s, t in $\mathcal{L}$, being subspaces, satisfies

$$\dim s \vee t = \dim s + \dim t - \dim s \wedge t ,$$

and thus the subadditivity of the rank function is satisfied. Hence, $\mathcal{L}$ forms a geometric lattice.   □

In order to relate the regions partitioned by a set of hyperplanes to the geometric lattice $\mathcal{L}$, we make use of the following fundamental result from combinatorial topology. A proof of this result is beyond our scope and is omitted here.

Lemma 2.13:   Consider a partition of n-dimensional Euclidean space by a set of hyperplanes, and for $k = 0, \ldots, n$, let α_k be the number of k-dimensional faces of the partition. Then,

$$\sum_{k=0}^n (-1)^k \alpha_k = (-1)^n.$$ □

The value of the above sum $(-1)^n$ is referred to as the Euler number.

Theorem 2.17:   The number of regions in $\mathcal{R}^n$ partitioned by a set of hyperplanes (all passing through the origin) is given by

$$\sum_{s \in \mathcal{L}} |\mu(s, I)| \ ,$$

where μ is the Möbius function of the geometric lattice $\mathcal{L}$ generated by the intersections of the hyperplanes.

Proof: Observe that every k-dimensional face in the partition corresponds to a region of exactly one k-dimensional member of $\mathcal{L}$. For any $s \in \mathcal{L}$, let β_s be the number of regions in s. Then

$$\alpha_k = \sum_{s \in \mathcal{L}, \ dim \ s=k} \beta_s \ ,$$

and the equation in Lemma 2.13 becomes

$$\sum_{s \in \mathcal{L}} (-1)^{dim \ s} \beta_s = (-1)^n \ .$$

Since Lemma 2.13 holds for any Euclidean space, in particular for every $t \in \mathcal{L}$, it follows from the above equation (by substituting the whole space by t) that

$$\sum_{s \in \mathcal{L}, \ s \preceq t} (-1)^{dim \ s} \beta_s = (-1)^{dim \ t} \ .$$

Now applying the Möbius inversion formula of Theorem 2.14 to the above equation, we obtain for every $t \in \mathcal{L}$,

$$(-1)^{dim \ t} \beta_t = \sum_{s \in \mathcal{L}, \ s \preceq t} (-1)^{dim \ s} \mu(s, t) \ .$$

In particular, setting t to be the entire space $\mathcal{R}^n$ (i.e., the maximal element I in the geometric lattice), we get

$$(-1)^n \beta_I = \sum_{s \in \mathcal{L}} (-1)^{dim \ s} \mu(s, I) \ .$$

Since $\mathcal{L}$ is a geometric lattice, by part (b) of Theorem 2.16, the sign of $\mu(s, I)$ is $(-1)^{n - dim\ s}$. Hence, the total number of regions can be expressed as

$$\beta_I = \sum_{s \in \mathcal{L}} |\mu(s, I)| \ .$$

$\square$

Now consider the special case where we have 2^n hyperplanes of the type $\mathbf{w}^T X_i = 0$, with direction vectors $X_i \in \{1, -1\}^n$, and the corresponding geometric lattice $\hat{\mathcal{L}}$. Theorem 2.17 allows us to compute the number of linear threshold functions over all 2^n vectors in $\{1, -1\}^n$ by computing the values of the Möbius function of lattice $\hat{\mathcal{L}}$. Although evaluating $\mu(s, I)$ for every $s \in \hat{\mathcal{L}}$ is difficult, one can derive an effective lower bound on the sum in Theorem 2.17 by estimating the cardinality of $\hat{\mathcal{L}}$. Lemma 2.14 yields a lower bound on the number of different subspaces generated by all possible intersections of hyperplanes with direction vectors in $\{1, -1\}^n$. Since its derivation requires involved calculations, we refer the proof to the original source [90].

Lemma 2.14:   Let $p \leq n(1 - 10/\log n)$. Let vectors $\mathbf{v}_1, \ldots, \mathbf{v}_p$ be chosen at random from $\{1, -1\}^n$, and S be the subspace spanned by them. The probability that there is a $\mathbf{u} \in \{1, -1\}^n$ such that $\mathbf{u}$ is in S but $\mathbf{u} \neq \pm\mathbf{v}_i$ for $i = 1, \ldots, p$ is

$$4\binom{p}{3}(\frac{3}{4})^n + O((\frac{7}{10})^n) \ .$$

$\square$

Now we are ready to the prove the main result of this section.

Proof of Theorem 2.13:   Consider the lattice $\hat{\mathcal{L}}$ that is generated by the intersection of some subset of the hyperplanes with direction vectors in $\{1, -1\}^n$. Each element in $\hat{\mathcal{L}}$ is an orthogonal complement to the subspace spanned by some subset of vectors in $\{1, -1\}^n$. Choose from $\{1, -1\}^n$ all subsets of cardinality $p_n = n - 10n/\log n$, and from each pair of opposite subsets we keep one subset. By Lemma 2.14, almost all subspaces spanned by the chosen subsets of vectors do not contain vectors from $\{1, -1\}^n$ different from those that generate them. Thus, the number of different subspaces generated by these subsets of vectors is asymptotically equal to

$$\frac{1}{2}\binom{2^n}{n(1 - 10/\log n)} \ , \tag{2.25}$$

and their orthogonal complements are the elements of the lattice $\hat{\mathcal{L}}$. More-over, it follows from part (a) of Theorem 2.16 that $\mu(s, I) \neq 0$ for all $s \in \hat{\mathcal{L}}$. Since μ has integer values, the sum in Theorem 2.17 is bounded below by the cardinality of $\hat{\mathcal{L}}$. As the regions in $\mathcal{R}^n$ partitioned by the hyperplanes with direction vectors in $\{1, -1\}^n$ form a one-to-one correspondence with the lin-ear threshold functions defined over these vectors, we have from Eqn. (2.25) that for sufficiently large n,

$$\log_2 N_n > n^2 (1 - 10/\log n) \ .$$

Together with the upper bound of Corollary 2.1, Theorem 2.13 follows.   □

2.7.4   Number of Weights in Universal Networks

Consider a set of n vectors (the dimension of the space is irrelevant). Suppose we wish to design a feedforward network that is capable of computing any binary function over these points. The question we address is: How many connections (i.e., weights) must such a network have?

Theorem 2.18:   Any feedforward network that is universal with respect to n input vectors must have a total number of weights

$$S \geq \frac{n}{1 + \log n} \ .$$

Proof: We first establish an upper bound for the number of binary func-tions that can be computed by a feedforward network with S weights.

Denote by S_i the number of weights and threshold of the ith LTE in the network. Thus, $S = \sum_i S_i$.

Now fix the input weight vectors for all LTEs in the network except for the ith LTE. Let $\{X_1', X_2', \ldots, X_n'\}$ denote the set of input patterns to the ith LTE corresponding to the set of n input vectors to the network. Note that these inputs cannot be changed by varying the input weights to the ith LTE since there is no feedback allowed in the network.

The number of different functions that the network can compute by *only* varying the input weights to the ith LTE is thus less than or equal to the number of different binary functions that the ith LTE can compute over the set $\{X_1', X_2', \ldots, X_n'\}$. But from Theorem 2.11, we know that the number of such functions $\leq C(n, S_i)$.

Hence, an upper bound on the number of functions that the network can compute is

$$\max_{(\sum_i S_i = S)} \prod_i C(n, S_i).$$

Consider the following upper bound for $C(n, S_i)$:

$$C(n, S_i) = 2 \sum_{j=0}^{S_i-1} \binom{n-1}{j} \leq 2 \sum_{j=0}^{S_i-1} n^j \leq 2n^{S_i}.$$

Hence, the total number of functions computable by the network with S weights is less than

$$\max_{(\sum_i S_i = S)} \prod (2n)^{S_i} \; = \; (2n)^S.$$

Now for the network to be able to compute all 2^n possible functions, one must have

$$2^n \; \leq \; (2n)^S.$$

This yields the result of the theorem, i.e., $S \geq \dfrac{n}{1 + \log_2 n}$.   □

In Chapter 3, we show the above lower bound to be tight (see Exercise 3.9).

2.8   Large Integer Weights are Sufficient

The size of the weights of an LTE plays an important role in determining the complexity of its hardware implementations. In this section, we show that if the inputs are bounded integers (e.g., in Boolean threshold circuits), then it suffices to have exponentially large (in the number of inputs) integer weights. In Chapter 6, we shall show that exponentially large weights in a threshold circuit can be reduced to polynomially bounded integer weights by incurring a small increase in the depth and a polynomial factor increase in the size.

Theorem 2.19:   Any n-input LTE can be assumed to have integer weights w_i such that $|w_i| \leq (n+1)! b^n$, where b is a bound on the magnitude of the integer inputs.

Proof: If the LTE has m input vectors in $\mathcal{R}^n$ (where, without loss of generality, $m \geq n + 1$), then we can assume (see Remarks 2.1 and 2.2) that there are m vectors $X_1, \ldots, X_m$, $X_i \in \mathcal{R}^{n+1}$, such that the weight vector, $\mathbf{w} \in \mathcal{R}^{n+1}$, of the LTE satisfies $\mathbf{w}^T X_i \geq 1$ for $i = 1, \ldots, m$. By assumption, the elements of X_i are integers and their magnitudes are bounded above by b. We also assume that $X_1, \ldots, X_m$ span the whole space $\mathcal{R}^{n+1}$. If this assumption is not met, then one can show that some of the weights can be always made 0 (see Exercise 2.3).

Let $\mathbf{w}_0$ be a weight vector such that it satisfies a *maximum* number of the constraints with equality. That is, there exist s input vectors $X_{i_1}, X_{i_2}, \ldots, X_{i_s}$ such that

$$\begin{bmatrix} X_{i_1}^T \\ X_{i_2}^T \\ \vdots \\ X_{i_s}^T \end{bmatrix} \mathbf{w}_0 = A\mathbf{w}_0 = \begin{bmatrix} 1 \\ 1 \\ \vdots \\ 1 \end{bmatrix}$$

and no other weight vector satisfies more than s of the constraints with equality. Note that for any other X_j, $j \notin \{i_1, i_2, \ldots, i_s\}$, $X_j^T \mathbf{w}_0 > 1$.

We claim that $\text{rank}(A) = n + 1$, i.e., $\mathbf{w}_0$ is uniquely determined by the above equations. Let us suppose that our claim is not true. Then, there exists a non-zero vector $\mathbf{u} \in \mathcal{R}^{n+1}$ that lies in the right null-space of matrix A. Consider a new weight vector $\mathbf{w}_\epsilon = \mathbf{w}_0 + \epsilon \mathbf{u}$. One can then show the following:

1. $X_{i_1}^T \mathbf{w}_\epsilon = X_{i_2}^T \mathbf{w}_\epsilon = \cdots = X_{i_s}^T \mathbf{w}_\epsilon = 1$ for any ϵ.

2. There exists at least one input vector X_k, $k \notin \{i_1, i_2, \ldots, i_s\}$, such that $X_k^T \mathbf{u} \neq 0$. This is because by assumption $\{X_1, \ldots, X_m\}$ span $\mathcal{R}^{n+1}$. Thus, one can choose ϵ appropriately such that for some input vector, $X_j^T \mathbf{w}_\epsilon = 1$, $j \notin \{i_1, i_2, \ldots, i_s\}$, and for other vectors $X_l^T \mathbf{w}_\epsilon \geq 1$, $l \notin \{i_1, i_2, \ldots, i_s, j\}$.

3. The weight vector $\mathbf{w}_\epsilon$ thus satisfies at least $s + 1$ of the constraints with equality, which is a contradiction. Hence, $\text{rank}(A) = n + 1$.

The above arguments imply that the elements of $\mathbf{w}_0$ are solutions of an $(n + 1) \times (n + 1)$ linear system of equations in which the coefficients are integers and bounded above in magnitude by b. Hence, the elements of $\mathbf{w}_0$ can be written as

$$\mathbf{w}_{0i} = \frac{\sum_{j=1}^{n+1} \det(M_{ij})}{\det(M)},$$

where M_{ij}'s are $n \times n$ matrices and M is an $(n+1) \times (n+1)$ matrix. One can easily show that $|\det(M_i)| \leq n! b^n$. Hence, by scaling the entries of $\mathbf{w}_0$ by $|\det(M)|$, we conclude that there exist integer weights bounded above in magnitude by $(n+1)! b^n$.   □

If the inputs are binary-valued, then $b = 1$ and the above theorem implies that each of the weights of the LTE can be represented by at most $O(n \log n)$ bits.

Corollary 2.3:   Any LTE with n binary inputs can be represented as an LTE with integer weights w_i, where $|w_i| \leq 2^{O(n \log n)}$.   □

Exercises

2.1:   [*A variation of the perceptron learning algorithm*]
During the [ADD] step in the perceptron learning algorithm, instead of setting $\mathbf{w}_{l+1} = \mathbf{w}_l + X_i$, set $\mathbf{w}_{l+1} = \mathbf{w}_l + l \cdot X_i$. Show that this modified algorithm will also converge when the input vectors are linearly separable.

2.2:   [*Unate functions and linear threshold functions*]
A Boolean function $f(x_1, \ldots, x_n)$ is *unate* if it has an SOP representation such that for each variable x_i, the SOP form contains either x_i or $\overline{x_i}$, but not both.

(a) Show that every linear threshold function is unate.
    Show that $f(x_1, x_2, x_3, x_4) = (x_1 \wedge x_2) \vee (x_3 \wedge x_4)$ is not linearly separable. Hence, conclude that the class of Boolean linear threshold functions is a proper subset of the class of unate functions.

(b) An SOP expression for a Boolean function f is *irredundant* if the removal of any of the product terms makes the expression not equal to f.

Show that a unate function f has exactly one irredundant SOP expression. Thus, conclude that any linear threshold function has a unique irredundant SOP representation.

2.3: [*Input matrix with full row rank*]
Consider an n-input threshold gate with weight vector $\mathbf{w} \in \mathcal{R}^{n+1}$ (i.e., $\mathbf{w}$ includes the threshold value of the gate as discussed in Remark 2.1). Let $\{X_1, \ldots, X_m\}$ $(m \geq n + 1)$, $X_i \in \mathcal{R}^{n+1}$, be the set of augmented input vectors to the gate, i.e., $x_{i(n+1)} = -1$ for $i = 1, \ldots, m$. Let $f \in \mathcal{R}^m$ be the vector comprising the outputs, i.e., the ith coordinate of f (denoted as $(f)_i$) equals $\text{sgn}(\mathbf{w}^T X_i)$. Let $\text{sgn}(P)$, $P \in \mathcal{R}^m$, be defined as the vector whose i^{th} coordinate is $\text{sgn}((P)_i)$. Then

$$f = \text{sgn}(D^T \mathbf{w}),$$

where $D = [X_1 \cdots X_m]$ denotes the augmented input matrix. Show that if the row rank of D is $(n + 1) - k$, $k \geq 1$, (i.e., D is not of full row rank), then one can set k of the input weights to 0, and yet have the same output function (by appropriately choosing the rest of the input weights).

In many instances in this book, we assume that the augmented input matrix D has full row rank.

2.4: [*Linearly nonseparable sets with exponentially many separable vectors*]
For any $n \geq 2$, construct a set $\{X_1, \ldots, X_{2^{n-2}+2}\}$ of $2^{n-2} + 2$ vectors in $\mathcal{R}^n$ as follows:

Let $\{X_1, \ldots, X_{2^{n-2}}\}$ be the set of 2^{n-2} distinct vectors in $\{1, -1\}^n$ such that $x_{i1} = x_{i2} = 1$ (i.e., the first two entries equal 1) for each X_i, $i = 1, \ldots, 2^{n-2}$. Also set

$$\begin{aligned} X_{2^{n-2}+1} &= [1 \quad -1 \quad 1 \quad \cdots \quad 1]^T \\ X_{2^{n-2}+2} &= [-1 \quad 1 \quad -1 \quad \cdots \quad -1]^T . \end{aligned}$$

(a) Show that in the above construction, $\{X_1, \ldots, X_{2^{n-2}}\}$ is the set of separable vectors, and $\{X_{2^{n-2}+1}, X_{2^{n-2}+2}\}$ is the set of nonseparable vectors.

(b) Using part (a), show that in Example 2.8, $\{X_1, X_2, X_3, X_4\}$ is the set of separable vectors, and $\{X_5, X_6\}$ is the set of nonseparable vectors.

Also, determine a linearly separable subset of maximum cardinality for the set of vectors in Example 2.8.

(c) For the function $f(x_1, x_2, x_3, x_4) = (x_1 \wedge x_2) \vee (x_3 \wedge x_4)$, list the sixteen different input vectors and the corresponding outputs, and preprocess the input vectors according to Remarks 2.1 and 2.2. Show that the resultant set of input vectors is linearly nonseparable, and that there are 10 separable vectors and six nonseparable vectors.

2.5:   [*Asummability and linear separability*]
Given $f : \{1, 0\}^n \rightarrow \{0, 1\}$, let $S_1 \subseteq \{0, 1\}^n$ and $S_0 \subseteq \{0, 1\}^n$ be the sets of inputs for which f equals 1 and 0, respectively. The function f is said to be m-summable if there exist subsets of the inputs $S_1' \subseteq S_1$, $S_0' \subseteq S_0$ and constants $p_i, r_i > 0$ such that

$$\sum_{X_i \in S_1'} p_i X_i = \sum_{X_j \in S_0'} r_j X_j,$$

and $\sum p_i = \sum r_i = m$. f is said to be *asummable* if it is not m-summable for any $m \geq 2$.

Using Theorem 2.3, show that a Boolean function f is linearly separable if and only if it is asummable.

2.6:   [*Linearly separable subsets of maximum cardinality*]

(a) Show that the output set V in the heuristic algorithm presented in Section 2.5.1 is linearly separable.

(b) Show that in the heuristic algorithm presented in Section 2.5.1, there always exists a choice of vectors that can be deleted (i.e., X_k's) so that the resulting output set V is a linearly separable subset of $\{X_1, X_2, \ldots, X_m\}$ of maximum cardinality.

2.7:   [*A generalized version of Problem 2.4 is also NP-complete*]
Problem 2.4 can be restated as follows: Given a set of m vectors, $S = \{X_1, X_2, \ldots, X_m\}$ $(X_i \in \mathcal{R}^d)$, determine a subset $S' \subseteq S$ of maximum cardinality such that $\mathbf{w}^T X_i > 0$ for all $X_i \in S'$, and for some $\mathbf{w} \in \mathcal{R}^d$. Define a more general problem, P', as follows: Given two sets (together comprising m vectors), $S_1 = \{X_1, \ldots, X_k\}$ and $S_2 = \{X_{k+1}, \ldots, X_m\}$ $(X_i \in \mathcal{R}^d)$,

determine subsets $S_1' \subseteq S_1$ and $S_2' \subseteq S_2$, such that $S_1' \cup S_2'$ is of maximum cardinality, and for some $\mathbf{w} \in \mathcal{R}^d$ and $t \in \mathcal{R}$: (1) $\mathbf{w}^T X_i - t > 0$ for all $X_i \in S_1'$, and (2) $\mathbf{w}^T X_i - t \leq 0$ for all $X_i \in S_2'$.

Show that given an instance of Problem 2.4 with m vectors, one can construct an instance of P' with $2m + 1$ vectors, such that solving such an instance of P' will yield a solution to the given instance of Problem 2.4. Hence, conclude that P' is also NP-complete.

2.8:   Complete the proof of Theorem 2.10 by showing that any vector not in the set $A' = \{Y_{k+1}, \ldots, Y_m\}$ is a nonseparable vector.

2.9:   [*An upper bound on the number of functions computable by depth-d threshold circuits*]

(a) Consider a threshold gate with $S(n)$ $(S(n) < 2^n)$ inputs, where each input is a Boolean function of n variables: $x_1, \ldots, x_n$. Use Corollary 2.1 to show that the total number of Boolean functions computable by the given gate is $\leq 2^{nS(n)+1}$.

(b) Let $F_n(S(n), 2)$ be the number of n-variable Boolean functions computable by a depth-2 threshold circuit with fan-in at most $S(n)$, where $n \leq S(n) << 2^n$ (e.g., consider $S(n) = O(n^c)$ for some constant $c > 0$). Show that

$$F_n(S(n), 2) \leq \binom{2^{n^2}}{S(n)} 2^{nS(n)+1} \leq 2^{O(n^2 S(n))}.$$

(c) Let $F_n(S(n), d)$ be the number of n-variable Boolean functions computable by depth-d threshold circuits with fan-in bounded by $S(n)$, where $2^n >> S(n) \geq n$. Show that

$$F_n(S(n), d+1) \leq \binom{F_n(S(n), d)}{S(n)} 2^{nS(n)+1}$$

$$\leq (F_n(S(n), d))^{S(n)} 2^{nS(n)+1}.$$

Hence, conclude that $F_n(S(n), d) \leq 2^{O(n^2 S(n)^{d-1})}$.

2.10:   [*Self-dual linear threshold functions*]

Let $f(x_1, \ldots, x_n) = \text{sgn}(\sum_{i=1}^{n} w_i x_i - t)$ be a self-dual linear threshold function. Show that one can appropriately choose a threshold t' such that $f(x_1, \ldots, x_n) = \text{sgn}(\sum_{i=1}^{n} w_i x_i - t')$ and $t' = (1 + \sum_{i=1}^{n} w_i)/2$. Hence, show that there always exists a choice of threshold for a self-dual function such that Eqn. (2.17) is satisfied.

2.11:   [*An alternative approach for deriving a lower bound on the number of linearly separable Boolean functions*]

(a) Given two n-variable functions $f(x_1, \ldots, x_n)$ and $g(x_1, \ldots, x_n)$, let

$$h(x_1, \ldots, x_n, x_{n+1}) = f(x_1, \ldots, x_n) \wedge x_{n+1} \bigvee g(x_1, \ldots, x_n) \wedge \overline{x_{n+1}}.$$

Show that $h(x_1, \ldots, x_n, x_{n+1})$ is a linear threshold function if and only if $f(x_1, \ldots, x_n)$ and $g(x_1, \ldots, x_n)$ are linear threshold functions, which can have the same weight vector (but may have different threshold values).

(b) Let

$$h(x_1, \ldots, x_n, x_{n+1}) = f(x_1, \ldots, x_n) \wedge x_{n+1} \bigvee g(x_1, \ldots, x_n) \wedge \overline{x_{n+1}},$$

and

$$h'(x_1, \ldots, x_n, x_{n+1}) = f'(x_1, \ldots, x_n) \wedge x_{n+1} \bigvee g'(x_1, \ldots, x_n) \wedge \overline{x_{n+1}}.$$

Show that $h(\cdot)$ and $h'(\cdot)$ are different functions if and only if $f(\cdot) \neq f'(\cdot)$ and/or $g(\cdot) \neq g'(\cdot)$.

Let $N(n)$ be the number of linear threshold functions of up to n variables. Also, let $s_n(f)$ be the number of linear threshold functions of up to n variables that can have the same weight vector with a given linear threshold function $f(\cdot)$. Then, show that

$$N(n+1) = \sum_{i=1}^{N(n)} s_n(f_i),$$

where f_i's are distinct threshold functions of up to n variables.

(c) Using Lemma 2.8, show that for any linear threshold function f, of up to n variables,

$$s_n(f) \geq 2^n + 1.$$

Hence, show that $N(n+1) \geq (2^n + 1)N(n)$. Now, using the fact that $N(6) > 2^{23}$ [86], show that $N(n) > 2^{\frac{n(n-1)}{2}+8}$.

Note that the lower bound stated in Lemma 2.10 applies to the class of positive self-dual linearly threshold functions, which is a proper subset of the class of general linear threshold functions.

2.12: [*Möbius function and the zeta function*]
For a finite poset P, verify that the Möbius function $\mu(x, y)$ and the zeta function $\zeta(x, y)$ are inverses of each other in the incidence algebra of P.

2.13: [*Duality of Möbius inversion formula*]
Given a finite poset P, let P^* denote the poset obtained by inverting the order of P, and let μ and μ^* be the Möbius functions of P and P^* respectively.

(a) Let $f(x)$ be a function defined for all x in P. Suppose there is an element q such that $f(x)$ vanishes unless $x \preceq q$. Show that if

$$g(x) = \sum_{x \preceq y} f(y) \, ,$$

then

$$f(x) = \sum_{x \preceq y} \mu(x, y) g(y) \, .$$

(b) Conclude from part (a) that $\mu^*(x, y) = \mu(y, x)$.

2.14: [*A tighter upper bound on the magnitude of the weights of an LTE for binary inputs*]
Consider an n-input LTE with inputs restricted to $\{1, -1\}^n$.

(a) Following the proof of Theorem 2.19, show that the weights (and the threshold) of the LTE can be written as

$$w_i = \frac{\sum_{j=1}^{n+1} \det(H_{ij})}{\det(H)},$$

where H_{ij}'s are $n \times n$ matrices with entries in $\{1, -1\}$ and H is an $(n+1) \times (n+1)$ matrix with entries in $\{1, -1\}$.

(b) Show that the determinant of an $n \times n$ matrix whose entries are in $\{1, -1\}$ is (i) bounded above in magnitude by $n^{n/2}$, and (ii) divisible by $2^{(n-1)}$.

(c) Hence, show that the weights of the LTE can be assumed to be integers with magnitudes bounded above by $2^{-(n-1)}(n+1)n^{n/2}$.

In Chapter 6, constructions of linear threshold functions that require integer weights of magnitude close to this upper bound are presented.

2.15:   [*Most threshold functions need exponentially large weights*]
Show that there are at most $2^{O(n \log n)}$ linearly separable Boolean functions with polynomially bounded integer weights. Thus, show that most linearly separable Boolean functions require exponentially large integer weights.

Bibliographic Notes

The Linear Threshold Element model was first introduced by McCulloch and Pitts in 1943 [77]. The basic properties of LTEs presented in Section 2.2 can be found in several books including [61, 70, 86]. The perceptron learning algorithm discussed in Section 2.3 is due to Rosenblatt [109]. Our presentation of the material, however, has been influenced by the work of Minsky and Papert [81]. Because of the emphasis of this book on the design and analysis of threshold circuits, we have not discussed several other important learning algorithms, including the Least Mean Square (LMS) algorithm developed by Widrow and Hoff [144]. A detailed discussion of such algorithms can be found in [145, 146].

The contents of Sections 2.4–2.6 on the analysis and learning of linearly nonseparable training sets are taken from Roychowdhury, Siu, and Kailath [112]. While learning and convergence properties of perceptrons are well understood for linearly separable input vectors, [112] presents the first known results on the structure of linearly nonseparable training sets (see Sections 2.4 and 2.5) and on the behavior of the perceptron learning algorithm when the set of input vectors is linearly nonseparable (see Section 2.6). An indirect form of Theorem 2.3 can be found in the early literature on thresh-old logic (see for example, [86]) for the special case of Boolean vectors, i.e., when entries of X_i's are restricted to binary values 0 and 1. An alternative proof for Theorem 2.6 can be found in [37]. Exercise 2.7 was suggested by K. S. Van Horn (private communication).

It is shown in [53] that it is co-NP-complete to decide whether a Boolean function given in a disjunctive normal form is linearly separable. (It is pointed out in [53] that an incomplete proof of NP-completeness for the same problem is given in [101].) For other work related to the learning of linearly nonseparable vectors, see Hassoun and Song [49] and Singleton [121]. Similarly, several complexity theoretic results for problems in learning are derived in Blum and Rivest [16], Judd [63], Vergis, Steiglitz, and Dickinson [138], and the collection of papers in Roychowdhury, Siu, and Orlitsky [113].

The upper bound result (Theorem 2.11) on the number of functions com-putable by an LTE was first derived in [117]. Several proofs of this result have appeared in the literature and can be found in Cover [31], Nechiporuk [87], and Winder [147]. The proof provided in Section 2.7 follows the ge-ometric approach developed in [31]. The lower bound on the number of linearly separable Boolean functions described in Section 2.7.2 can be found in [85]. The tighter lower bound stated in Theorem 2.13 is due to [150]. Our presentation of the results from the theory of posets and the theory of geometric lattices is taken from Rota [110] and Zaslavsky [151]. Proof of Lemma 2.13 can be found in Buck [23]. The result on the number of con-nections in a universal network is from [32]. A related result by Nechiporuk on the number of nodes in a universal network is presented in [87] and is discussed in Chapter 3. The results on the size of integer weights for an LTE can be found in several papers including [40, 86, 105].

Chapter 3

Computing Symmetric Functions

3.1 Introduction

As discussed in Chapter 1, many of the arithmetic circuits presented in later chapters are based on the efficient designs of threshold circuits for computing symmetric functions. In this chapter, we present efficient constructions of threshold circuits for the class of symmetric functions. Moreover, we show that the design of threshold circuits for symmetric functions presented here is almost optimal in terms of gate count. We also show that any Boolean function can be considered as a generalization of symmetric functions and that we can extend the construction of threshold circuits for symmetric functions to general Boolean functions. We conclude this chapter by presenting a lower bound on the size of depth-2 threshold circuits computing the Parity function.

Definition 3.1: (**Symmetric Function**) A Boolean function $f : \{0,1\}^n \to \{0,1\}$ is said to be *symmetric* if

$$f(x_1, \ldots, x_n) = f(x_{(1)}, \ldots, x_{(n)})$$

for any permutation $(x_{(1)}, \ldots, x_{(n)})$ of $(x_1, \ldots, x_n)$, or equivalently, f depends only on the sum of its input values $\sum_{i=1}^{n} x_i$. □

115

A common example of a symmetric function is the *Parity* function, for which the output is 1 if the sum of its inputs is odd, and 0 otherwise. Other examples of symmetric functions that will be of interest to us are Majority and Exact_k.

Definition 3.2: **(Majority and Exact$_k$)** The n-variable functions $MAJ_n(X)$ and $EX_k^n(X)$ where $X = (x_1, ..., x_n) \in \{0,1\}^n$, and $k = 1, ..., n$, are defined as

$$MAJ_n(X) = \begin{cases} 1 & \text{if } \sum_{i=1}^{n} x_i \geq \dfrac{n}{2}, \\ 0 & \text{otherwise.} \end{cases} \tag{3.1}$$

$$EX_k^n(X) = \begin{cases} 1 & \text{if } \sum_{i=1}^{n} x_i = k, \\ 0 & \text{otherwise.} \end{cases} \tag{3.2}$$

We refer to the family of functions $\{MAJ_n\}$ in Eqn. (3.1) and $\{EX_k^n\}$ in Eqn. (3.2) as Majority and Exact_k, respectively. $\square$

We first present a depth-2 threshold circuit for computing any n-variable symmetric function with at most $n+1$ gates. Afterwards, we reduce the gate count of the depth-2 construction to $\lceil n/2 \rceil + 1$ gates by applying the 'telescopic' technique. We then generalize this technique and give a construction of a depth-3 threshold circuit with $O(\sqrt{n})$ gates that computes a general n-variable symmetric function. Finally, based on a counting argument, we present a lower bound result showing that the $O(\sqrt{n})$ gate count cannot be reduced significantly even if we allow threshold circuits of any depth.

The results of this chapter also demonstrate that threshold circuits are computationally more powerful than AND/OR circuits. In Chapter 9, we shall show that Parity cannot be computed by constant-depth polynomial-size circuits of AND/OR gates. On the other hand, we show in this chapter that all symmetric functions have efficient implementations in small-depth threshold circuits.

3.2   A Depth-2 Construction

Theorem 3.1: Let $f(x_1, x_2, ..., x_n)$ be a symmetric function. Then f can be computed using a depth-2 threshold circuit with at most $n+1$ gates and $O(n^2)$ connections.

Proof: Since f only depends on the sum of its inputs $\sum\limits_{i=1}^{n} x_i$, there exists a set of s disjoint subintervals in $[0, n]$, say $[k_1, \tilde{k}_1], [k_2, \tilde{k}_2], ..., [k_s, \tilde{k}_s]$, where k_j's, $\tilde{k}_j$'s are integers, $k_{j+1} > \tilde{k}_j + 1$, and $k_j \le \tilde{k}_j$, such that $f(x_1, x_2, ..., x_n) = 1$ if and only if for some j,

$$\sum_{i=1}^{n} x_i \in [k_j, \tilde{k}_j] .$$

Our circuit consists of two levels. In the first level, there are $2s$ threshold gates computing for $j = 1, \ldots, s$,

$$y_{k_j} = \mathrm{sgn}(\sum_{i=1}^{n} x_i - k_j)$$

and

$$\tilde{y}_{k_j} = \mathrm{sgn}(\tilde{k}_j - \sum_{i=1}^{n} x_i) .$$

In the second level, the output gate computes

$$f(x_1, x_2, ..., x_n) = \mathrm{sgn}(\sum_{j=1}^{s} (y_{k_j} + \tilde{y}_{k_j}) - s - 1) .$$

To show that the output gate gives the correct value of the function, note the following:

If for $j = 1, \ldots, s$, $\sum\limits_{i=1}^{n} x_i \notin [k_j, \tilde{k}_j]$, then $y_{k_j} + \tilde{y}_{k_j} = 1$ for all j. Thus,

$$\mathrm{sgn}(\sum_{j=1}^{s}(y_{k_j} + \tilde{y}_{k_j}) - s - 1) = \mathrm{sgn}(s - s - 1) = 0.$$

On the other hand, if $\sum\limits_{i=1}^{n} x_i \in [k_j, \tilde{k}_j]$ for some $j \in \{1, \ldots, s\}$, then

$y_{k_j} + \tilde{y}_{k_j} = 2$ and $y_{k_i} + \tilde{y}_{k_i} = 1$ for $i \ne j$. Thus, $\mathrm{sgn}(\sum\limits_{i=1}^{s}(y_{k_i} + \tilde{y}_{k_i}) - s - 1) =$

$\mathrm{sgn}(s + 1 - s - 1) = 1.$

Since s is at most $\lceil n/2 \rceil$, there are at most $2\lceil n/2 \rceil + 1$ threshold gates in the circuit. If $s < \lceil n/2 \rceil$, we obtain a depth-2 threshold circuit for f of size at most $n + 1$. If $s = \lceil n/2 \rceil$, then applying the same construction to the negation of f yields a depth-2 threshold circuit of at most $2\lfloor \frac{n}{2} \rfloor + 1 \le n + 1$

gates. In this case, we can easily modify the threshold circuit for $\overline{f}$ to yield a threshold circuit for f of the same size (see Exercise 3.2). Moreover, since each gate in the circuit has fan-in at most n, the number of connections is $O(n^2)$. □

In fact, the proof of the above theorem implies a stronger result, stated as follows.

Theorem 3.2:   Let $f(x_1, x_2, ..., x_n)$ be a symmetric function. Then f can be represented as a sum of $O(n)$ $\widehat{LT}_1$ functions, i.e., linear threshold functions with polynomially bounded integer weights.

Proof: Using the same notation as in Theorem 3.1, we only have to observe that

$$f(x_1, x_2, ..., x_n) = \sum_{j=1}^{s}(y_{k_j} + \tilde{y}_{k_j}) - s \ .$$

□

We next present an ingenious technique called the 'telescopic' technique. We shall make repeated use of this technique in order to obtain a threshold circuit with $O(\sqrt{n})$ gates for any symmetric function.

Lemma 3.1:   Let the integer interval $[0, n]$ be divided into $k + 1$ subintervals $[b_0, b_1 - 1], [b_1, b_2 - 1], \ldots, [b_{k-1}, b_k - 1], [b_k, n]$, where $b_0 = 0 < b_1 < b_2 < \cdots < b_k < n$. Let $y_i = \text{sgn}(\sum_{j=1}^{n} x_j - b_i)$, for all $i = 1, \ldots, k$. Then

$$\sum_{j=1}^{k}(a_j - a_{j-1})y_j = a_m$$

if $\sum_{j=1}^{n} x_j \in [b_m, b_{m+1} - 1]$, where, $a_0 = 0$, and $a_1, \cdots, a_k$ are arbitrary real numbers.

Proof: Recall that $y_i = \text{sgn}(\sum_{j=1}^{n} x_j - b_i)$ equals 1 if and only if $\sum_{j=1}^{n} x_j \geq b_i$, otherwise it equals 0. Hence, if $\sum_{j=1}^{n} x_j \in [b_m, b_{m+1} - 1]$ (assume for now $m \geq 1$), then $y_1 = \cdots = y_m = 1$, and $y_{m+1} = \cdots = y_k = 0$. Hence,

$$\sum_{j=1}^{k}(a_j - a_{j-1})y_j = \sum_{j=1}^{m}(a_j - a_{j-1}) \ = \ a_m \ .$$

Note that if $\sum_{j=1}^{n} x_j \in [b_0, b_1 - 1]$, i.e., $m = 0$, then $y_i = 0$, for all $i = 1, \ldots, k$,

and hence, $\sum_{j=1}^{k} (a_j - a_{j-1}) y_j = 0 = a_0$.   □

Using Lemma 3.1, we can construct a depth-2 threshold circuit with at most $\lceil \frac{n}{2} \rceil + 1$ gates that computes a general symmetric function.

Theorem 3.3:   Any symmetric function of n variables can be computed by a depth-2 threshold circuit with at most $\lceil \frac{n}{2} \rceil + 1$ gates.

Proof: Let $f(X) : \{0,1\}^n \rightarrow \{0,1\}$ be a general symmetric function of n variables. Define a set of integers s_i and S_i with $0 \leq s_i \leq S_i \leq n$ for $i = 1, \ldots, r$, and $S_i + 1 < s_{i+1}$ for $i < r$, such that $f(X) = 1$ if and only if for some i,

$$s_i \leq \sum_{j=1}^{n} x_j \leq S_i.$$

The first layer of the circuit consists of r threshold gates, computing $y_i = \text{sgn}(\sum_{j=1}^{n} x_j - s_i)$ for each i, $1 \leq i \leq r$. The output gate in the second layer computes

$$z = \text{sgn}(\sum_{j=1}^{r} (S_j - S_{j-1}) y_j - \sum_{j=1}^{n} x_j).$$

Let us verify that this circuit gives us the correct output for the function. Note that for some m, $\sum_{j=1}^{n} x_j \in [s_m, s_{m+1} - 1]$. It follows from Lemma 3.1 that

$$\sum_{j=1}^{r} (S_j - S_{j-1}) y_j = S_m,$$

where we define $S_0 = 0$. If $f(X) = 1$, then $s_m \leq \sum_{j=1}^{n} x_j \leq S_m$ and hence,

$z = \text{sgn}(S_m - \sum_{j=1}^{n} x_j) = 1$. On the other hand, if $f(X) = 0$, then $S_m <$

$\sum_{j=1}^{n} x_j \leq s_{m+1} - 1$ and hence, $z = \text{sgn}(S_m - \sum_{j=1}^{n} x_j) = 0$.

The number of threshold gates in the circuit is $r + 1$. Since r can be at most $\lceil \frac{n}{2} \rceil$, the result follows.   □

3.3   A Depth-3 Construction

By generalizing the telescopic technique to a depth-3 construction, we can reduce the gate count significantly.

Theorem 3.4:   Any symmetric function of n variables can be computed by a depth-3 threshold circuit with $2\sqrt{n} + O(1)$ threshold gates.

Proof: Recall that we can define f by giving a set of integers s_i and S_i, $i = 1, \ldots, r$, with $s_i \leq S_i < s_{i+1}$ such that $f(X) = 1$ if and only if for some i,

$$s_i \leq \sum_{j=1}^{n} x_j \leq S_i.$$

We can always divide the interval $[0, n]$ into d consecutive subintervals, $[s_{1_1}, s_{2_1} - 1]$, $[s_{2_1}, s_{3_1} - 1]$, $\ldots$, $[s_{d_1}, n]$, so that each subinterval (except possibly the last one) contains the same number l of the integers s_i and S_i, where $l \leq \lceil \frac{n}{2d} \rceil$. The i-th subinterval will contain integers $s_{i_1} \leq S_{i_1} < s_{i_2} \leq S_{i_2} < \ldots < s_{i_l} \leq S_{i_l}$. In the following arguments, we assume that the last subinterval also contains l s_i's and l S_i's to avoid the use of cumbersome notations. The readers can easily modify the arguments for the general case.

The first layer of our circuit consists of d threshold elements, each computing the value z_i, where

$$z_i = \mathrm{sgn}(\sum_{j=1}^{n} x_j - s_{i_1})$$

for $i = 1, \ldots, d$.

For each $k = 1, \ldots, l$ we define two telescopic sums as follows:

$$
\begin{aligned}
T_k &= S_{1_k} z_1 + (S_{2_k} - S_{1_k})z_2 + (S_{3_k} - S_{2_k})z_3 + \ldots + (S_{d_k} - S_{d-1_k})z_d, \\
t_k &= s_{1_k} z_1 + (s_{2_k} - s_{1_k})z_2 + (s_{3_k} - s_{2_k})z_3 + \ldots + (s_{d_k} - s_{d-1_k})z_d.
\end{aligned}
$$

Clearly, T_k and t_k are linear combinations of the outputs from the first layer.

The second layer of our circuit will consist of $2l$ threshold elements, each utilizing the integer T_k or t_k as a 'threshold' value and computing the value Q_k or q_k, where

$$Q_k = \mathrm{sgn}(T_k - \sum_{j=1}^{n} x_j)$$

and

$$q_k = \text{sgn}(\sum_{j=1}^{n} x_j - t_k).$$

Finally, the output threshold element in the third layer computes

$$f(X) = \text{sgn}(\sum_{k=1}^{l} 2(Q_k + q_k) - 2l - 1) .$$

Now we show that our circuit gives the correct outputs on all inputs $X = (x_1, \ldots, x_n)$. Suppose $\sum_{j=1}^{n} x_j$ lies within the m-th interval, i.e., $\sum_{j=1}^{n} x_j \in [s_{m_1}, s_{(m-1)_1} - 1]$. By Lemma 3.1, the telescopic sums T_k, and t_k ($k = 1, \cdots, l$) assume the following values:

$$T_k = S_{m_k},$$

$$t_k = s_{m_k} .$$

By definition, $f(X) = 1$ if and only if for some k,

$$s_{m_k} \le \sum_{j=1}^{n} x_j \le S_{m_k} .$$

Let us assume that $f(X) = 1$ for the given input X; this implies that there exists an i such that $s_{m_i} \le \sum_{j=1}^{n} x_j \le S_{m_i}$. In the second layer, $\sum_{j=1}^{n} x_j$ is compared with $T_k = S_{m_k}$ and $-t_k = -s_{m_k}$ for each k, and the outputs are Q_k and q_k, respectively. Since, $s_{m_i} \le \sum_{j=1}^{n} x_j \le S_{m_i}$, the values of the outputs of the second layer (Q_k, q_k) can be seen as follows:

$$Q_k + q_k = \begin{cases} 2 & \text{if } k = i, \\ 1 & \text{if } k \ne i. \end{cases}$$

Thus, the output threshold element gives $\text{sgn}(\sum_{k=1}^{l} 2(Q_k + q_k) - 2l - 1) = \text{sgn}(2l + 2 - 2l - 1) = 1$.

Similarly, if $f(X) = 0$, i.e., for no k, $s_{m_k} \leq \sum_{j=1}^{n} x_j \leq S_{m_k}$, then $Q_k + q_k = 1$ for all k, and the output threshold element computes

$$\text{sgn}(\sum_{k=1}^{l} 2(Q_k + q_k) - 2l - 1) = \text{sgn}(2l - 2l - 1) = 0.$$

Hence, our depth-3 threshold circuit gives the correct outputs on all inputs $X = (x_1, \ldots, x_n)$.

To determine the size of our circuit, note that the first layer consists of d elements, the second layer $2l \leq \lceil n/d \rceil$ elements, and the third layer one output element. So all together, at most $d + \lceil n/d \rceil + 1$ threshold elements are required. To complete the proof, take $d = \lceil \sqrt{n} \rceil$ to minimize the size and thus obtain a depth-3 circuit of size $2\sqrt{n} + O(1)$.   □

Example 3.1:   We illustrate the above procedure by realizing the Parity function of 11 variables. Recall that the output of a Parity function is 1 if $\sum_{i=1}^{11} x_i$ is odd. Hence, $s_1 = S_1 = 1$, $s_2 = S_2 = 3$, $s_3 = S_3 = 5$, $s_4 = S_4 = 7$, $s_5 = S_5 = 9$, and $s_6 = S_6 = 11$. In the constructive proof of Theorem 3.4, set $s_{1_1} = 1$, $s_{2_1} = 5$, and $s_{3_1} = 9$, i.e., choose $d = 3$. Thus, each subinterval has two of the s_i's in it. In other words, $s_{1_1} = S_{1_1} = 1 < s_{1_2} = S_{1_2} = 3$; $s_{2_1} = S_{2_1} = 5 < s_{2_2} = S_{2_2} = 7$; and $s_{3_1} = S_{3_1} = 9 < s_{3_2} = S_{3_2} = 11$. Based on these choices, the realization of the function is shown in Fig. 3.1.

Note that the depth-3 realization requires eight threshold gates; in comparison, an efficient depth-2 realization, using the construction of Theorem 3.3, would require seven gates. However, as the number of variables increases, the efficiency of the depth-3 construction becomes apparent. For example, realizing the Parity function of 36 variables by the procedure outlined in the proof of Theorem 3.3 would require 19 gates, whereas the depth-3 construction would require only 13 gates.   □

Remark 3.1:   From our proof of Theorem 3.4, it is clear that the size of the threshold circuit computing a symmetric function f actually depends on r_f, the number of different integers s_i's (or equivalently S_i's) required for defining the function. Recall from our definition of the integers s_i's and S_i's: $f(X) = 1$ if and only if $s_i \leq \sum_{j=1}^{n} x_j \leq S_i$ for some i. On closer observation,

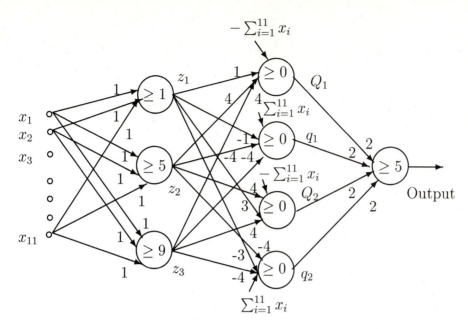

Figure 3.1 A depth-3 network for computing the Parity of 11 variables.

the proof of Theorem 3.4 shows that our construction of the threshold circuit computing f has size $2\sqrt{r_f} + O(1)$. For example, for the Parity function of n variables, $r_f = \lceil n/2 \rceil$, and the gate count is $O(\sqrt{n})$. $\qquad\square$

3.4 Generalized Symmetric Functions and General Boolean Functions

We extend our results in the preceding section to a more general class of Boolean functions. Recall that a symmetric function depends only on the sum of its input variables $\sum_{j=1}^{n} x_j$. We have shown that a symmetric function can be computed by threshold circuits of small size and depth. A natural generalization is to consider the class of functions that depends only on a weighted sum of its variables. In order for this class of functions to have polynomial-size circuits, we also require that the weighted sum be polynomially bounded.

Definition 3.3: **(Generalized Symmetric Function)** A Boolean function $f : \{0,1\}^n \to \{0,1\}$ is a *generalized symmetric* function if there is a function $\hat{f} : Z \to \{0,1\}$ such that

$$f(x_1, \ldots, x_n) = \hat{f}(\sum_{j=1}^{n} w_j x_j),$$

where the w_j's are positive integers and $\sum_{j=1}^{n} w_j = O(n^c)$ for some constant $c > 0$. □

Any generalized symmetric function can be considered to be a symmetric function of $\sum_{j=1}^{n} w_j$ variables by repeating each variable x_j w_j times . For example, if $f(x_1, x_2, x_3) = \hat{f}(2x_1 + x_2 + 3x_3)$, then f can be considered as a symmetric function f_s of six variables, where $f(x_1, x_2, x_3) = f_s(x_1, x_1, x_2, x_3, x_3, x_3)$. Theorem 3.4 thus implies the following result.

Corollary 3.1: Let $f(X)$ be a generalized symmetric function of n variables such that $f(X) = \hat{f}(\sum_{j=1}^{n} w_j x_j)$, then $f(X)$ can be computed by a depth-3 threshold circuit comprising $O(\sqrt{\sum_{j=1}^{n} w_j})$ gates. □

Remark 3.2: Since we assume the negations of the input variables, $\bar{x}_i$'s, to be available as inputs to the circuit, there is no loss of generality in assuming the w_j's in Definition 3.3 to be positive (Exercise 3.3). □

In later chapters, we shall make frequent use of the following result about generalized symmetric functions.

Theorem 3.5: Let $f_i(X)$'s be n-variable generalized symmetric functions and $g(X) = \sum_{i=1}^{N} w_i f_i(X)$, where the w_i's and N are polynomially bounded integers in n. Then $g(X)$ can be represented as a sum of polynomially many $\widehat{LT}_1$ functions.

Proof: Since each $f_i(X)$ can be considered as a symmetric function of $O(n^c)$ variables for some constant $c > 0$, it follows from Theorem 3.2 that we can represent each f_i as

$$f_i(X) = \sum_{j=1}^{O(n^c)} \ell_{ij}(X) ,$$

where $\ell_{ij}(X) \in \widehat{LT}_1$. Now substituting this expression for each f_i into the expression for $g(X)$, we conclude that

$$g(X) = \sum_{i=1}^{N} \sum_{j=1}^{O(n^c)} w_i \ell_{ij}(X) ,$$

which is a sum of polynomially many $\widehat{LT}_1$ functions.   □

We often refer to the following special case of Theorem 3.5, where each f_i is a Parity function of a subset of the input variables.

Corollary 3.2:   Let $S \subset \{0,1\}^n$ with $|S| = O(n^c)$ and

$$h(X) = \sum_{\alpha \in S} w_\alpha \left(\bigoplus_{\{i : \alpha_i = 1\}} x_i \right),$$

where $\alpha = (\alpha_1, \ldots, \alpha_n)$, and w_α's are polynomially bounded integers. Then $h(X)$ can be represented as a sum of polynomially many $\widehat{LT}_1$ functions. □

We now determine how many threshold gates (allowing unbounded fan-in) are sufficient to compute a general Boolean function. As a consequence of Theorem 3.4, we have the following theorem.

Theorem 3.6:   Any Boolean function of n variables can be computed by a depth-3 threshold circuit with $O(2^{n/2})$ threshold gates.   □

Proof: Note that the sum $\sum_{j=1}^{n} 2^j x_j$ is distinct for different values of $X = (x_1, \ldots, x_n) \in \{0,1\}^n$. Thus, *any* Boolean function could be considered as a function of an (exponentially large) weighted sum of its input variables, or equivalently, a *symmetric* function of $2^n - 1$ variables (by repeating each variable x_j 2^{j-1} times). The result follows from Theorem 3.4.   □

We next address the following question: Can one design more size-efficient threshold circuits for computing symmetric functions than the one presented in Theorem 3.4? It turns out that the upper bound presented in Theorem 3.4 on the size of depth-3 threshold circuit computing a general symmetric function is quite close to a lower bound that can be derived for the more general case where no restrictions are imposed on the weights and the circuit depth. This lower bound result follows from a lower bound on the number of threshold gates in an unrestricted depth circuit that computes a general Boolean function. The latter lower bound result follows directly from Theorem 2.18 that we proved in Chapter 2.

Theorem 3.7:   If a general Boolean function of n variables can be computed by a threshold circuit with $S(n)$ gates (where both the depth and the weights of the circuit are unrestricted), then $S(n) = \Omega(2^{n/2}/\sqrt{n})$.

Proof: It can be shown (see Exercise 3.8) that there is a threshold circuit C_U of N gates and

$$\sum_{i=1}^{N}(n+i) = O(nN + N^2)$$

number of weights with the following property: Given *any* threshold circuit $\hat{C}$ with n inputs and N or less gates, there always exists a choice of the weights in C_U such that the resultant circuit computes the same function as $\hat{C}$.

It follows that a threshold circuit with n input variables and $S(n)$ gates can have at most

$$\sum_{i=1}^{S(n)}(n+i) = O(nS(n) + S^2(n))$$

number of weights. Now the total number of input vectors to a circuit computing a general Boolean function is 2^n. By Theorem 2.18, the number of weights in the circuit must be at least $\Omega(2^n/n)$. Hence,

$$nS(n) + S^2(n) = \Omega(2^n/n),$$

which implies that

$$S(n) = \Omega(2^{n/2}/\sqrt{n}) .\qquad\qquad \square$$

Corollary 3.3:   If a general symmetric function of n variables can be computed by a threshold circuit with $\Gamma(n)$ gates (where both the depth and the weights of the circuit are unrestricted), then $\Gamma(n) = \Omega(\sqrt{n/\log n})$.

Proof: Suppose that any symmetric function of n variables can be computed by a threshold circuit with at most $\Gamma(n)$ gates. Since any n-variable Boolean function is equivalent to a symmetric function of $2^n - 1$ variables (see Theorem 3.6), one can compute any Boolean function using $\Gamma(2^n - 1)$ threshold gates. Now Theorem 3.7 implies that $\Gamma(2^n - 1) = \Omega(2^{n/2}/\sqrt{n})$, or equivalently, $\Gamma(n) = \Omega(\sqrt{n/\log n})$. ☐

In [87], constant-depth threshold circuits of size $O(2^{n/2}/n)$, which compute general Boolean functions, have been derived. Using such constructions and threshold circuits for computing the sum of n 1-bit numbers (as presented in Chapter 5), one can derive size-optimal constant-depth threshold circuits for computing general symmetric functions.

3.5 Hyperplane Cuts of a Hypercube

In Section 3.2, we have shown that the n-variable Parity function PAR_n can be computed by a depth-2 threshold circuit with $(\frac{n}{2} + 1)$ gates. Using techniques of rational approximation, we shall show in Chapter 7 that if a depth-2 threshold circuit computing PAR_n has weights that are polynomially bounded integers, then the circuit size (number of gates) must be at least almost linear $(\Omega(n/\log^2 n))$. Here we derive a lower bound $\Omega(\sqrt{n})$ on the size of any depth-2 threshold circuit with arbitrary weights computing PAR_n. The proof of this result employs the theory of posets, which is also used in Chapter 2 to derive a lower bound on the number of linear threshold functions. For a summary of the terminologies of posets, see Section 2.7.3.

Recall that given any set $\hat{S}$, the class of all subsets of $\hat{S}$ is a poset under set inclusion ($\subseteq$). A collection $\mathcal{C}$ of subsets of $\hat{S}$ is an antichain if for any two distinct sets A_i, A_j in $\mathcal{C}$, we have $A_i \not\subseteq A_j$. Sperner's Lemma, as stated in Lemma 3.2 below, gives a tight bound on the maximum cardinality of any such antichain.

Lemma 3.2: (**Sperner's Lemma**) Let $\mathcal{C}$ be an antichain of subsets of an n-element set $\hat{S}$. Then

$$|\mathcal{C}| \le \binom{n}{\lceil \frac{n}{2} \rceil}.$$

Proof: Given any subset A of $\hat{S}$, we say that a permutation π of elements in $\hat{S}$ *begins with A* if the first $|A|$ members of π are the elements of A in some order. Observe that there are $|A|! \, (n-|A|)!$ permutations that begin with A. If a permutation begins with A_i and A_j, then either $A_i \subseteq A_j$ or $A_j \subseteq A_i$. Thus, permutations beginning with different sets in $\mathcal{C}$ are distinct. Hence,

$$\sum_{A \in \mathcal{C}} |A|! \, (n-|A|)! \le n! \ .$$

If we let m_k denote the number of members of $\mathcal{C}$ of size k, then

$$\sum_{k} k! \, (n-k)! \, m_k \le n! \ .$$

Therefore,

$$\sum_{k} \frac{m_k}{\binom{n}{k}} \le 1 \ .$$

Since $\binom{n}{k} \le \binom{n}{[\frac{n}{2}]}$ for all k, we have

$$\begin{aligned}
|\mathcal{C}| &= \sum_{k} m_k \\
&= \binom{n}{[\frac{n}{2}]} \sum_{k} \frac{m_k}{\binom{n}{[\frac{n}{2}]}} \\
&\le \binom{n}{[\frac{n}{2}]} \sum_{k} \frac{m_k}{\binom{n}{k}} \\
&\le \binom{n}{[\frac{n}{2}]} \ .
\end{aligned}$$

$\square$

The upper bound stated in Sperner's Lemma can be achieved by taking $\mathcal{C}$ to be the collection of all subsets of cardinality $[\frac{n}{2}]$. In the proof above, each permutation $\pi = s_{i_1} \cdots s_{i_n}$ of elements in $\hat{S}$ has a one-to-one correspondence to a maximal chain $\hat{C}_\pi = \{S_{i_1}, \ldots, S_{i_n}\}$ in the poset of subsets of $\hat{S}$, where $S_{i_k} = \{s_{i_1}, \ldots, s_{i_k}\}$. Given any subset A of $\hat{S}$, a permutation π begins with A if and only if the corresponding maximal chain $\hat{C}_\pi$ contains A. Moreover, each maximal chain contains at most one member of $\mathcal{C}$. Since the total number of maximal chains in the poset is $n!$, $|A|! \, (n-|A|)! \, /n!$ thus represents the probability that a randomly chosen maximal chain will contain A, and this probability is the same for all subsets A with the same cardinality. These observations lead to the following generalization of Sperner's Lemma to more general posets.

Lemma 3.3: Let S be a finite poset with unique minimal and maximal elements O and I, and a rank function r, such that for each $k \geq 0$, all elements of rank k

(i) are covered by the same number a_k of elements of rank $(k+1)$, and

(ii) cover the same number b_k of elements of rank $(k-1)$.

If m_k is the number of elements of rank k in S, then any antichain $\mathcal{A}$ in S has at most M elements, where $M = \max_k m_k$.

Proof: For each k, let S_k be the subset of elements of S of rank k. For $x \in S$, let q_x be the number of maximal chains in S that contains x. If N denotes the number of all maximal chains of S, then $p_x = q_x/N$ is the probability that a randomly chosen maximal chain will contain x.

First, note that for any $x \in S_k$, there are $b_k \cdots b_1$ maximal chains in the interval $[O, x]$ and $a_k \cdots a_{n-1}$ maximal chains in the interval $[x, I]$, where $n = r(I)$. Thus, $p_x = b_k \cdots b_1 a_k \cdots a_{n-1}/N$, and this probability is the same for all $x \in S_k$.

Second, for a given k, each maximal chain of S contains exactly one element of S_k. Addition of probabilities for mutually exclusive events then gives

$$1 = \sum_{x \in S_k} p_x = m_k p_x .$$

Thus, $p_x = 1/m_k \geq 1/M$.

Finally, note that each maximal chain of S contains at most one element of $\mathcal{A}$, since $\mathcal{A}$ is an antichain. Therefore, the probability that a given maximal chain contains some element of $\mathcal{A}$ is $\sum_{x \in \mathcal{A}} p_x$, which does not exceed 1. Hence,

$$1 \geq \sum_{x \in \mathcal{A}} p_x \geq \sum_{x \in \mathcal{A}} 1/M = |\mathcal{A}|/M ,$$

or $|\mathcal{A}| \leq M$. $\qquad\square$

An n-dimensional hypercube $\hat{H}_n$ is a graph with vertices in $\{0, 1\}^n$ such that x and y in $\{0, 1\}^n$ are adjacent if they differ in exactly one coordinate. Consider a stratified depth-2 threshold circuit C_n (see Exercise 3.15) computing the n-variable Parity function PAR_n, where we assume without loss of generality that the argument in the linear threshold function for each gate is never zero. Then, each threshold gate in the first layer corresponds to an $(n-1)$-dimensional hyperplane passing through (cutting) the edges of $\hat{H}_n$. Since PAR_n has different values on each pair of adjacent vertices in $\hat{H}_n$, each edge in $\hat{H}_n$ must be cut by some of these hyperplanes. So a lower bound on the number of $(n-1)$-dimensional hyperplanes that cut all the edges of $\hat{H}_n$ also yields a lower bound on the size of any stratified depth-2 threshold circuit (with no restriction on its weights) computing PAR_n. Such a lower bound result can then be applied to non-stratified depth-2 threshold circuits.

We first derive an upper bound on the maximum number of edges in $\hat{H}_n$ that can be cut by a hyperplane using the generalization of Sperner's Lemma (Lemma 3.3). The underlying idea is to impose a partial order on the edges of the hypercube. Without loss of generality, we can assume each cutting hyperplane to have the following form:

$$\sum_{i=1}^{n} w_i x_i = 1, \text{ or } \vec{w} \cdot \vec{x} = 1 , \tag{3.3}$$

where $0 \leq x_i \leq 1$, $i = 1, \ldots, n$. Since we are concerned with the maximum number of edges of the hypercube that can be cut by a hyperplane defined above, we can further assume that the coefficients w_i's are non-negative (Exercise 3.14).

Let E_n be the set of edges of the hypercube $\hat{H}_n$. We denote an edge $e \in E_n$ by $(\vec{x}_i, \vec{x}_j)$, where $\vec{x}_i, \vec{x}_j \in \{0,1\}^n$ are the adjacent vertices on which e is incident, and $\vec{x}_i \leq \vec{x}_j$ (i.e., the inequality holds for the corresponding coordinates in $\vec{x}_i$ and $\vec{x}_j$). Let $e = (\vec{x}_i, \vec{x}_j)$ and $e' = (\vec{x'}_i, \vec{x'}_j)$ be any two edges in E_n. Then a partial order $\preceq$ on E_n can be defined as follows: $e \preceq e'$ if $\vec{x}_j \leq \vec{x'}_i$. We can also define a rank function r on E_n: for $e = (\vec{x}_i, \vec{x}_j) \in E$, $r(e)$ is the number of 1's in $\vec{x}_i$. Let $\tilde{S}$ be the set of edges in $\hat{H}_n$ cut by any hyperplane P defined by Eqn. (3.3) with all coefficients w_i's being non-negative. With the partial order defined on E_n, we obtain the following result.

Lemma 3.4: $\tilde{S}$ is an antichain in E_n.

Proof: Suppose on the contrary that P cuts the edges $e = (\vec{x}_i, \vec{x}_j)$ and $e' = (\vec{x'}_i, \vec{x'}_j)$, with $e \preceq e'$. Then $\vec{w} \cdot \vec{x}_i < 1$, $\vec{w} \cdot \vec{x}_j > 1$, $\vec{w} \cdot \vec{x'}_i < 1$, and $\vec{w} \cdot \vec{x'}_j > 1$. Since $e \preceq e'$, we have $\vec{x'}_i - \vec{x}_j = \vec{z} \in \{0,1\}^n$. However, by assumption, all coefficients w_i's defining the hyperplane P are positive. Thus, $1 < \vec{w} \cdot \vec{x}_j \leq \vec{w} \cdot (\vec{x}_j + \vec{z}) = \vec{w} \cdot \vec{x'}_i$, which contradicts $\vec{w} \cdot \vec{x'}_i < 1$. $\square$

Lemma 3.5:   The maximum number of edges in E_n that can be cut by any hyperplane is bounded above by

$$(n - \lceil \tfrac{n}{2} \rceil) \binom{n}{\lceil \tfrac{n}{2} \rceil}.$$

Proof: First observe that for any $k > 0$, all elements of E_n of rank k are covered by the same number of elements of rank $(k+1)$ and cover the same number of elements of rank $(k-1)$. By Lemma 3.3, we conclude that the set $\tilde{S}$ has at most $M = \max_k m_k$ elements, where m_k is the number of elements of rank k in E_n. m_k achieves its maximum when $k = \lceil \tfrac{n}{2} \rceil$ and is given by the expression stated in the lemma. The result follows.   □

Theorem 3.8:   Any depth-2 threshold circuit computing PAR_n must have $\Omega(\sqrt{n})$ gates.

Proof: First consider a stratified depth-2 threshold circuit computing PAR_n. Then, the hyperplanes corresponding to the threshold gates in the first layer of the circuit must cut all the edges of the n-dimensional hypercube $\hat{H}_n$. Since the total number of edges in $\hat{H}_n$ is $n2^n$, by Lemma 3.5 the number of such hyperplanes must be at least

$$\frac{n2^n}{(n - \lceil \tfrac{n}{2} \rceil)\binom{n}{\lceil \tfrac{n}{2} \rceil}} = \Omega(\sqrt{n}).$$

By Exercise 3.15, if PAR_n can be computed by a non-stratified depth-2 threshold circuit with N gates in the first layer, then PAR_{n-1} can be computed by a stratified depth-2 threshold circuit with $2N$ gates in the first layer. Hence, the theorem follows.   □

Exercises

3.1:   [*An alternative definition of symmetric function*]
Show that a function $f(x_1, \cdots, x_n)$ is invariant under any permutation of the inputs $x_1, \cdots, x_n$ if and only if $f(X)$ is determined by the sum of the inputs $\sum_{i=1}^{n} x_i$. This result partly justifies Definition 3.1.

3.2:   [*Threshold circuits for a function and its negation*]
Given a threshold circuit C_f for a function f, show that one can modify the circuit to yield a threshold circuit $C_{\overline{f}}$ for $\overline{f}$ with the same depth and size as C_f, where $\overline{f}$ is the negation of f.

3.3:   [*Weighted sum and generalized symmetric functions*]
Using the fact that $\overline{x}_i = 1 - x_i$ for a Boolean variable $x_i \in \{0, 1\}$, and assuming that $\overline{x}_i$'s are also considered as input variables, show that the class of generalized symmetric functions remain the same if the w_j's in Definition 3.3 are allowed to be negative.

3.4:   [*Parity and depth-2 AND-OR circuits*]
Show that if the Parity function of n variables is computed by a depth-2 AND-OR circuit, then at least $(2^{n-1} + 1)$ gates are needed. Also show that $(2^{n-1} + 1)$ gates are sufficient.

3.5:   [*Majority gates and threshold gates*]
Show that any linear threshold function of n variables with polynomially bounded integer weights (i.e., $\widehat{LT}_1$ function) is a projection of a majority function with polynomially many (in n) variables.

3.6:   [*Majority and Exact functions are complete for the class of symmetric functions under AC^0 reduction*]
Given a class of functions $\mathcal{F}$, a function $f(X) \in \mathcal{F}$ is said to be complete for $\mathcal{F}$ under AC^0 reduction, if every function in $\mathcal{F}$ is computable by a constant-depth polynomial-size circuit with AND/OR gates and additional gates that compute f.

(a) Show that the functions Majority and Exact_k (for some k) are complete for the class of symmetric functions under AC_0 reduction.

(b) Give an explicit realization of Majority using a depth-2 circuit comprising only Exact_k gates.

3.7:   [*Spectral coefficients of symmetric functions*]
Show that the spectral coefficients of a symmetric function $f(X)$ satisfy the following property: $a_{\alpha_1} = a_{\alpha_2}$ if $|\alpha_1| = |\alpha_2|$, where $|\alpha|$ is the number of 1's in α.

3.8:   [*Networks with a maximum number of weights*]
Consider an n-input threshold circuit comprising N gates, where each gate is labeled by a unique integer i, $1 \le i \le N$. The set of inputs to the ith gate consists of all the n input variables, and the outputs of the gates with labels less than i. Thus, the circuit depth is N and the fan-in of the ith gate equals $(n + i - 1)$. Such a circuit is invoked in the proof of Theorem 3.7.

Show that there always exists a choice of the weights of the above circuit such that the resultant circuit computes the same function as *any* other threshold circuit with N or less gates and n inputs.

3.9:   [*Tight lower bound on the number of weights*]
It is known that any Boolean function of n variables can be computed by a threshold circuit comprising $2^{n/2}/\sqrt{n}$ gates (compare with Theorem 3.7). Use this fact to show that the asymptotic lower bound stated in Theorem 2.18 is tight.

3.10:   [*Computing symmetric functions with bounded fan-in AND-OR circuits of linear edge complexity*]

  (a) Show that the sum of three k-bit integers can be reduced to the sum of two $(k + 1)$-bit integers using constant-depth circuits with $O(k)$ bounded fan-in AND/OR gates.

  (b) Use the above result to construct bounded fan-in AND-OR circuits of $O(n)$ gates and $O(\log n)$-depth that compute the sum of n 1-bit integers.

  (c) Hence, conclude that any symmetric function of n variables can be computed by a bounded fan-in AND-OR circuit of $O(\log n)$-depth, $O(n)$ gates and edges.

3.11:   [*Computing general Boolean functions in depth-3 AND-OR circuits*]
Let $f(x_1, ..., x_n) : \{0,1\}^n \rightarrow \{0,1\}$ be a general Boolean function. Let $m = \lceil n/2 \rceil$. For any $\vec{\gamma} = (\gamma_1, ..., \gamma_m) \in \{0,1\}^m$, let $p_{\vec{\gamma}}(x_1, ..., x_m)$ equal 1 if $\vec{\gamma} = (x_1, ..., x_m)$, and equal 0 otherwise. Note that p can be realized by either an AND gate or an OR gate. For any $\vec{\beta} = (\beta_{m+1}, ..., \beta_n) \in \{0,1\}^{n-m}$, let $\vec{\beta}' = (\overline{\beta}_{m+1}, ..., \overline{\beta}_n)$ denote the negation of $\vec{\beta}$.

(a) For any $\vec{\alpha} = (\alpha_1, ..., \alpha_m) \in \{0,1\}^m$, show that

$$f(\vec{\alpha}, x_{m+1}, ..., x_n) = \bigwedge_{\{\vec{\beta} \in \{0,1\}^{n-m} | f(\vec{\alpha}, \vec{\beta}') = 0\}} p_{\vec{\beta}}(x_{m+1}, ..., x_n) \ .$$

(b) Show that f can be expressed as

$$f(x_1, ..., x_n) = \bigvee_{\vec{\alpha} \in \{0,1\}^m} p_{\vec{\alpha}}(x_1, ..., x_m) f(\vec{\alpha}, x_{m+1}, ..., x_n).$$

(c) Use the above results to construct a depth-3 AND-OR circuit for a general Boolean function with at most $O(2^{n/2})$ gates.

3.12:   [*Depth-3 threshold circuits for Inner Product Mod 2*]
Show that the Inner Product Mod 2 function of n variables

$$IP(X, Y) = \bigoplus_{i=1}^{n/2} (x_i \wedge y_i)$$

can be computed by a depth-3 threshold circuit comprising at most $\lceil \frac{3n}{4} \rceil + 1$ gates.

3.13:   [*Uniqueness of antichains that achieve Sperner's bound*]
Let S be a given set of n elements and P be the poset of all subsets of S under set inclusion.

(a) Show that if n is even, there is only one maximum-sized antichain, namely, the collection of all subsets of cardinality $n/2$.

(b) Let $n = 2m + 1$ be odd, and $\mathcal{A}$ be a maximum-sized antichain in P.

   (i) Show that all sets in $\mathcal{A}$ must be of size m or $(m + 1)$.
   (ii) Suppose X, Y are subsets of size m, $(m + 1)$, respectively, and that $X \subset Y$. Show that either X or Y must be in $\mathcal{A}$.
   (iii) Suppose $\mathcal{A}$ contains some but not all of the subsets of $(m + 1)$ elements. Then, without loss of generality, we can assume for some i, $E = \{x_1, ..., x_{m+1}\} \in \mathcal{A}$ but $F = \{x_i, ..., x_{m+i}\} \notin \mathcal{A}$. Argue that there must be an integer $j < i$ with $E^* = \{x_j, ..., x_{m+j}\} \in \mathcal{A}$ but $F^* = \{x_{j+1}, ..., x_{m+j+1}\} \notin \mathcal{A}$, and thus $E^* \cap F^* \notin \mathcal{A}$.

(iv) Show that part (ii) implies that either $E^* \cap F^*$ or F^* must be in $\mathcal{A}$, and this contradicts the result of part (iii). Hence, conclude that $\mathcal{A}$ indeed consists of subsets all of the same size.

3.14:   [*Hypercube edges cut by hyperplanes with non-negative coefficients*]
Let m be the number of edges in the hypercube $\hat{H}_n$ that are cut by a hyperplane $\sum\limits_{i=1} w_i x_i = 1$. Show that there is a hyperplane $\sum\limits_{i=1} \tilde{w}_i x_i = 1$ with all $\tilde{w}_i$'s non-negative that cuts m edges in $\hat{H}_n$.

3.15:   [*Stratified vs. non-stratified depth-2 threshold circuits for Parity*]
Let C_n be any depth-2 threshold circuit computing the Parity of n inputs. Denote its node complexity by $N(n)$ and its edge complexity by $E(n)$.

(a) Let x_i be any input variable. Let C' be a circuit derived from C_n by assigning the value 1 to x_i. Let C'' be a circuit derived from C_n by assigning the value 0 to x_i and negating the output threshold and the weights of all incoming edges to the output node. Show that circuits C' and C'' compute the same function, which is the negation of the Parity of the rest $(n-1)$ inputs.

(b) Let $\mathrm{sgn}(\sum w_i f'_i - t)$ and $\mathrm{sgn}(\sum -w_i f''_i + t)$ be the output nodes of C' and C'', respectively. Obtain a new circuit $\hat{C}$ by merging the two output nodes of C' and C'' such that the output node of $\hat{C}$ computes $\mathrm{sgn}(\sum w_i f_i - w_i f''_i)$. Show that $\hat{C}$ computes the same function as C' and C''.

(c) Conclude from the result of part (b) that if PAR_n is computable by a non-stratified depth-2 threshold circuit with N gates in the first layer, then PAR_{n-1} can be computed by a stratified (i.e., there are no direct edges from the input variables to the output node) depth-2 threshold circuit with $2N$ gates in the first layer.

Bibliographic Notes

The construction of depth-2 threshold circuits with $n+1$ gates for computing n-variable symmetric functions is shown in Muroga [84]. The telescopic technique used in reducing the gate count of the depth-2 construction (Theorem 3.3) is from Minnick [80]. The reduction in the size of threshold circuits computing symmetric functions from $O(n)$ to $O(\sqrt{n})$ by increasing the depth from 2 to 3 is shown in Siu, Roychowdhury, and Kailath [127]. Similar techniques are used in Redkin [107] in the derivation of threshold circuits for computing general Boolean functions. The result in [107] is similar to the one in Theorem 3.6.

The lower bound result stated in Theorem 3.7 is derived in Nechiporuk [87], which also establishes a tight (within a constant factor of $\sqrt{2}$) upper bound. The result of Nechiporuk is refined in Lupanov [73, 74]. Techniques similar to Nechiporuk are used in Cover [32] to derive a lower bound on the number of weights in a feedforward network of linear threshold elements that can compute any binary function over N vectors. This result of Cover [32] is proved in Chapter 2 (Theorem 2.18).

The result on the maximum number of edges in a hypercube that can be cut by any hyperplane (Lemma 3.5) is shown in O'Neil [92], which remains the best-known result, and from which the lower bound on the size of depth-2 threshold circuits for Parity (Theorem 3.8) easily follows. In Chapter 7, we derive a stronger lower bound on the circuit size where the weights of the threshold circuits are assumed to be polynomially bounded integers. There are many proofs of Sperner's Lemma in the literature, and ours is due to Lubell [72]. The generalization to Sperner's Lemma (Lemma 3.3) is due to Baker [9]. Anderson [7] provides an excellent account of Sperner's Lemma and its applications.

Chapter 4

Depth-Efficient Arithmetic Circuits

4.1 Introduction

In this chapter, we derive small-depth polynomial-size threshold circuits that compute common arithmetic functions. We emphasize here the minimization of the depth of threshold circuits rather than the size. All the threshold circuits derived here have polynomial size and polynomially bounded integer weights, and many of the circuits are shown to be depth optimal.

In Section 4.2, we consider two basic functions: Comparison and Addition. These two functions will be used later in the realizations of many of the arithmetic functions. First, using harmonic analysis, we show the existence of depth-2 polynomial-size threshold circuits. The key concept underlying these depth-2 circuits shows that the functions Comparison and Addition can be closely approximated by sparse polynomials. Then, using techniques from the theory of error-correcting codes, we also give explicit constructions of such sparse polynomials and the resulting depth-2 threshold circuits computing the functions.

In Section 4.3, we derive threshold circuits for Multiplication (computing the product of two n-bit integers). We first show that Multiple Sum (computing the sum of n n-bit integers) can be computed by depth-3 threshold circuits. By a simple reduction, we then show that Multiplication can be computed by depth-4 threshold circuits. Based on the construction for

137

Multiple Sum, we also derive depth-efficient threshold circuits for Division (computing the quotient of two n-bit integers) and related problems. In particular, we show in Section 4.4 that Division of two n-bit integers and Powering can be computed by depth-4 threshold circuits.

The results for Comparison are used in Section 4.5 to derive depth-3 threshold circuits for the sorting of a set of n n-bit integers. In Section 4.6, we also show that any polynomial-size threshold circuit computing Sorting or Division must have depth at least 3.

The presentation of depth-optimal threshold circuits for Multiple Sum, Multiplication, and Division will be postponed until Chapter 6, where the issues of depth-weight tradeoffs in threshold circuits are also considered. In Chapter 6, we shall integrate the techniques developed here with other advanced combinatorial methods to construct depth-optimal circuits.

4.2   Depth-2 Threshold Circuits for Comparison and Addition

This section concerns two basic functions: Comparison and Addition. For the convenience of presenting the harmonic analysis and approximation techniques, we adopt the $\{1, -1\}$ notations. In other words, we define a Boolean function as $f : \{1, -1\}^n \to \{1, -1\}$. Also recall that $\widehat{LT}_d$ denotes the class of Boolean functions that can be computed by depth-d polynomial-size threshold circuits with polynomially bounded integer weights.

4.2.1   Existence Proof via Harmonic Analysis

The function Comparison has been introduced in Section 1.4. Let us restate its definition in terms of $\{1, -1\}$ notation here.

Definition 4.1:   Let $X = (x_1, \ldots, x_n)$, $Y = (y_1, \ldots, y_n) \in \{1, -1\}^n$. The function Comparison is defined as

$$COMP_n(X, Y) = \text{sgn}(\sum_{i=1}^{n} 2^i(x_i - y_i) + 1).$$ □

Before we show that Comparison can be computed by depth-2 polynomial-size threshold circuits, we first derive a more general result. Recall from the discussions on spectral representation (see Section 1.5) that every Boolean function $f(X) : \{1, -1\}^n \rightarrow \{1, -1\}$ can be represented as $f(X) = \sum_{\alpha \in \{0,1\}^n} a_\alpha X^\alpha$, and that its spectral norm is defined as $\|f\|_{\mathcal{F}} = \sum_{\alpha \in \{0,1\}^n} |a_\alpha|$. In what follows, we show that any Boolean function with polynomially bounded L_1 spectral norm can be closely approximated by a *sparse polynomial*. A polynomial of n variables is said to be *sparse* if the number of monomials is polynomially bounded in n.

Lemma 4.1:   Let $f(X) : \{1, -1\}^n \rightarrow \{1, -1\}$ such that $\|f\|_{\mathcal{F}} \leq n^c$ for some c. Then, for any $k > 0$, there exists a sparse polynomial

$$F(X) = \frac{1}{N} \sum_{\alpha \in S} w_\alpha X^\alpha$$

such that

$$|F(X) - f(X)| \leq n^{-k},$$

where w_α and N are integers, $S \subset \{0, 1\}^n$, the size of S, w_α, and N are all bounded by a polynomial in n. Hence, $f(X) \in \widehat{LT}_2$.

Proof: The proof is based on probabilistic arguments. It suffices to show that there exists a random sparse polynomial of the above form $F(X)$ such that $|F(X) - f(X)| \leq n^{-k}$ with probability > 0. Let $f(X) = \sum_{\alpha \in \{0,1\}^n} a_\alpha X^\alpha$ and $L_1 = \|f\|_{\mathcal{F}} = \sum_{\alpha \in \{0,1\}^n} |a_\alpha|$. Let $\lceil L_1 \rceil$ be the least integer not smaller than L_1. Note that $\lceil L_1 \rceil < \|f\|_{\mathcal{F}} + 1$ and is polynomially bounded. For $\alpha \in \{0, 1\}^n$, let $p_\alpha = \dfrac{|a_\alpha|}{\lceil L_1 \rceil}$. We define independent identically distributed random variables $Z_i(X)$ such that

$$Z_i(X) = \begin{cases} \operatorname{sgn}(a_\alpha) X^\alpha & \text{with prob. } p_\alpha, \ \alpha \in \{0, 1\}^n, \\ 0 & \text{with prob. } 1 - \sum_{\alpha \in \{0,1\}^n} p_\alpha. \end{cases}$$

Note that the expected value of $Z_i(X)$ is

$$
\begin{aligned}
E[Z_i(X)] &= \sum_{\alpha \in \{0,1\}^n} p_\alpha \operatorname{sgn}(a_\alpha) X^\alpha \\
&= \sum_{\alpha \in \{0,1\}^n} a_\alpha X^\alpha / \lceil L_1 \rceil = \frac{f(X)}{\lceil L_1 \rceil},
\end{aligned}
$$

and the variance is

$$
\begin{aligned}
Var[Z_i(X)] &= E[Z_i^2(X)] - E^2[Z_i(X)] \\
&= \frac{\|f\|_{\mathcal{F}}}{\lceil L_1 \rceil} - \frac{1}{\lceil L_1 \rceil^2} = \Omega(1).
\end{aligned}
$$

Therefore, by the *Central Limit Theorem* and for sufficiently large n, we have

$$Pr\{|\sum_{i=1}^{N}(Z_i(X) - \frac{f(X)}{\lceil L_1 \rceil})| > \sqrt{nN}\}$$

$$= Pr\{|\frac{\lceil L_1 \rceil}{N}(\sum_{i=1}^{N} Z_i(X)) - f(X)| > n^{-k}\}$$

$$= O(e^{-n})$$

$$< 2^{-n}$$

for $N = O(\lceil L_1 \rceil^2 n^{2k+1})$, which is polynomially bounded. Now, take $F(X) = \frac{\lceil L_1 \rceil}{N}\sum_{i=1}^{N} Z_i(X)$. By the union bound, we obtain

$$Pr\{|F(X) - f(X)| > n^{-k} \text{ for some } X \in \{1, -1\}^n \} < 1.$$

Equivalently,

$$Pr\{|F(X) - f(X)| \leq n^{-k} \text{ for all } X \in \{1, -1\}^n \} > 0.$$

We can rewrite $F(X) = \frac{1}{N}\sum_{\alpha \in S} w_\alpha X^\alpha$, where w_α and the cardinality of S are bounded by N. Observe that

$$f(X) = \text{sgn}(F(X)) = \text{sgn}(\sum_{\alpha \in S} w_\alpha X^\alpha).$$

Since each monomial X^α is a symmetric function and thus can be written as a sum of polynomially many $\widehat{LT}_1$ functions, it follows that $f(X) \in \widehat{LT}_2$ (see Theorem 3.5 and Corollary 3.2).   □

As a consequence of Lemma 4.1, we obtain the following lemma.

Lemma 4.2:   Let $f(X) : \{1, -1\}^n \to \{1, -1\}$ such that $\|f\|_{\mathcal{F}} \leq n^c$ for some c. Then, for any $k > 0$, there exists a linear combination of $\widehat{LT}_1$ functions

$$F(X) = \frac{1}{N}\sum_{i=1}^{s} w_j t_j(X)$$

such that

$$|F(X) - f(X)| \leq n^{-k},$$

where $t_j(X) \in \widehat{LT}_1$, and s, w_j's, and N are integers bounded by a polynomial in n.

Proof: The proof follows immediately from Lemma 4.1 by rewriting every monomial X^α in $F(X)$ as a sum of polynomially many $\widehat{LT}_1$ functions. □

Theorem 4.1: Comparison $\in \widehat{LT}_2$.

Proof: It suffices to show that Comparison has polynomially bounded L_1 spectral norms. We write a recursion for the spectral representation of $COMP_n(X, Y)$. If C_n is the polynomial corresponding to the function of $X = (x_1, \ldots, x_n)$ and $Y = (y_1, \ldots, y_n)$, it is easy to see that

$$C_n = \frac{x_n - y_n}{2} + \frac{1 + x_n y_n}{2} C_{n-1}.$$

This shows that the L_1 spectral norm increases by 1 when n is increased by 1. So if we denote the L_1 spectral norm by $\|\cdot\|_{\mathcal{F}}$ as before, then with $\|C_1\|_{\mathcal{F}} = 2$, we have by induction that $\|C_n\|_{\mathcal{F}} = n + 1$ and thus is polynomially bounded. Hence, Comparison $\in \widehat{LT}_2$. □

Remark 4.1: It follows trivially by taking $-Y$ that

$$\tilde{C}(X, Y) = \text{sgn}(\sum_{i=1}^{n} 2^i (x_i + y_i) + 1) \in \widehat{LT}_2 .$$

□

Similar techniques also apply to Addition.

Definition 4.2: Given two n-bit integers X and Y, Addition is defined as the problem of computing the $(n + 1)$-bit sum $X + Y$. □

Theorem 4.2: Addition $\in \widehat{LT}_2$.

Proof: It suffices to show that the L_1 spectral norm of each bit of the sum is polynomially bounded. Let $X_1 = x_{1_{n-1}} x_{1_{n-2}} \cdots x_{1_0}$, $X_2 = x_{2_{n-1}} x_{2_{n-2}} \cdots x_{2_0}$ be the two n-bit binary numbers whose sum is to be computed. Let the input variables to Addition be arranged as $(x_{1_{n-1}}, x_{2_{n-1}}, \ldots, x_{1_0}, x_{2_0})$. Let $s = s_n s_{n-1} \cdots s_0$ denote the resulting $(n + 1)$-bit sum and c_k denote the kth carry bit, for $k = 1, \ldots, n$. Then, $s_n = c_n$, $s_k = \text{XOR}(x_{1_k}, x_{2_k}, c_k)$ for $1 \le k \le n - 1$, $s_0 = \text{XOR}(x_{1_0}, x_{2_0})$. It follows that $\|s_k\|_{\mathcal{F}} = \|x_{1_k} x_{2_k} c_k\|_{\mathcal{F}} = \|c_k\|_{\mathcal{F}}$. Hence, it suffices to show that c_k has a polynomially bounded L_1 spectral norm.

As in the proof of Theorem 4.1, we write a recursion for the spectral representation of the nth-carry bit c_n as a function of $(x_{1_{n-1}}, x_{2_{n-1}}, \ldots, x_{1_0}, x_{2_0})$. Then, if $\tilde{C}_n$ is the polynomial corresponding to the nth-carry bit, we have

$$\tilde{C}_n = \frac{x_{1_{n-1}} + x_{2_{n-1}}}{2} + \frac{1 - x_{1_{n-1}} x_{2_{n-1}}}{2} \tilde{C}_{n-1}.$$

This shows that the L_1 spectral norm increases by 1 when n is increased by 1. The same calculation as in the case of Comparison gives $\|\tilde{C}_n\|_{\mathcal{F}} = n + 1$ for all $n \geq 1$. It follows that the L_1 spectral norm of each bit of the resultant sum is polynomially bounded and hence, Addition $\in \widehat{LT}_2$. $\square$

4.2.2 Explicit Constructions Based on Error-Correcting Codes

In the preceding section, we have seen that any Boolean function with polynomially bounded spectral norm can be approximated by a sparse polynomial (Lemma 4.1) and thus can be computed by a depth-2 threshold circuit. Since the proof of Lemma 4.1 is based on probabilistic methods, for a general Boolean function with polynomially bounded spectral norm, the result does not yield an explicit construction of the corresponding sparse polynomial approximation. For some specific functions, such as Comparison and Addition, we can in fact provide explicit constructions of the approximating polynomials and thus the corresponding depth-2 threshold circuits.

The construction of the sparse polynomials for Comparison and Addition will be based on some known results from the theory of error-correcting codes. We first introduce some terminologies and review some basic concepts.

Let $G = (g_{ij})_{1 \leq i \leq n, 1 \leq j \leq N}$ be the $(n \times N)$ generator $0, 1$ matrix of a linear block code over $GF(2)$, where N is polynomially bounded in n. The rows of G are linearly independent and every code word $c \in \{0, 1\}^N$ can be written as $c = uG$ for some $u = (u_1, \ldots, u_n) \in \{0, 1\}^n$. Associate each u with a unique $X_u = (x_1, x_2, \ldots, x_n) \in \{1, -1\}^n$ such that $x_i = -1$ if $u_i = 1$, and $x_i = 1$ if $u_i = 0$. Moreover, for each jth column in the generator matrix, consider the monomial $\prod\limits_{\{i : g_{ij} = 1\}} x_i$. If $c = uG$, then

$$\prod_{\{i : g_{ij} = 1\}} x_i = \begin{cases} 1 & \text{if } c_j = 0, \\ -1 & \text{if } c_j = 1. \end{cases} \tag{4.1}$$

Now suppose that in the linear block code, the Hamming weight (i.e., the number of 1's) of each non-zero code word is between $(1-\epsilon)\frac{N}{2}$ and $(1+\epsilon)\frac{N}{2}$ for some small $\epsilon > 0$. In other words, the Hamming weight of any non-zero code word is sufficiently close to half the length of the code word. Define a polynomial of N monomial terms as follows:

$$P_G(X) = \sum_{j=1}^{N} \prod_{i;g_{ij}=1} x_i,$$

where $X = (x_1, \ldots, x_n) \in \{1, -1\}^n$. Observe from Eqn. (4.1) that $|P_G(X_u)|$ is the absolute value of the difference between the number of 0's and the number of 1's in the code word $c = uG$. It follows from the above assumption on the block code that for each $X_u \neq (1, 1, \ldots, 1)$ (i.e., the corresponding $u \neq (0, 0, \ldots, 0)$),

$$|P_G(X_u)| \leq (1+\epsilon)\frac{N}{2} - (1-\epsilon)\frac{N}{2} = \epsilon N.$$

Moreover, when $X_u = (1, 1, \ldots, 1)$, we have $P_G(1, 1, \ldots, 1) = N$.

Examples of linear block codes with properties described above can be found in [6] for any $\alpha \geq 2$, $N = \Theta(n^{\alpha+\delta})$ and $\epsilon = O(n/\sqrt{N})$, for some $\delta \geq 0$. We summarize the above discussions and the results that will be useful for us as the following lemma.

Lemma 4.3: Let $X = (x_1, \ldots, x_n) \in \{1, -1\}^n$. For every fixed $\alpha > 2$, there exists a polynomial $P(X)$ of N monomials, where $N = \Theta(n^{\alpha+\delta})$ and $\delta \geq 0$, such that $P(1, 1, \ldots, 1) = N$, and for $X \neq (1, 1, \ldots, 1)$, $|P(X)| \leq \epsilon N$ where $\epsilon = O(n/\sqrt{N})$. □

Recall that

$$COMP_n(X, Y) = \text{sgn}\{\sum_{i=1}^{n} 2^i(x_i - y_i) + 1\},$$

where $X = (x_1, \ldots, x_n)$ and $Y = (y_1, \ldots, y_n)$. It is easy to see that $COMP_n(X, Y) = 1$ if and only if

1. $X = Y$, or

2. There exists an i, $1 \leq i \leq n$ such that $x_i = 1$ and $y_i = -1$, and $x_j = y_j$ for each j where $i < j \leq n$.

It is the above structure of Comparison and the existence of the polynomial described in Lemma 4.3 that enables us to construct a sparse polynomial approximation to Comparison.

Theorem 4.3:   Let $q_i(X,Y) = P(x_n y_n, x_{n-1} y_{n-1}, \ldots, x_{i+1} y_{i+1}, 1, \ldots, 1)$, and $q_n(X,Y) = N$, where $P(\cdot)$ is the polynomial (of n variables) described in Lemma 4.3 with $N = \Theta(n^{4+2k})$. Let

$$\tilde{C}(X,Y) = \frac{1}{2N}[2q_0(X,Y) + \sum_{i=1}^{n}(x_i - y_i)q_i(X,Y)].$$

Then, for all X and Y in $\{1, -1\}^n$,

$$|COMP_n(X,Y) - \tilde{C}(X,Y)| = O(n^{-k}).$$

Proof: First, observe that $q_i(X,Y) = N$ when $x_j = y_j$ for $n \geq j > i$, and $|q_i(X,Y)| \leq \epsilon N$ otherwise. There are two cases to be considered in proving Theorem 4.3:

Case 1: $COMP_n(X,Y) = 1$.
If $X = Y$, then $\tilde{C}(X,Y) = 1 = COMP_n(X,Y)$. Otherwise, there exists an l, $1 \leq l \leq n$ such that $x_l = 1$ and $y_l = -1$, and $x_j = y_j$ for each j where $l < j \leq n$. Then, $(x_l - y_l)q_l(X,Y) = 2N$ and $|(x_i - y_i)q_i(X,Y)| \leq 2\epsilon N$ for all $i < l$. Hence,

$$\frac{1}{2N}(2n\epsilon N + 2N) \geq \tilde{C}(X,Y) \geq \frac{1}{2N}(-2n\epsilon N + 2N).$$

Equivalently, $|COMP_n(X,Y) - \tilde{C}(X,Y)| \leq n\epsilon = O(n^2/\sqrt{N})$.

Case 2: $COMP_n(X,Y) = -1$.
In this case, there exists an l, $1 \leq l \leq n$ such that $x_l = -1$ and $y_l = 1$, and $x_j = y_j$ for each j where $l < j \leq n$. Thus,

$$\frac{1}{2N}(2n\epsilon N - 2N) \geq \tilde{C}(X,Y) \geq \frac{1}{2N}(-2n\epsilon N - 2N).$$

Again, we have $|COMP_n(X,Y) - \tilde{C}(X,Y)| \leq n\epsilon = O(n^2/\sqrt{N})$.

So in both cases, we have shown that the sparse polynomial $\tilde{C}(X,Y)$ (with $O(nN)$ monomial terms) gives an approximation to Comparison with an error at most $O(n^2/\sqrt{N})$. The proof is completed by substituting $N = \Theta(n^\alpha)$ for some $\alpha \geq 4 + 2k$. $\square$

The construction of the sparse polynomial approximation to Addition is similar to Comparison. If we adopt the convention that 0, 1 are respectively encoded by 1, -1, then the kth sum bit, s_k, of two n-bit integers $X = (x_{n-1}, \ldots, x_0) \in \{1, -1\}^n$ (x_{n-1} the most significant bit) and $Y = (y_{n-1}, \ldots, y_0) \in \{1, -1\}^n$ can be expressed as $x_k y_k c_k$, where c_k is the kth carry bit in $\{1, -1\}$ notation. So, if $\hat{C}_k(X, Y)$ is a sparse polynomial approximation to c_k, then $x_k y_k \hat{C}_k(X, Y)$ is the corresponding sparse polynomial for s_k with the same approximation error. Without loss of generality, we only give the construction for the last carry bit c_n, where we denote as $C_n(X, Y)$ to indicate its dependence on the values of X and Y.

Theorem 4.4:   Let

$$X = (x_{n-1}, \ldots, x_0) \in \{1, -1\}^n$$

and

$$Y = (y_{n-1}, \ldots, y_0) \in \{1, -1\}^n.$$

Also, let

$$\hat{q}_i(X, Y) = P(-x_{n-1}y_{n-1}, -x_{n-2}y_{n-2}, \ldots, -x_{i+1}y_{i+1}, 1, \ldots, 1),$$

where $P(\cdot)$ is the polynomial (of n variables) described in Lemma 4.3 with $N = \Theta(n^{4+2k})$. Define $g(x_i, y_i) = \frac{1}{4}(1 - x_i - y_i + x_i y_i)$ and

$$\hat{C}(X, Y) = \frac{1}{N}[N - 2\sum_{i=0}^{n-1} g(x_i, y_i)\hat{q}_i(X, Y)].$$

Then, for all X and Y,

$$|C_n(X, Y) - \hat{C}(X, Y)| = O(n^{-k}).$$

Proof: First, observe that $\hat{q}_i(X, Y) = N$ if $x_j \neq y_j$ (i.e. $x_j y_j = -1$), for $n \geq j > i$, and $|\hat{q}_i(X, Y)| \leq \epsilon N$ otherwise. Moreover, $g(x_i, y_i)$ equals 1 only if $(x_i, y_i) = (-1, -1)$, and equals 0 otherwise. In other words, when $g(x_i, y_i) = 1$, a carry bit is generated by x_i and y_i.

Similar to the proof of Theorem 4.3, there are two cases to be considered:

Case 1: $C_n(X, Y) = -1$, i.e., there is a carry to bit n.
In this case, we must have a carry bit generated ($g(x_i, y_i) = 1$) and propagated. Consider the largest index i where the carry is generated, i.e.,

$0 \leq i < n$ such that $x_i = -1$ and $y_i = -1$ (where the carry is generated), and $x_j \neq y_j$ (i.e. $x_j y_j = -1$) for each j where $i < j < n$. Then, $g(x_i, y_i)\hat{q}_i(X, Y) = N$ and $|g(x_j, y_j)\hat{q}_j(X, Y)| \leq \epsilon N$ for all $j < i$. Hence,

$$\frac{1}{N}(N + 2n\epsilon N - 2N) \geq \tilde{C}(X, Y) \geq \frac{1}{N}(N - 2n\epsilon N - 2N).$$

Equivalently, $|C_n(X, Y) - \hat{C}(X, Y)| \leq 2n\epsilon = O(n^2/\sqrt{N})$.

Case 2: $C_n(X, Y) = 1$, i.e., there is no carry to bit n.

For each $i < n$, there are two subcases:

1. There is no carry generation, i.e., $g(x_i, y_i) = 0$. Then $\hat{C}(X, Y) = 1$.

2. There is a carry generation but no carry propagation. Then there exists a j, $i \leq j \leq n - 1$ such that $x_j = y_j = 1$. It follows from the properties of the polynomial $\hat{q}_i(X, Y)$ that $|\hat{q}_i(X, Y)| \leq \epsilon N$.

Hence,

$$\frac{1}{N}(N + 2n\epsilon N) \geq \hat{C}(X, Y) \geq \frac{1}{N}(N - 2n\epsilon N).$$

Again, we have

$$|C_n(X, Y) - \hat{C}(X, Y)| \leq 2n\epsilon = O(n^2/\sqrt{N}).$$

The proof of Theorem 4.4 is completed by substituting $N = \Theta(n^{\hat{\alpha}})$ for some $\hat{\alpha} \geq 4 + 2k$. $\quad\square$

4.3   Multiple Sum and Multiplication

In the preceding section, we have adopted the $\{1, -1\}$ notation for the convenience of presenting the techniques of harmonic analysis. In this section, we convert back to the more natural $\{0, 1\}$ notation and construct threshold circuits for the functions Multiple Sum and Multiplication. First, we give formal definitions of these two functions.

Definition 4.3:   Given n n-bit integers, $z_i = \sum_{j=0}^{n-1} z_{i,j} 2^j$, $i = 1, \ldots, n$, $z_{i,j} \in \{0, 1\}$, Multiple Sum is the problem of computing the $(n + \log n)$-bit sum $\sum_{i=1}^{n} z_i$ of the n integers. $\quad\square$

Definition 4.4: Given 2 n-bit integers, $X = \sum_{j=0}^{n-1} x_j 2^j$ and $Y = \sum_{j=0}^{n-1} y_j 2^j$, we define Multiplication to be the problem of computing the $(2n)$-bit product of X and Y. $\square$

To prove the result in this section, we make use of a restricted version of Theorem 3.5, as stated below.

Lemma 4.4: Let $X = (x_{n-1}, x_{n-2}, \ldots, x_0) \in \{0,1\}^n$, and $f : \{0,1\}^n \to \{0,1\}$ be a Boolean function of X that depends only on a polynomially bounded weighted sum of the input variables. Then, $f(X)$ can be represented as a sum of polynomially many $\widehat{LT}_1$ functions in X. Hence, $f(X)$ can be computed by a depth-2 threshold circuit, i.e., $f(X) \in \widehat{LT}_2$. $\square$

Theorem 4.5: Multiple Sum $\in \widehat{LT}_3$.

Proof: The underlying idea of computing the sum of many large numbers is to reduce the multiple sum to the sum of only two numbers. This technique, which we call the 'block-save' technique, is a generalization of the traditional 'carry-save' technique.

Block-save technique: The main difficulty in computing the sum of many numbers is to compute the carry bits in parallel. The traditional carry-save technique reduces the sum of three numbers to the sum of two numbers in one step, where one of the resultant two numbers only consists of the carry bits.

Let the original three numbers be $X = x_{n-1}\ x_{n-2}\ \cdots\ x_0$, $Y = y_{n-1}\ y_{n-2}\ \cdots\ y_0$, $Z = z_{n-1}\ z_{n-2}\ \cdots\ z_0$ (all numbers are in binary representation). The ith bit of each number can be added to give $x_i + y_i + z_i = 2c_{i+1} + w_i$, where the c_{i+1} is the carry bit generated by the sum of the ith bits x_i, y_i, and z_i. Thus, $X + Y + Z = C + W$, where $C = c_n\ c_{n-1}\ \cdots\ c_0$ with $c_0 = 0$ and $W = w_{n-1}\ w_{n-2}\ \cdots\ w_0$. Note that $c_{i+1} = x_i y_i \vee y_i z_i \vee z_i x_i$ and $w_i = x_i \oplus y_i \oplus z_i$, hence the c_{i+1}'s and w_i's can all be computed in parallel in one step. For example, let $X = 1001$, $Y = 0111$, and $Z = 1101$. Then $W = 0011$ and $C = 11010$.

The block-save technique generalizes this idea and reduces the sum of n $O(n)$-bit numbers to the sum of two $O(n)$-bit numbers in one step. We now present this block-save idea for the case where each of the n numbers has exactly n bits.

Denote each of the n-bit numbers by $x_i = x_{i_{n-1}} x_{i_{n-2}} \cdots x_{i_0}$, for $i = 1, \ldots, n$. For simplicity, assume $\log n$ and $N = n/\log n$ to be integers. Partition each binary number x_i into N consecutive blocks $\tilde{x}_{i_0}, \tilde{x}_{i_1}, \ldots, \tilde{x}_{i_{N-1}}$ of $(\log n)$-bits each so that

$$x_i = \sum_{j=0}^{N-1} \tilde{x}_{i_j} \cdot 2^{\log n \cdot j},$$

where $0 \le \tilde{x}_{i_j} < 2^{\log n}$. Hence, the total sum (after rearranging the indices of summation) becomes

$$S = \sum_{i=1}^{n} x_i = \sum_{j=0}^{N-1} \left(\sum_{i=1}^{n} \tilde{x}_{i_j} \right) \cdot 2^{\log n \cdot j}.$$

Observe that for each $j = 0, \ldots, N-1$, the block sum

$$\tilde{s}_j = \sum_{i=1}^{n} \tilde{x}_{i_j} < \sum_{i=1}^{n} 2^{\log n} = 2^{2 \log n},$$

and thus $\tilde{s}_j$ can be represented in $2\log n$ bits. Therefore, each block sum $\tilde{s}_j$ can be expressed as

$$\tilde{s}_j = \tilde{c}_{j+1} 2^{\log n} + \tilde{w}_j,$$

where $0 \le \tilde{c}_{j+1} < 2^{\log n}$ is the $(\log n)$-bit carry of the sum $\tilde{s}_j$ with $\tilde{c}_0 = 0$, and $\tilde{w}_j < 2^{\log n}$. Thus,

$$\begin{aligned} S &= \sum_{j=0}^{N-1} \left(\sum_{i=1}^{n} \tilde{x}_{i_j} \right) \cdot 2^{\log n \cdot j} \\ &= \sum_{j=0}^{N-1} \tilde{s}_j 2^{\log n \cdot j} \\ &= \sum_{j=0}^{N-1} \tilde{c}_{j+1} 2^{(j+1)\log n} + \sum_{j=0}^{N-1} \tilde{w}_j 2^{j \log n}. \end{aligned}$$

Since both $\tilde{c}_{j+1}$ and $\tilde{w}_j$ are $< 2^{\log n}$, the binary representation of

$$\sum_{j=0}^{N-1} \tilde{c}_{j+1} 2^{(j+1)\log n}$$

is simply the concatenation of the bits $\tilde{c}_N \, \tilde{c}_{N-1} \cdots \tilde{c}_0$. Similarly, concatenating the bits $\tilde{w}_{N-1} \cdots \tilde{w}_0$ gives the binary representation of $\sum_{j=0}^{N-1} \tilde{w}_j 2^{j \log n}$. Hence, we have shown the reduction from the sum of n n-bit numbers to the sum of two $O(n)$-bit numbers. Furthermore, we can compute all $\tilde{c}_j$'s and $\tilde{w}_j$'s in parallel to obtain the resultant two numbers.

It remains to be shown how to compute the $(2 \log n)$-bit representation of each block sum $\tilde{s}_j = \tilde{c}_{j+1} 2^{\log n} + \tilde{w}_j$. Since each number of the summands is of the form $\sum_{k=0}^{\log n - 1} 2^k x_k < n$, the total sum is a polynomially bounded weighted sum of the $n \log n$ variables x_{i_l}'s, where $1 \leq i \leq n$ and $(j-1) \log n < l \leq j \log n$. Thus, each bit of $\tilde{c}_{j+1}$ and $\tilde{w}_j$ is a function of a polynomially bounded weighted sum of $n \log n$ variables. In other words, each bit of $\tilde{c}_{j+1}$ and $\tilde{w}_j$ is a generalized symmetric function (see Section 3.4), where the weighted sum the function depends on is polynomially bounded. Hence, it follows from Lemma 4.4 that each bit of the block sum $\tilde{s}_j$ can be represented as a sum of polynomially many linear threshold functions in the variables x_{i_j}'s.

Figure 4.1 illustrates the block-save technique by showing the computation of four 16-bit integers.

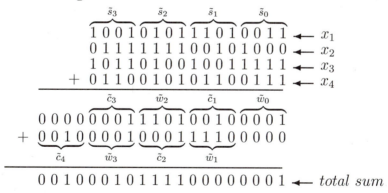

Figure 4.1 An example illustrating the computation of a multiple sum using the block-save technique.

Now it is clear how to compute the sum of n $O(n)$-bit numbers with a depth-3 threshold circuit. First, use the block-save technique to reduce this multiple sum to the sum of two $O(n)$-bit numbers. By Lemma 4.4, each bit of the resultant two numbers is a linear combination of polynomially many linear threshold functions computed in the first layer. Then, take this linear combination as inputs to the depth-2 threshold circuit (by Theorem 4.2) for computing the sum of the resultant two numbers. Thus, only three layers are needed.   □

Theorem 4.6:   Multiplication $\in \widehat{LT}_4$.

Proof: To compute the product of two n-bit numbers, $x = x_{n-1}x_{n-2}\cdots x_0$, $y = y_{n-1}y_{n-2}\cdots y_0$, the first layer of our circuit outputs the n $2n$-bit numbers $z_i = z_{i_{2n-1}}z_{i_{2n-2}}\cdots z_{i_0}$ for $i = 0, \ldots, n-1$, where

$$z_i = \underbrace{0\cdots 0}_{n-i}(x_{n-1}\wedge y_i)(x_{n-2}\wedge y_i)\cdots(x_0 \wedge y_i)\underbrace{0\cdots 0}_{i}.$$

Another three layers compute the sum of all z_i's as shown before. Thus, the product of 2 n-bit numbers can be computed in four layers.   □

4.4   Division and Related Problems

In this section, we derive small-depth threshold circuits for Division and related problems. First, we present formal definitions of these problems and introduce some number-theoretic notions.

Suppose one wants to compute the quotient of two integers. Since some quotient in binary representation might require infinitely many bits, a Boolean circuit can compute only the most significant bits of the quotient. If a number has both finite and infinite binary representation (e.g., $0.1 = 0.0111...$), we always express the number in its finite binary representation.

Definition 4.5:   Let X and $Y \geq 1$ be two n bit integers. Let $X/Y = \sum_{i=-\infty}^{n-1} z_i 2^i$ be the quotient of X divided by Y. $\mathrm{DIV}_k(X/Y)$ is the problem of computing X/Y *truncated* to the $(n+k)$-bit number, i.e.,

$$\mathrm{DIV}_k(X/Y) = \sum_{i=-k}^{n-1} z_i 2^i \ .   □$$

In particular, $\mathrm{DIV}_0(X/Y)$ is $\lfloor X/Y \rfloor$, which is the greatest integer $\leq X/Y$.

The above problem is to find the quotient of two integers with truncation, which is a commonly adopted definition of the division problem. Similarly, the "rounded" quotient can be solved using the techniques developed in this chapter. The circuit for division will be developed through a number of steps, each of which would involve implementations of other arithmetic functions such as Powering and Exponentiation.

Definition 4.6:   Let X be an n-bit number. Powering is the problem of computing the n^2-bit representation of X^n. $\square$

Definition 4.7:   Let Y be an $O(\log n)$-bit number (i.e., $Y \leq n^k$ for some constant k), and $c \geq 0$ be a fixed integer. Exponentiation is the problem of computing the $O(n^k)$-bit representation of c^Y. $\square$

Remark 4.2:   Since we are only concerned with functions that can be computed by polynomial-size circuits, it is necessary that the outputs of the function be representable by a polynomial number of bits. The assumption in the function Exponentiation that the exponent Y is polynomially bounded is simply to guarantee that c^Y can be represented in a polynomial number of bits. $\square$

We need some number-theoretic terminologies in presenting our results.

Definition 4.8:   Let x, y, and m be positive integers. We write

$$x \equiv y \pmod{m}$$

if $(x - y)$ is divisible by m, and $x \bmod m$ to denote the unique integer z such that $0 \leq z < m$ and $x \equiv z \pmod{m}$. $\square$

We shall make use of the following results from number theory.

Fermat's Theorem
Let p be a prime number. Then, for any positive integer a not divisible by p, $a^{p-1} \equiv 1 \pmod{p}$. $\square$

Chinese Remainder Theorem
Let p_i for $i = 1, \ldots, n$ be relatively prime numbers, and $P_n = \prod_{i=1}^{n} p_i$. Let $0 \leq Z < P_n$, $q_i = P_n/p_i$, $\tilde{q}_i$ to be an integer such that $q_i \tilde{q}_i = 1 \pmod{p_i}$, and $m_i = q_i \cdot \tilde{q}_i$. (Note that $\tilde{q}_i$ exists since q_i and p_i are relatively prime.) Further, let $r_i \equiv Z \pmod{p_i}$. Then,

$$Z = (\sum_{i=1}^{n} r_i \cdot m_i) \bmod P_n .$$ $\square$

Example 4.1:   Let $Z = 14$, $p_1 = 3$, $p_2 = 5$. Then, $P_n = 15$, $q_1 = 5$, $q_2 = 3$, $\tilde{q}_1 = 2$, $\tilde{q}_2 = 2$, $m_1 = 10$, $m_2 = 6$, $r_1 = 2$, $r_2 = 4$. So $Z = (r_1 m_1 + r_2 m_2) \bmod P_n = 44 \bmod 15 = 14$.   □

The Chinese Remainder Theorem enables us to represent a number in a 'mixed-radix' system (as opposed to the traditional 'fixed-radix' system), where the r_i is the 'coefficient' with respect to the 'radix' m_i. This representation is useful because sometimes it is much easier to compute the coefficients r_i's than Z itself, and the r_i's can be computed in parallel.

4.4.1   Exponentiation and Powering Modulo a 'Small' Number

Another way of interpreting the Chinese Remainder Theorem is that a positive integer $Z < P_n = \prod_{i=1}^{n} p_i$ is uniquely determined by $r_i \equiv Z \pmod{p_i}$, the values of Z modulo the primes p_i's. When we apply this result to obtain a small-depth threshold circuit for division, the primes p_i's are 'small' (i.e., polynomially bounded) and known in advance. We now show that an input n-bit integer modulo a 'small' number can be computed by a small-depth circuit.

Theorem 4.7:   Let $X = (x_{n-1}, x_{n-2}, \ldots, x_0) \in \{0,1\}^n$ denote an input n-bit integer, and m be a fixed positive integer bounded by a polynomial in n. Then, $X \bmod m$ can be computed by a depth-2 threshold circuit.

Proof: It can be verified that if $a \equiv \tilde{a} \pmod{m}$, $b \equiv \tilde{b} \pmod{m}$, then $a + b \equiv \tilde{a} + \tilde{b} \pmod{m}$ and $a \cdot b \equiv \tilde{a} \cdot \tilde{b} \pmod{m}$. Therefore, $\sum_{i=0}^{n-1} x_i 2^i \bmod m = [\sum_{i=0}^{n-1} x_i (2^i \bmod m)] \bmod m$. Now $(2^i \bmod m) < m$ and is polynomially bounded in n, thus each bit in the output depends only on a polynomially bounded weighted sum of the input bits. It follows then from Lemma 4.4 that $X \bmod m$ can be computed by a depth-2 threshold circuit.   □

Before we show that Powering can be computed by a depth-4 threshold circuit, we first show how to compute powering and exponentiation modulo a small (i.e., polynomially bounded) prime number.

Theorem 4.8:   Let $X = (x_{n-1}, x_{n-2}, \ldots, x_0) \in \{0,1\}^n$ be an input n-bit integer, p be a prime number bounded by a polynomial in n, and c

be a positive integer not divisible by p. Then, each bit of $X^n \bmod p$ and $c^X \bmod p$ can be represented as a sum of polynomially many $\widehat{LT}_1$ functions in X. Hence, $X^n \bmod p$ and $c^X \bmod p$ can both be computed by depth-2 threshold circuits.

Proof: Observe that

$$
X^n \bmod p \;=\; (\sum_{i=0}^{n-1} x_i 2^i)^n \bmod p
$$

$$
=\; [\sum_{i=0}^{n-1} x_i (2^i \bmod p)]^n \bmod p \;.
$$

Similarly, since p is a prime and c not divisible by p, by Fermat's Theorem, $(c^{p-1} \equiv 1 \bmod p)$ and $(c^X \bmod p) = (c^{X \bmod (p-1)} \bmod p)$. Thus, each bit of $X^n \bmod p$ depends only on $[\sum_{i=0}^{n-1} x_i (2^i \bmod p)]$, and $c^X \bmod p$ depends only on $[\sum_{i=0}^{n-1} x_i (2^i \bmod (p-1))]$, both being a polynomially bounded weighted sum of the input bits. It follows from Lemma 4.4 that each bit of $X^n \bmod p$ and $c^X \bmod p$ can be represented as a sum of polynomially many linear threshold functions in X and, hence, can be computed by a depth-2 threshold circuit.

□

Remark 4.3:   First, notice that if c is divisible by p, then trivially $c^X \equiv 0$ (mod p). Also, it is important to note that in Theorem 4.8, the numbers p and c are known in advance. The only input variables are the $X = (x_{n-1}, x_{n-2}, \ldots, x_0)$. We make crucial use of this fact in the proof of Theorem 4.8. Also notice that $c^X \bmod p$ can be represented in $O(\log n)$ bits whereas c^X requires exponentially many bits (since X is exponentially large). We restrict ourselves in the Exponentiation function c^Y to input Y that is polynomially bounded in n.   □

Theorem 4.9:   Exponentiation $\in \widehat{LT}_2$.

Proof: Let Y be an input positive number $\leq n^k$, and c be a fixed positive integer. Since each bit of the $O(n^k)$-bit representation of c^Y depends only on Y, which is polynomially bounded, it follows from Lemma 4.4 that c^Y can be computed by a depth-2 threshold circuit.   □

4.4.2   Powering in Depth-4

Recall that Multiplication of two n-bit numbers can be computed by a depth-4 threshold circuit. We proved this result by reducing Multiplication to the sum of n $O(n)$-bit numbers. It is not clear if the function Powering can be similarly reduced to a multiple sum of polynomially many $O(n^k)$-bit numbers. Intuitively, this function seems 'harder' in terms of circuit complexity than Multiplication. Using the techniques we developed in the preceding sections and the Chinese Remainder Theorem, we show that as in the case of Multiplication, Powering only requires at most four layers of threshold gates to compute. In order to prove this result, we shall establish several lemmas (Lemmas 4.5 to 4.8), which are useful in reducing the circuit depth.

Suppose $f(X) = \text{sgn}(\sum_{i=1}^{m} w_i t_i(X))$. If each $t_i(X)$ can be computed by a depth-k threshold circuit, then clearly $f(X)$ can be computed by a depth-$(k + 1)$ threshold circuit. The following lemma states that we can reduce the depth by one if each $t_i(X)$ can be *closely approximated* by a linear combination of outputs from polynomially many depth-$(k - 1)$ threshold circuits.

Lemma 4.5:   Suppose

$$f(X) = \text{sgn}\left(\sum_{i=1}^{m} w_i t_i(X)\right),$$

where $\sum_{i=1}^{m} w_i t_i(X) \neq 0$ for all inputs X, w_i's are integers, and $\sum_{i=1}^{m} |w_i| \leq n^c$. If for each i,

$$\left| t_i(X) - \frac{1}{N} \sum_{j=1}^{m_i} \tilde{w}_{i_j} \tilde{t}_{i_j}(X) \right| < n^{-c},$$

where m_i's, $\tilde{w}_{i_j}$'s, and N are integers bounded by a polynomial in n, and each $\tilde{t}_{i_j}$ can be computed by a depth-$(k - 1)$ threshold circuit, then $f(X)$ can be computed by a depth-k threshold circuit.

Proof: Let $r_i(X) = \dfrac{1}{N} \sum_{j=1}^{m_i} \tilde{w}_{i_j} \tilde{t}_{i_j}(X)$. By assumption, $t_i(X) = r_i(X) + \epsilon_i$ where $|\epsilon_i| < n^{-c}$. Then,

$$\text{sgn}\left(\sum_{i=1}^{m} w_i r_i(X)\right) = \text{sgn}\left(\sum_{i=1}^{m} w_i(t_i(X) - \epsilon_i)\right)$$

$$= \text{sgn} \left(\sum_{i=1}^{m} w_i t_i(X) - \epsilon \right),$$

where $\epsilon = \sum_{i=1}^{m} w_i \epsilon_i$. Note that

$$|\epsilon| = \left| \sum_{i=1}^{m} w_i \epsilon_i \right| < n^{-c} \sum_{i=1}^{m} |w_i| \leq 1.$$

Since by assumption, $\sum_{i=1}^{m} w_i t_i(X)$ is a non-zero integer for all X, therefore

$$f(X) = \text{sgn} \left(\sum_{i=1}^{m} w_i t_i(X) \right)$$

$$= \text{sgn} \left(\sum_{i=1}^{m} w_i r_i(X) \right)$$

$$= \text{sgn} \left(\sum_{i=1}^{m} w_i \left(\frac{1}{N} \sum_{j=1}^{m_i} \tilde{w}_{i_j} \tilde{t}_{i_j}(X) \right) \right).$$

By assumption, each $\tilde{t}_{i_j}(X)$ can be computed by a depth-$(k-1)$ threshold circuit. Hence, $f(X) = \text{sgn} \left(\sum_{i=1}^{m} w_i (\sum_{j=1}^{m_i} \tilde{w}_{i_j} \tilde{t}_{i_j}(X)) \right)$ can be computed by a depth-k threshold circuit. □

Remark 4.4: By Lemma 4.2, each bit of the sum of two n-bit numbers can be *closely approximated* by a linear combination of outputs from polynomially many linear threshold gates. It follows from our reduction technique that each bit of the sum of n $O(n)$-bit numbers can be closely approximated by a linear combination of outputs from polynomially many depth-2 threshold circuits. □

The following lemma, which will be useful later, generalizes the result of Lemma 4.4 to any Boolean function that has a fixed number of the functions stated in the assumptions of Lemma 4.4 as inputs.

Lemma 4.6: For a fixed integer $d \geq 1$, let $g(X) = h(f_1(X), \ldots, f_d(X))$, where the functions f_i satisfy the assumptions stated in Lemma 4.4. Then, $g(X)$ can be represented as a sum of polynomially many $\widehat{LT}_1$ functions in X.

Proof: By assumption, one can write each $f_i(X) = \tilde{f}(\sum_{j=0}^{n-1} w_{ij} x_j)$, where w_{ij} are integers bounded by a polynomial in n. Let N be the maximum of $\sum_{j=0}^{n-1} |w_{ij}|$ over all $i = 1, \ldots, d$. Now observe that for each

$i = 1, \ldots, d$, the sum $\sum_{j=0}^{n-1} w_{ij} x_j$ can be determined from the value $\sum_{i=0}^{d-1} (N+1)^i (\sum_{j=0}^{n-1} w_{ij} x_j)$. But this sum is also polynomially bounded for all X. Thus, $g(X)$ is a function that depends only on a polynomially bounded weighted sum of the input variables. Our claim follows from Lemma 4.4. $\square$

We make use of the following specific form of Lemma 4.6. As usual, the symbol $\wedge$ denotes the logic AND function.

Corollary 4.1:   Let $g(X) = f_1(X) \wedge f_2(X)$, where $f_1(X)$ and $f_2(X)$ are functions computed by depth-k threshold circuits with inputs X. Then, $g(X)$ can be represented as a sum of outputs from polynomially many depth-k threshold circuits with inputs X.

Proof: Simply observe that each $f_i(X)$ depends on a polynomially bounded weighted sum of the outputs from depth-$(k-1)$ threshold circuits. Then, from Lemma 4.6, we can write $g(X)$ as a sum of polynomially many linear threshold functions of the outputs from depth-$(k-1)$ threshold circuits. $\square$

We need a slight generalization of Lemma 4.5 in our proof of Theorem 4.10. Lemma 4.7 states that if t_1 and t_2 can be closely approximated by a polynomially bounded sum of outputs from depth-k threshold circuits, so can their product $t_1 \wedge t_2$.

Lemma 4.7:   Suppose for $i = 1, 2$, and for every $c > 0$, there exist integers m_i, w_{i_j}, and N that are bounded by a polynomial in n, such that for all inputs X,

$$|t_i(X) - \frac{1}{N} \sum_{j=1}^{m_i} w_{i_j} t_{i_j}(X)| = O(n^{-c}),$$

where each t_{i_j} can be computed by a depth-k threshold circuit. Then there exist integers $\tilde{m}$, $\tilde{w}_j$'s, and $\tilde{N}$ that are bounded by a polynomial in n such that

$$|t_1(X) \wedge t_2(X) - \frac{1}{\tilde{N}} \sum_{j=1}^{\tilde{m}} \tilde{w}_j \tilde{t}_j(X)| = O(n^{-c}),$$

where each $\tilde{t}_j$ can be computed by a depth-k threshold circuit.

Proof: For $i = 1, 2$, let $t_i(X) = \frac{1}{N} \sum_{j=1}^{m_i} w_{i_j} t_{i_j}(X) + \epsilon_i$ where $|\epsilon_i| = O(n^{-c})$ by assumption. Since each $t_i(X)$ is either 0 or 1, thus $|\frac{1}{N} \sum_{j=1}^{m_i} w_{i_j} t_{i_j}(X)| =$

$1 + O(n^{-c})$. Moreover,

$$t_1(X) \wedge t_2(X) = t_1(X)t_2(X) = \frac{1}{N^2} \sum_{j=1}^{m_1} \sum_{l=1}^{m_2} w_{1_j} w_{2_l} t_{1_j}(X) t_{2_l}(X) + O(n^{-c}).$$

Now by Corollary 4.1, each $t_{1_j}(X)t_{2_l}(X) = t_{1_j}(X) \wedge t_{2_l}(X)$ can be represented as a sum of outputs from polynomially many depth-k threshold circuits. The conclusion follows by substituting this sum into the preceding expression for $t_1(X) \wedge t_2(X)$.   □

Remark 4.5:   Clearly, we can apply Lemma 4.7 repeatedly so that the result can be generalized to the product of a fixed number of $t_i(X)$'s. In other words, the product of a fixed number of $t_i(X)$'s can also be closely approximated by a polynomially bounded sum of outputs from depth-k threshold circuits.   □

Recall that each threshold gate in our circuit is only allowed to have polynomially bounded integer weights. Lemma 4.8 states that a linear threshold gate with arbitrary (even real-value) weights can be *closely approximated* by a linear combination of outputs from polynomially many depth-2 threshold circuits. We shall present the proof of the lemma in Chapter 6.

Lemma 4.8:   Let $f(X)$ be a linear threshold function (of n variables) whose weights can be arbitrary real numbers. Then for any $k > 0$, there exists a linear combination of functions $t_j(X)$ computable by depth-2 threshold circuits (with polynomially bounded integer weights),

$$F(X) = \frac{1}{N} \sum_{j=1}^{s} w_j t_j(X),$$

such that

$$|F(X) - f(X)| \leq n^{-k},$$

where s, w_j's, and N are integers bounded by a polynomial in n.   □

Theorem 4.10:   Powering $\in \widehat{LT}_4$.

Proof: Let X denote an input n-bit integer. We want to compute the n^2-bit representation of $Z = X^n$. Our main tool will be the Chinese Remainder

Theorem. It allows us first to compute Z modulo small prime numbers in parallel, and then combine the results using small-depth threshold circuits.

Let p_i denote the ith prime number and let $\pi(k)$ denote the number of primes $\leq k$. Let

$$P_n = \prod_{i=1}^{\pi(n^2)} p_i$$

be the product of all primes $\leq n^2$. Let

$$q_i = P_n/p_i, \quad \tilde{q}_i = q_i^{-1} \bmod p_i, \quad \text{and} \quad m_i = q_i \cdot \tilde{q}_i.$$

$\tilde{q}_i$ exists since p_i and q_i are relatively prime. Observe that by the *Prime Number Theorem*, P_n is greater than 2^{n^2} for sufficiently large n. Moreover, $(Z \bmod P_n) = Z$ since $Z < 2^{n^2} < P_n$. We compute $(Z \bmod P_n)$ with a small-depth threshold circuit via the Chinese Remainder Theorem, and the steps are as follows:

1. For $i = 1, \ldots, n^2$, compute in parallel the values $r_i = Z \bmod p_i$.

2. $\tilde{Z} = \sum_{i=1}^{\pi(n^2)} r_i \cdot m_i$.

3. $Z = (Z \bmod P_n) = (\tilde{Z} \bmod P_n)$.

It is important to note that the m_i's are known in advance, and thus Step 2 is actually the computation of Multiple Sum. Also note that $r_i \leq n^2$ and $m_i \leq P_n$. Therefore, $\tilde{Z} \leq \sum_{i=1}^{\pi(n^2)} n^2 P_n \leq n^4 \cdot P_n$. Hence, $Z = (\tilde{Z} \bmod P_n) = \tilde{Z} - k \cdot P_n$ for some k, where $0 \leq k \leq n^4$. For each $k \in \{0, \ldots, n^4\}$, let

$$
\begin{aligned}
EQ_k(Z) &= \operatorname{sgn}(\tilde{Z} - k \cdot P_n) + \operatorname{sgn}((k+1)P_n - \tilde{Z} - 1) - 1 \\
&= \begin{cases} 1 & \text{if } Z = (\tilde{Z} \bmod P_n) = \tilde{Z} - k \cdot P_n, \\ 0 & \text{otherwise.} \end{cases}
\end{aligned}
$$

Let z_{jk} be the jth bit of $\tilde{Z} - k \cdot P_n$. Then the jth bit of Z is

$$\bigvee_{0 \leq k \leq n^4} (EQ_k(Z) \wedge z_{jk}).$$

Now let us calculate the number of layers needed to compute each jth bit of Z. We can compute the values r_i in Step 1 as a sum of the outputs from polynomially many linear threshold gates in the first layer by Theorem 4.8.

By expressing $(-k \cdot P_n)$ in 2's complement representation, $(\tilde{Z} - k \cdot P_n)$ becomes a multiple sum (with input variables r_i's). It follows from Remark 4.4 that each z_{jk} can be closely approximated by a sum of outputs from polynomially many depth-2 threshold circuits whose inputs are the variables r_i. Similarly, Lemma 4.8 implies that the functions $EQ_k(Z)$ can also be closely approximated by a sum of outputs from polynomially many depth-2 threshold circuits whose inputs are the variables r_i. Thus, $EQ_k(Z)$ and z_{jk} can be closely approximated by a sum of the outputs from depth-3 threshold circuits whose inputs are the variables X. Now Lemma 4.7 implies that $(EQ_k(Z) \wedge z_{jk})$ can also be closely approximated by a sum of the outputs from depth-3 threshold circuits. Then it follows from Lemma 4.5 that the output $\bigvee_{0 \leq k \leq n^4} (EQ_k(Z) \wedge z_{jk})$ can be computed by a depth-4 threshold circuit. $\qquad \Box$

Remark 4.6: Note that for any *fixed* c, this result also implies that we can compute X^{n^c} using a depth-4 polynomial-size threshold circuit.

4.4.3 Multiple Product in Depth-5

Definition 4.9: Given n n-bit numbers z_i, $i = 1, \ldots, n$. Multiple Product is the problem of computing the n^2-bit representation of $\prod_{i=1}^{n} z_i$. $\qquad \Box$

We now show that this function can be computed by depth-5 threshold circuits.

Theorem 4.11: Multiple Product $\in \widehat{LT}_5$.

Proof: We use the same notation as in the proof of Theorem 4.10. Let $Z = \prod_{j=1}^{n} z_j$. Again, the proof is similar to the proof of Theorem 4.10, except that we need to show how to compute each $r_i = Z \bmod p_i$. In the case of Powering, each $r_i = (X^n \bmod p_i)$ can be expressed as a sum of polynomially many linear threshold functions. To prove Theorem 4.11, it suffices to show that in the case of Multiple Product, each $r_i = \prod_{j=1}^{n} z_j \bmod p_i$ can be expressed as a sum of the outputs from polynomially many depth-2 threshold circuits. It then follows that we can compute Multiple Product with a depth-5 threshold circuit.

Let Z_{p_i} be the set of all integers modulo p_i. Since each p_i is a prime number, Z_{p_i} is a finite field. Let g_i be a generator of Z_{p_i}, i.e., every integer $\tilde{Z} \equiv g_i^{\tilde{e}_i} \pmod{p_i}$ for some $\tilde{e}_i \in \{0, \ldots, p_i-1\}$. Note that $r_i = (\prod_{j=1}^n z_j \bmod p_i) = \prod_{j=1}^n (z_j \bmod p_i) \bmod p_i$. The idea is to compute $e_{i,j}$, the exponent of $(z_j \bmod p_i)$ with respect to the generator g_i, i.e., $(z_j \bmod p_i) = (g_i^{e_{i,j}} \bmod p_i)$. Then

$$r_i = \prod_{j=1}^n z_j \bmod p_i = \prod_{j=1}^n g_i^{e_{i,j}} \bmod p_i = g_i^{(\sum_{j=1}^n e_{i,j})} \bmod p_i.$$

Now observe that each bit of $e_{i,j}$ is a function of $(z_j \bmod p_i)$, which is a polynomially bounded weighted sum of the bits in z_j. By Lemma 4.4, $e_{i,j}$ can be represented as a sum of polynomially many linear threshold functions in the input bits of z_j. Since each $e_{i,j} < p_i$, $\sum_{j=1}^n e_{i,j} < np_i$, and is polynomially bounded. Moreover, each r_i only depends on $\sum_{j=1}^n e_{i,j}$. Again Lemma 4.4 implies that r_i can be represented as a sum of polynomially many linear threshold functions in the input bits of $e_{i,j}$. It follows that r_i can be expressed as a sum of the outputs from polynomially many depth-2 threshold circuits, with inputs z_j. $\square$

4.4.4 Division in Depth-4

We now show that $DIV_k(X/Y)$ can be computed by a depth-4 threshold circuit. First, we need a lemma that is a slight generalization of the result stated in Theorem 4.8.

Lemma 4.9:   Let $N, p_i, m_i, a_i, b_i, c_i, K$ be fixed integers, where N and p_i are positive integers bounded by a polynomial in n, and a_i, b_i, and c_i are integers with polynomially bounded number of bits. Let x and y be the input integers with polynomially bounded number of bits. Let $Z = \sum_{j=1}^N a_j x(b_j + c_j y)^j$. Let $r_i \equiv Z \pmod{p_i}$ where r_i is polynomially bounded. Then each bit $\tilde{z}_l$ of

$$\tilde{Z} = \sum_{i=1}^N r_i \cdot m_i - K$$

can be closely approximated by a linear combination of outputs from depth-3 threshold circuits. That is, for any $k > 0$, there exists a linear combination of functions t_j computable by depth-3 threshold circuits (with polynomially bounded integer weights) such that

$$|\tilde{z}_l - \frac{1}{M} \sum_{j=1}^{s} w_j t_j| \leq n^{-k},$$

where s, w_j's and M are polynomially bounded integers.

Proof: We first show that each term $z_{i,j} = a_j x(b_j + c_j y)^j \bmod p_i$ can be represented as a sum of polynomially many linear threshold functions with inputs x and y. Note that $z_{i,j} = \{a_j(x \bmod p_i)[(b_j + c_j y) \bmod p_i]^j\} \bmod p_i$. Let $x \equiv \sum_l \hat{w}_l x_l \pmod{p_i}$, and $(b_j + c_j y) \equiv \sum_l \tilde{w}_l y_l \pmod{p_i}$ (the weights are polynomially bounded). Let $M_y = \sum_l |\tilde{w}_l| + 1$. Then, each bit of $z_{i,j}$ can be expressed as a function of $(M_y \sum_l \hat{w}_l x_l) + (\sum_l \tilde{w}_l y_l)$, which is still a polynomially bounded weighted sum of the input bits of x and y. By Lemma 4.4, each bit of $z_{i,j}$ can be represented as a sum of polynomially many linear threshold functions with input bits x and y.

Now $\tilde{Z} = \sum_{i=1}^{N}(\sum_{j=1}^{N} z_{i,j}) \cdot m_i - K$, which is simply a multiple sum, since m_i's are known in advance. From Remark 4.4, each bit of this sum can be closely approximated by a linear combination of outputs from polynomially many depth-2 threshold circuits with inputs $z_{i,j}$. Since each bit of $z_{i,j}$ can be represented as a sum of polynomially many linear threshold functions with input bits x and y, the lemma follows. $\square$

Theorem 4.12: Let x and $y > 0$ be two n-bit integers. Then $\mathrm{DIV}_k(x/y) \in \widehat{LT}_4$.

Proof: Note that $\mathrm{DIV}_k(x/y) = 2^{-k}\mathrm{DIV}_0(2^k x/y)$, so it suffices to prove our claim for the case $k = 0$. The resulting threshold circuits for the general case, where k is polynomial in n, have the same depth and the size will increase by a polynomial factor.

The underlying idea is to compute an *over-approximation* $\tilde{a}$ to x/y such that $x/y \leq \tilde{a} \leq x/y + 2^{-(n+1)}$. We claim that $\lfloor \tilde{a} \rfloor = \lfloor x/y \rfloor$. Clearly, the claim is true if x/y is an integer. Suppose x/y is not an integer. Then

$$x = \lfloor x/y \rfloor y + r,$$

where r is an integer and $0 < r < y < 2^n$. Thus,

$$x/y - \lfloor x/y \rfloor = r/y \geq 2^{-n},$$

and

$$\lfloor x/y \rfloor + 1 - x/y = 1 - r/y \geq 2^{-n}.$$

In other words,

$$\lfloor x/y \rfloor + 2^{-n} \leq x/y \leq \lfloor x/y \rfloor + 1 - 2^{-n}.$$

Since by assumption, $0 \leq \tilde{a} - x/y \leq 2^{-(n+1)}$, we must have

$$\lfloor x/y \rfloor < \tilde{a} < \lfloor x/y \rfloor + 1.$$

Hence, $\lfloor \tilde{a} \rfloor = \lfloor x/y \rfloor$.

Since x/y is equal to the product of x and y^{-1}, it is enough to get an over-approximation $\tilde{y}^{-1}$ of y^{-1} with error $\leq 2^{-(2n+1)}$. Then, using Lemma 4.9, we can compute $q = x \cdot \tilde{y}^{-1}$ with error $\leq x2^{-(2n+1)} \leq 2^{-(n+1)}$ with a small-depth threshold circuit.

To construct an over-approximation of y^{-1}, let $j \geq 1$ be the integer such that $2^{j-1} \leq y < 2^j$. Note that $|1 - y2^{-j}| \leq \frac{1}{2}$ and we can express y^{-1} as a series expansion

$$y^{-1} = 2^{-j} \cdot (1 - (1 - y2^{-j}))^{-1} = 2^{-j} \sum_{i=0}^{\infty} (1 - y2^{-j})^i.$$

If we put

$$\tilde{y}^{-1} = 2^{-j} \Big(\sum_{i=0}^{2n+1} (1 - y2^{-j})^i + (1 - y2^{-j})^{2n+1} \Big),$$

then the difference

$$0 \leq (\tilde{y}^{-1} - y^{-1}) \leq 2^{-(2n+1)}.$$

Since $x < 2^n$, we have

$$0 \leq (x\tilde{y}^{-1} - xy^{-1}) < 2^{-(n+1)}.$$

Suppose for the moment that we can find the integer $j \geq 1$ such that $2^{j-1} \leq y < 2^j$. Now we can rewrite

$$x\tilde{y}^{-1} = \frac{1}{2^{j(2n+2)}} \Big(\sum_{i=0}^{2n+1} 2^{j(2n+1-i)} x(2^j - y)^i + x(2^j - y)^{2n+1} \Big).$$

Using Lemma 4.9, we can proceed as in the proof of Theorem 4.10 to construct a small-depth threshold circuit for $x\tilde{y}^{-1}$.

Let $Z_j = \sum_{i=0}^{2n+1} 2^{j(2n+1-i)} x(2^j - y)^i + x(2^j - y)^{2n+1}$ Then, $x\tilde{y}^{-1} = \frac{1}{2^{j(2n+2)}} Z_j$, a shifting of the bits in Z_j. Let N be a sufficiently large integer such that the product of the first N primes $\prod_{i=1}^{N} p_i = P_N > Z_j$ for all $j = 1, \ldots, n$. It is easy to see that all p_i's are polynomially bounded (in n). Let

$$q_i = P_N/p_i, \quad \tilde{q}_i = q_i^{-1} \bmod p_i, \quad \text{and} \quad m_i = q_i \cdot \tilde{q}_i.$$

Again we compute Z_j via the Chinese Remainder Theorem as follows:

1. For $i = 1, \ldots, N$, compute in parallel the values $r_{i,j} = Z_j \bmod p_i$.

2. $\tilde{Z}_j = \sum_{i=1}^{N} r_{i,j} \cdot m_i$.

3. $Z_j = (Z_j \bmod P_N) = (\tilde{Z}_j \bmod P_N)$.

Note that $\tilde{Z}_j \le n^\alpha P_N$ for some $\alpha > 0$. Hence, $Z_j = (\tilde{Z}_j \bmod P_N) = \tilde{Z}_j - kP_n$ for some k, where $0 \le k \le n^\alpha$. For each $k \in \{0, \ldots, n^\alpha\}$, let

$$
\begin{aligned}
EQ_k(Z_j) &= \operatorname{sgn}\{\tilde{Z}_j - kP_N\} + \operatorname{sgn}\{(k+1)P_N - \tilde{Z}_j - 1\} - 1 \\
&= \begin{cases} 1 & \text{if } Z_j = (\tilde{Z}_j \bmod P_N) = \tilde{Z}_j - kP_N, \\ 0 & \text{otherwise.} \end{cases}
\end{aligned}
$$

Let $\sum_l z_{j,k,l} 2^l = \frac{1}{2^{j(2n+2)}} (\tilde{Z}_j - kP_N)$. If $EQ_{k^*}(Z_j) = 1$, i.e., $(\tilde{Z}_j \bmod P_N) = \tilde{Z}_j - k^* P_N$, then

$$\operatorname{DIV}_0(x/y) = \sum_{l=0}^{n-1} z_{j,k^*,l} 2^l.$$

Thus, the ith bit of $\operatorname{DIV}_0(x/y)$ can be computed as

$$\bigvee_{1 \le k \le n^\alpha} (EQ_k(Z_j) \wedge z_{j,k,i}).$$

The preceding expression is based on the assumption that we can find the unique integer $j \ge 1$ such that $2^{j-1} \le y < 2^j$. We can compute such integer j in parallel without increasing the depth of the circuit. To see this, for each $j \in \{1, \ldots, n\}$, let

$$
\begin{aligned}
I_j &= \operatorname{sgn}\{y - 2^{j-1}\} + \operatorname{sgn}\{2^j - y - 1\} - 1 \\
&= \begin{cases} 1 & \text{if } 2^{j-1} \le y < 2^j, \\ 0 & \text{otherwise.} \end{cases}
\end{aligned}
$$

Then, the ith bit of $DIV_0(x/y)$ is

$$\bigvee_{1 \leq j \leq n} \bigvee_{1 \leq k \leq n^\alpha} (I_j \wedge EQ_k(Z_j) \wedge z_{j,k,i}).$$

Note that by Lemma 4.9, each of the $z_{j,k,i}$, $EQ_k(Z_j)$, and I_j can be closely approximated by a linear combination of outputs from polynomially many depth-3 threshold circuits, with inputs x and y. Remark 4.5 following Lemma 4.7 implies that each $(I_j \wedge EQ_k(Z_j) \wedge z_{j,k,i})$ can also be closely approximated by a linear combination of outputs from polynomially many depth-3 threshold circuits. Hence, it follows from Lemma 4.5 that the final result can be computed by a depth-4 threshold circuit. $\square$

4.5   Sorting in Depth-3 Threshold Circuits

Given n n-bit numbers, we show in this section that the problem of sorting the numbers can be carried out in depth-3 threshold circuits.

Definition 4.10:   Let $X_1, X_2, \ldots, X_n$ be n input n-bit integers. Sorting$(X_1, X_2, \ldots, X_n)$ is the problem of computing the list of integers sorted in nondecreasing order, $X_{i_1} \leq X_{i_2} \leq \ldots \leq X_{i_n}$. $\square$

Theorem 4.13:   Sorting $\in \widehat{LT}_3$.

Proof: Let $z_i = z_{i_n} z_{i_{n-1}} \cdots z_{i_1}$, for $i = 1, \ldots, n$, denote the input binary numbers. Define

$$c_{ij} = \begin{cases} 1 & \text{if } z_i > z_j \text{ or } (z_i = z_j \text{ and } i \geq j), \\ 0 & \text{otherwise.} \end{cases}$$

If we let $p_i = \sum_{j=1}^n c_{ij}$, then p_i is the position of z_i in the sorted list. Let $L_m(p_i) = \text{sgn}\{p_i - m\}$; then $L_m(p_i)$ equals 1 if $p_i \geq m$, and equals 0 otherwise. Similarly, let $l_m(p_i) = \text{sgn}\{m - p_i\}$; then $l_m(p_i)$ equals 1 if $p_i \leq m$, and equals 0 otherwise. Thus, the kth bit of the mth number in the sorted list is

$$\bigvee_{1 \leq i \leq n} (L_m(p_i) \wedge l_m(p_i) \wedge z_{ik}),$$

where $\vee$ and $\wedge$ respectively denote the OR and AND functions.

By Lemma 4.2, Comparison can be closely approximated by a sum of polynomially many linear threshold functions. It is easy to modify Comparison so that the each c_{ij} can also be closely approximated by a sum of polynomially many linear threshold functions. Hence, by Lemma 4.5, each $L_m(p_i)$ and $l_m(p_i)$ can be computed by a depth-2 threshold circuit. Now apply Corollary 4.1 twice, it follows that each $L_m(p_i) \wedge l_m(p_i) \wedge z_{ik}$ can be represented as a sum of outputs from polynomially many depth-2 threshold circuits. We need one more threshold gate to compute the output $\bigvee$ gate. Thus, all together only 3 layers are needed.   □

4.6   Lower Bounds for Division and Sorting

In this section, we show that any polynomial-size threshold circuit for Division or Sorting must have depth at least 3. This demonstrates that the sorting circuit derived in the preceding section has the minimum possible depth. In Chapter 6, we combine the techniques in this chapter with other advanced combinatorial techniques to prove that Division actually can be computed by depth-3 polynomial-size threshold circuits. Thus, the lower bound on the division circuit is the best possible.

The proofs of these lower bounds make use of the following result.

Theorem 4.14:   Any polynomial-size threshold circuit with polynomially bounded integer weights computing the function Inner Product Mod 2, i.e., $IP(x_1, \ldots, x_n, y_1, \ldots, y_n) = (x_1 \wedge y_1) \oplus \cdots \oplus (x_n \wedge y_n)$, must have depth at least 3.   □

A proof for Theorem 4.14 is presented in Chapter 8.

As a result of Theorem 4.14, we have the following corollary. The proof is left as an exercise.

Corollary 4.2:   Any polynomial-size threshold circuit with polynomially bounded integer weights computing the product of two n-bit integers must have depth at least 3.   □

Clearly, the square z^2 of an n-bit integer z can be computed using a circuit for Multiplication by setting both the input integers of the circuit to z. It turns out that computing the square of an n-bit integer is as difficult as Multiplication with respect to the depth of polynomial-size threshold circuits.

Corollary 4.3:   Any polynomial-size threshold circuit with polynomially bounded integer weights computing the square of an n-bit integer must have depth at least 3.

Proof: By Corollary 4.2, it suffices to show that Multiplication is a projection of squaring. Given any two n-bit integers x and y, let $z = x2^{2(n+1)} + y$. Then

$$z^2 = x^2 2^{4(n+1)} + xy 2^{2(n+1)+1} + y^2.$$

Note that each of xy and y^2 has length at most $2n$ bits. Thus, the product xy can be obtained from the binary representation of z^2 with the least significant bit at position $2(n+1)+1$ from the least significant bit of z^2. $\square$

The above corollary will be useful in proving the lower bound on the depth of threshold circuits computing Division.

Theorem 4.15:   Any polynomial-size threshold circuit with polynomially bounded integer weights for Division must have depth at least 3.

Proof: It suffices to show that the lower bound holds for a special case of Division, namely $DIV_k(1/y)$, for some k that is polynomially bounded. The idea is to show that the square of an integer z can be obtained as portions of the outputs of $DIV_n(1/y)$ (i.e., squaring is a projection of the reciprocal function). The lower bound then follows from Corollary 4.3.

Let $z = z_{n-1} \cdots z_0$ be an n-bit integer whose square z^2 is to be computed. Let $m = 4n$ and $M = 10n$. Clearly, the number $q = z2^{-m} + 2^{-M} = 0.q_{-1} \cdots q_{-M}$ is a projection of z, with $q_{-m+(n-1)} = z_{n-1}, \ldots, q_{-m} = z_0$, $q_{-M} = 1$, and the rest of q_i's being zeros. Consider the number $1 - q = 0.\hat{q}_{-1} \cdots \hat{q}_{-M}$. Then, $\hat{q}_{-M} = q_{-M} = 1$, and for $i < M$, $\hat{q}_{-i} = \overline{q}_{-i}$. Thus, $1 - q$ is also a projection of z.

Since $0 < q < 1$, we have

$$
\begin{aligned}
\frac{1}{(1-q)} &= 1 + q + q^2 + q^3 + \cdots \\
&= 1 + (z2^{-m} + 2^{-M}) + (z2^{-m} + 2^{-M})^2 \\
&\quad + (z2^{-m} + 2^{-M})^3 + \cdots \\
&= 1 + z2^{-4n} + z^2 2^{-8n} + r \ ,
\end{aligned}
$$

where

$$r = 2^{-M} + z2^{1-m-M} + 2^{-2M} + (z2^{-m} + 2^{-M})^3 + \cdots$$
$$\leq 2^{-10n} + 2^{1-13n} + 2^{-20n} + 2^{1-9n}.$$

For $n > 1$, since $z^2 2^{-8n} \leq 2^{-6n}$ and $r < 2^{-8n}$, z^2 appears as an isolated block in the binary representation of $(1-q)^{-1}$, with the least significant bit at position $-8n$. Thus, z^2 can be obtained as the least $2n$ significant bits of the binary representation of $DIV_{8n}(1/(1-q))$, where $(1-q)$ is a projection of z. By Corollary 4.3, any threshold circuit with polynomially bounded integer weights computing $DIV_{8n}(1/(1-q))$ must have depth at least 3. □

In Exercise 4.11, one is asked to follow the arguments developed in Section 4.4 and to show that the "rounded quotient" of two n-bit integers can be computed by depth-4 polynomial-size threshold circuits. The preceding lower bound result for Division which is defined as the truncated quotient also holds for the rounded quotient.

Corollary 4.4:   Any polynomial-size threshold circuit with polynomially bounded integer weights computing the rounded quotient of two n-bit integers must have depth at least 3.

Proof: The proof follows from the simple observation that the reciprocal $(1-q)^{-1}$ rounded to the least $8n$ significant bits is the same as the truncated reciprocal $DIV_{8n}(1/(1-q))$.   □

We now show that the circuit for Sorting derived in Theorem 4.13 is also depth optimal.

Theorem 4.16:   Any polynomial-size threshold circuit with polynomially bounded integer weights for Sorting must have depth at least 3.

Proof: We show that if we can sort $2n + 1$ integers of length $(\log n + 3)$-bits, then the Inner Product Mod 2 function $IP(x_1, \ldots, x_n, y_1, \ldots, y_n)$ can be computed by setting some of the inputs of Sorting to variables x_i, y_i, or constants. Let b(i) be the $\log n$-bit binary representation of the integer i, for $i = 0, \ldots, n - 1$. We choose the following input to sorting (each line represents one integer):

$$x_1 y_1 1 \cdots 1$$
$$x_2 y_2 1 \cdots 1$$
$$\cdots$$
$$x_n y_n 1 \cdots 1$$
$$110\mathrm{b}(0)$$
$$110\mathrm{b}(1)$$
$$\cdots$$
$$110\mathrm{b}(n-1)$$
$$110\mathrm{b}(n)$$

Note that all of the above $2n + 1$ integers have the same number of bits in their binary representation. We claim that the least significant bit of the $(n + 1)^{st}$ integer in the output list (in ascending order) always yields the value of the $IP(x_1, \ldots, x_n, y_1, \ldots, y_n)$ function.

To justify our claim, consider the position in the output list represented by the integer $110\mathrm{b}(k)$. Clearly, $110\mathrm{b}(j) > 110\mathrm{b}(k)$ if $j > k$, and $x_i y_i 1 \cdots 1 > 110\mathrm{b}(k)$ if and only if $(x_i, y_i) = (1, 1)$. Hence, if $x_i y_i = 1$ for exactly r many i's, then there are exactly $n - k + r$ many integers that are greater than $110\mathrm{b}(k)$. Moreover, $IP(x_1, \ldots, x_n, y_1, \ldots, y_n)$ is 1 if r is odd and 0 otherwise, and is the same as the least significant bit of $\mathrm{b}(r)$.

Choosing $k = r$ reveals that the integer that appears at position $n + 1$ in the sorting list is the integer $110\mathrm{b}(r)$. The least significant bit of this integer is identical with the output of $IP(x_1, \ldots, x_n, y_1, \ldots, y_n)$. Hence, a circuit for Sorting can be used for computing the Inner Product Mod 2 function. By Theorem 4.14, we conclude that any polynomial-size threshold circuit for Sorting has depth at least 3.   □

Exercises

4.1:   [*Lower bound for Multiplication*]
Use the fact that $IP_2 \notin \widehat{LT_2}$ to show that Multiplication is not in $\widehat{LT_2}$.

4.2:   [*Fermat's Theorem*]
Let p be a prime number and a be any positive integer not divisible by p.

(a) Show that for $j = 1, \ldots, p - 1$, there is a unique integer $r_j \in [1, p-1]$ such that $aj \equiv r_j \bmod p$.

(b) Show that $\displaystyle\prod_{j=1}^{p-1} aj \equiv \prod_{j=1}^{p-1} r_j \bmod p$.

(c) Conclude that $a^{p-1} \equiv 1 \bmod p$.

4.3:   [*Prime Number Theorem*]
Use the Prime Number Theorem to derive an upper bound on the Nth prime number p_N. Conclude that p_N is polynomially bounded if N is.

4.4:   [*Exact function and its spectral norm*]
Calculate the L_1 spectral norm of the $Exact_k$ function and show that it is not polynomially bounded.

4.5:   [*Threshold circuits for string equivalence testing*]
Consider two n-bit strings, $x_1, \ldots, x_n$ and $y_1, \ldots, y_n$. The two strings are defined to be *equivalent* if there exists a j, $1 \le j \le n-2$, such that $x_i = y_i$, for all $i \le j$, and $g(x_{j+1}, x_{j+2}, y_{j+1}, y_{j+2}) = 1$, where $g(\cdot)$ is a predetermined 4-variable Boolean function. Construct a depth-2 polynomial-size threshold circuit whose output is 1 if two given n-bit strings are equivalent, and is 0 otherwise.

4.6:   [*Threshold circuits for merging sorted sequences*]
Given two sorted sequences each comprising $\frac{n}{2}$ n-bit integers, design a threshold circuit that merges the two sequences to generate a single sorted sequence of n integers.

4.7:   [*Threshold circuits for computing polynomials, and square roots*]
Give an upper bound on the size and the depth of threshold circuits computing the following functions:

(a) $\displaystyle\sum_{i=0}^{c} a_i X^i$, where c is a constant, and a_i and X_i are n-bit integers.

(b) $\lfloor \sqrt{X} \rfloor$, where X is an n-bit positive integer.

4.8:   [*Threshold circuits for computing Maximum*]
Given a sequence of n n-bit integers, determine a threshold circuit for computing the maximum of the sequence. Is it more depth/size efficient than a threshold circuit for sorting?

4.9:   [*Computing Majority using Multiplication*]
Show that Majority is a projection of Multiplication. Hence, conclude that Multiplication is complete for the class of symmetric functions under AC^0 reductions.

4.10:   [*Computing Multiple Sum with n-bit adder*]
Derive a small-depth circuit for Multiple Sum using devices that compute the sum of n bits.

4.11:   [*The rounded quotient*]
Define the rounded quotient $RQ(x/y)$ of two n-bit integers x and y as follows: $RQ(x/y) = \lfloor x/y \rfloor$ if $x/y - \lfloor x/y \rfloor < 1/2$, and $RQ(x/y) = \lfloor x/y \rfloor + 1$ otherwise. Modify the arguments developed in Section 4.4 to show that $RQ(x/y)$ can also be computed by depth-4 polynomial-size threshold circuits.

Bibliographic Notes

It is implicit in Chandra, Stockmeyer, and Vishkin [26] that Multiplication and Sorting can be computed by constant-depth polynomial-size threshold circuits, but the issue of minimizing the depth of these circuits is not addressed. Siu and Bruck [122] and Hofmeister, Hohberg, and Köhling [56] independently derive depth-4 threshold circuits for Multiplication; the circuit size in [56] is smaller than the one in [122]. The existence of depth-2 threshold circuits for Addition and Comparison, and the existence of depth-4 threshold circuits for Sorting are shown in Siu and Bruck [123] using the techniques of harmonic analysis presented in this chapter. The result that a function with a polynomially bounded spectral norm can be written as the sign of a sparse polynomial is shown in Bruck and Smolensky [22]. Siu and Bruck [123] strengthen the result and show that a function with polynomially bounded spectral norm can in fact be closely approximated by a sparse polynomial. This result is used several times in the construction of the depth-efficient threshold circuits presented in this chapter. The explicit construction of the sparse polynomials for Comparison and Addition is taken from Alon and Bruck [6].

Beame, Cook and Hoover [12] apply the Chinese Remainder Theorem to derive $O(\log n)$-depth AND/OR circuits with bounded fan-in (NC^1 circuits) for Division. It is observed in Reif [108] that the construction in [12] can be modified to obtain constant-depth threshold circuits for Division and related problems. However, the issue of constructing minimal-depth threshold circuits for Division and related problems is not addressed in [12, 108]. Combining the techniques in [6], [12], and [123], Siu, Bruck, Kailath, and Hofmeister [124] improve the constants of the circuits in [12, 108] and provide explicit construction of depth-4 threshold circuits for Division and Powering, and depth-5 threshold circuits for Multiple Product. Minimal-depth threshold circuits for Multiplication and Division are derived in Siu and Roychowdhury [125]. Depth-3 threshold circuits for Sorting are derived in [124], where it is also shown that such circuits have optimal depth.

The fact that Inner Product Mod 2 and Multiplication cannot be computed by polynomial-size threshold circuits with polynomially bounded integer weights and depth less than 3 was first shown in Hajnal, Maass, Pudlák, Szegedy and Turán [48]. However, it is left unresolved in [48] whether Multiplication can be computed by depth-3 polynomial-size threshold circuits. The depth-4 threshold circuits for Multiplication derived in [56] and [122] are not depth-optimal. Using the results in [40] and [41], Siu and Roychowdhury [125] show that in fact one can construct depth-3 polynomial-size threshold circuits for Multiplication and Division, and depth-4 threshold circuits for Multiple Product. The proof that Division cannot be computed by polynomial-size threshold circuits with polynomially bounded integer weights and depth less than three can be found in Hofmeister and Pudlák [57, 55]. Our proof of this result (Theorem 4.15) is taken from Wegener [141], which presents a simpler derivation.

Chapter 5

Depth/Size Trade-offs

5.1 Introduction

In Chapter 4, we present depth-efficient threshold circuits for many common arithmetic functions. The underlying goal there is to minimize the circuit depth while allowing the circuit size to be polynomially bounded. In this chapter, we study the trade-offs between the circuit size and the circuit depth. We deal with two different but related complexity measures of circuit size: the *node complexity* and the *edge complexity*. The node complexity refers to the number of gates in the circuit, while the edge complexity refers to the number of connections.

In earlier chapters, we have almost exclusively used the term 'circuit size' to denote the node complexity of the circuit. In our circuit model, the edge complexity of a circuit is polynomially bounded if and only if its node complexity is also polynomially bounded. Thus, there is no confusion if we say a circuit has polynomial size, since the term 'polynomial size' applies to both complexity measures. However, a circuit might have, for example, $O(n)$ node complexity but $O(n \log n)$ edge complexity. We shall explicitly indicate the complexity measure we are referring to whenever it is needed to distinguish the two. Moreover, all threshold circuits presented in the rest of this chapter are assumed to have polynomially bounded integer weights. To avoid cumbersome notations, we also assume that all values have been rounded to their nearest integers.

In Section 5.2.1, we first present trade-off results on the node complexity and the circuit depth for Parity and Multiple Sum. We make repeated use of the telescopic technique developed in Chapter 3, where we reduce the node complexity of threshold circuit for a general symmetric function from $O(n)$ (in depth-2 circuit) to $O(\sqrt{n})$ (in depth-3 threshold circuit). In fact, we shall show that the telescopic technique is applicable to any symmetric function with a periodic structure.

We then study depth/size trade-off issues for the functions Comparison and Addition in Section 5.3. To underscore the computational power of threshold circuits, we show that any depth-2 circuit of AND/OR gates requires exponentially many gates to implement Comparison or Addition. On the other hand, in Chapter 4, these two functions are shown to be computable by depth-2 polynomial-size threshold circuits. We show in Section 5.3 that the size complexity for these two functions can be substantially reduced by using depth-3 threshold circuits. Furthermore, in Section 5.4, we present trade-off results on the edge complexity and the circuit depth for Addition using ideas of parallel prefix computation.

The fan-in of most of the circuits under study so far increases at least linearly $\Omega(n)$ with the input size. In practice, there is always a certain limit to the maximum fan-in (and fan-out) that each gate can accommodate. Therefore, these results are meaningful in practice only when the value of n does not exceed the maximum parameter for the feasible implementation of the physical circuit. Hence, it is interesting to impose some constraints on the fan-in (and fan-out) of the circuit and derive a trade-off between the circuit size and the circuit depth with the fan-in (and fan-out) as the parameters. In Section 5.5, we focus on such issues and discuss trade-offs between the node complexity and the circuit depth when the fan-in of the circuits is restricted. Whenever it is appropriate, we also point out trade-off results for restricted fan-in circuits in the early sections of the chapter.

5.2   Trade-offs between Node Complexity and Circuit Depth

5.2.1   Parity

Here we construct depth-d threshold circuits for computing the Parity function $PAR_n(X)$. Since our construction is a generalization of the depth-3 threshold circuit construction for a general symmetric function, the readers might find it helpful to review the telescopic technique presented in Section 3.3.

Theorem 5.1: For every $d < \log n$, $PAR_n(X)$ can be computed by a depth-$(d+1)$ threshold circuit with $O(dn^{1/d})$ gates.

Proof: Since $PAR_n(X)$ is a symmetric function, we can rewrite it as a function $\tilde{f}(z)$, where $z = \sum_{i=1}^{n} x_i$, and $(x_1, x_2, \ldots, x_n) \in \{0,1\}^n$ are the inputs. Then, $\tilde{f}$ has *period* 2 since $\tilde{f}(z+2) = \tilde{f}(z)$. We apply the telescopic technique (see Theorem 3.4) recursively in our construction. The underlying idea is that by using this periodic structure, one can reduce the Parity function of n variables to a Parity function of $n^{1-1/d}$ variables with one layer of $n^{1/d}$ threshold gates.

We use notations similar to those introduced in Theorem 3.4. Divide the interval $[0, n]$ into $n^{1/d}$ consecutive subintervals of the same length $n^{1-1/d}$. The first layer of the circuit consists of $n^{1/d}$ threshold gates each computing the value y_i, where for $i = 1, \ldots, n^{1/d}$,

$$z_i = \text{sgn}(\sum_{i=1}^{n} x_i - s_i),$$

and $s_i = (i-1)n^{1-1/d}$. Let

$$t = s_1 z_1 + (s_2 - s_1)z_2 + (s_3 - s_2)z_3 + \cdots + (s_{n^{1/d}} - s_{n^{1/d}-1})z_{n^{1/d}} .$$

If $\sum_{j=1}^{n} x_j$ lies within the mth interval, then

$$z_i = \begin{cases} 1 & \text{if } i \leq m, \\ 0 & \text{if } i > m, \end{cases}$$

and because of the 'telescoping' effect of the coefficients, $t = s_m$. From the definition of $PAR_n(X)$, the function will be 1 if and only if $(\sum_{j=1}^{n} x_j)$ is odd. Without loss of generality, assume $n^{1-1/d}$ is even. The following observation is crucial: $(\sum_{j=1}^{n} x_j)$ is odd if and only if $(\sum_{j=1}^{n} x_j - t)$ is odd. Since $(\sum_{j=1}^{n} x_j - t)$ lies between 0 and $n^{1-1/d}$, we have reduced the problem to the Parity function of $n^{1-1/d}$ variables. Applying the reduction $d-1$ times, each time increasing the depth by 1 and the gate count by $n^{1/d}$, we reduce the problem to the Parity function of $n^{1/d}$ variables. Finally, it takes another depth-2 circuit of $O(n^{1/d})$ gates to compute the function. So all together the threshold circuit has $O(dn^{1/d})$ gates and depth $d + 1$. This holds for all $d < \log n$.   □

Remark 5.1:   In Exercise 5.2, one is asked to show that the Parity function $PAR_n(X)$ can be computed by a depth-$(\log n)$ threshold circuit comprising $\log n$ gates. This is a special case of our general construction. In Chapter 7, we prove that the depth-d circuits for $PAR_n(X)$ derived in Theorem 5.1 are almost optimal in the number of threshold gates.   □

The construction of Theorem 5.1 makes crucial use of the periodicity of the function when it is viewed as a function of $\sum_{i=1}^{n} x_i$ within $[0, n]$. In fact, for a general symmetric function without the periodic structure, one cannot in general obtain the depth/size trade-off results as in the case of Parity. This is because Corollary 3.3 implies that a general n-variable symmetric function cannot be computed by a threshold circuit with gate count smaller than $O(\sqrt{n/\log n})$, even if there is no restriction on the depth.

In general, the telescopic technique used in the proof of Theorem 5.1 can be applied to any symmetric function with periodic structure. Let T be the period of any such symmetric function $f(\sum_{i=1}^{n} x_i)$, i.e., f has the same value at $(\sum_{i=1}^{n} x_i)$ and $(\sum_{i=1}^{n} x_i + T)$ ($T = 2$ for Parity). We can use the same technique recursively to reduce the function f of n variables to the same function of $n^{1-1/d} + O(1)$ variables with one layer of $n^{1/d} + O(1)$ threshold gates. We only have to make sure that in each reduction step, the length of each subinterval (except possibly the last subinterval) is a multiple of the period T.

Another periodic symmetric function that we shall study in more detail in later chapters is the Complete Quadratic $(CQ(X))$ function.

Definition 5.1:   The Complete Quadratic function $(CQ(X))$ of n variables equals the Parity function of all $\binom{n}{2}$ possible ANDs between pairs of variables:

$$CQ(x_1, x_2, \ldots, x_n) = \bigoplus_{(i,j),\, i \neq j} (x_i \wedge x_j) .   □$$

Theorem 5.2:   For every $d < \log n$, there exists a depth-$(d+1)$ threshold circuit with $O(dn^{1/d})$ gates that computes the function CQ(X).

Proof: An alternative definition of CQ in $\{0,1\}$ notation can be stated as follows (see Exercise 7.1):

$$CQ(X) = \begin{cases} 0 & \text{if no. of 1's in } X \bmod 4 = 0 \text{ or } 1, \\ 1 & \text{otherwise.} \end{cases}$$

Because of this periodic structure of the function, the construction is similar to that of Parity as in Theorem 5.1.   □

Note that the fan-in of the general depth-$(d+1)$ circuits for Parity stated in Theorem 5.1 is $\Omega(n)$. When the fan-in is restricted to, say, $m \ll n$, we need to modify the construction in order to obtain the trade-offs between the circuit depth and the number of threshold gates.

Theorem 5.3: For every $d > 0$, the Parity function $PAR_n(X)$ can be computed using a threshold circuit of $O(dnm^{1/d})$ edges, depth $O(d\log n/\log m)$ and fan-in bounded by m. Moreover, the number of threshold gates in the circuit is $O(dn/m^{1-1/d})$.

Proof: Theorem 5.1 implies that for every integer $d > 0$, one can compute the Parity of m inputs using a depth-$(d+1)$ threshold circuit of $O(dm^{1/d})$ gates, $O(dm^{1+1/d})$ edges, and fan-in bounded by $O(m)$. Consider an m-ary tree with n leaves in which each node corresponds to such a depth-$(d+1)$ threshold circuit. Then, the output gate of the root node computes the Parity of n inputs corresponding to the leaves of the m-ary tree. Moreover, the number of nodes in the tree is $O(n/m)$ and the number of levels is $O(\log n/\log m)$. Hence, the circuit has $O(n/m \cdot dm^{1/d}) = O(dn/m^{1-1/d})$ gates, $O(dnm^{1/d})$ edges and $O(d\log n/\log m)$ depth.   □

5.2.2 Multiple Sum

Next we show that Multiple Sum can be computed by a constant-depth threshold circuit with $O(n^{1+\epsilon})$ gates for any fixed $\epsilon > 0$. Recall that the depth-3 threshold circuit for Multiple Sum presented in Section 4.3 consists of two main steps: the sum of n n-bit numbers is first reduced to the sum of two $(n + \log n)$-bit numbers using one layer, and then the final sum is computed using another two layers. We use this technique as our basis and utilize the result stated in Theorem 5.4 to reduce the gate count of each step while keeping the depth constant. First, let us prove two results that will lead us to the trade-off results for Multiple Sum.

Lemma 5.1:   Let $S_1(X) = \sum_i w_i x_i \in [0, 2^l]$ and $0 \le k < l$. For any positive integer d, the $(k+1)^{th}$ bit (counted from the least significant bit) of $S_1(X)$ can be computed by a depth-$(d+1)$ threshold circuit with $O(2^{(l-k)/d})$ gates.

Proof: Note that the $(k+1)^{th}$ bit of $S_1(X)$ is 1 if $S_1(X) \in [j2^k, (j+1)2^k - 1]$ for $j = 1, 3, 5, \ldots, 2^{l-k} - 1$, and is 0 otherwise. Also assume that $(l-k)/d$ is an integer.

The underlying idea is as follows: Each layer of the circuit has $2^{(l-k)/d}$ threshold gates. We use each layer recursively to remove the most significant $(l-k)/d$ bits of the sum computed from the preceding layer. Since the original sum is $(l+1)$-bit long, the most significant bit of the reduced sum after d steps is the $(k+1)^{th}$ bit of the original sum.

More formally, the first layer of our circuit consists of $2^{(l-k)/d}$ threshold gates, each computing the value y_j, where for $j = 0, 1, \ldots, 2^{(1-k)/d} - 1$,

$$y_j = \text{sgn}\left(\sum_i w_i x_i - j2^{l-(l-k)/d}\right).$$

Let

$$t = \sum_{j=0}^{2^{(l-k)/d}-1} 2^{l-(l-k)/d} y_j \ .$$

If $j^* 2^{l-(l-k)/d} \le \sum_i w_i x_i < (j^* + 1)2^{l-(l-k)/d}$, then $t = j^* 2^{l(1-1/d)+k/d}$, and

$$y_j = \begin{cases} 1 & \text{if } j \le j^*, \\ 0 & \text{if } j > j^*. \end{cases}$$

Thus, $0 \le (S_1(X) - t) < 2^{l-(l-k)/d}$. Let $S_2(X) = S_1(X) - t$. Note that $S_1(X) \in [j2^k, (j+1)2^k - 1]$ if and only if $S_2(X) \in [j2^k, (j+1)2^k - 1]$ for $j = 1, 3, 5, \ldots, 2^{l-k} - 1$. In other words, the $(k+1)^{th}$ bit of $S_1(X)$ is that of $S_2(X)$. Now apply the above reduction recursively. In general, for $2 \le m \le d$, the mth layer of our threshold circuit consists of $2^{(l-k)/d}$ threshold gates, each computing the value y_j^m, where for $j = 0, 1, \ldots, 2^{(1-k)/d} - 1$,

$$y_j^m = \text{sgn}\left(S_{m-1}(X) - j2^{l-m(l-k)/d}\right) ,$$

and

$$S_m(X) = S_{m-1}(X) - \sum_{j=0}^{2^{(1-k)/d}-1} 2^{l-m(l-k)/d} y_j \ .$$

The reduction is applied d times, each time increasing the depth by 1 and the gate count of the circuit by $2^{(l-k)/d}$. At the $(d+1)^{th}$ layer, the output gate computes $\text{sgn}(S_d(X) - 2^k)$, which is the $(k+1)^{th}$ significant bit of $S_1(X)$. $\square$

Corollary 5.1:   Given $\epsilon > 0$, and n $(\log n)$-bit numbers, each bit of their $2\log n$-bit sum can be computed by a constant-depth threshold circuit with $O(n^\epsilon)$ gates.

Proof: Let $\epsilon > 0$ be arbitrary and let the sum be represented by $s_{2\log n - 1}$ $s_{2\log n - 2} \cdots s_0$, with s_0 being the least significant bit. We use Lemma 5.1 to show that each s_k can be computed by a constant-depth threshold circuit with $O(n^\epsilon)$ gates. Choose d such that $\epsilon > 1/d$. We consider two cases. For $k = 0, \ldots, \log n - 2$, s_k depends only on a sum $\in [0, \ldots, 2^{\log n + k}]$. In this case, put $l = (\log n + k)$ in Lemma 5.1, it follows that each s_k can be computed by a constant-depth circuit with $O(2^{(l-k)/d}) = O(n^\epsilon)$ threshold gates. Now consider $k = \log n - 1, \ldots, 2\log n$. Since the total sum $\in [0, 2^{2\log n}]$, put $l = 2\log n$ in Lemma 5.1, and in this case $(l - k) \leq \log n$ and, thus, $O(2^{(l-k)/d}) = O(n^\epsilon)$ threshold gates suffice. $\square$

Theorem 5.4:   For any fixed $\epsilon > 0$, Multiple Sum can be computed by constant-depth threshold circuits with $O(n^{1+\epsilon})$ gates.

Proof: First apply the block-save technique (see Section 4.3) to reduce the sum of n n-bit numbers to the sum of two $O(n + \log n)$-bit numbers. The reduction consists of computing $O(n/\log n)$ block sums of n $(\log n)$-bit numbers. By Corollary 5.1, each bit of the block sum can be computed by a constant-depth threshold circuit with $O(n^\epsilon)$ gates. Thus, the entire block-save reduction step can be computed by constant-depth threshold circuits with $O(n^{1+\epsilon})$ gates. Now the sum of the remaining two $(n + \log n)$-bit numbers can be shown to be computable by a constant-depth threshold circuit with $O(n^{1+\epsilon})$ gates. In fact, we shall prove a stronger result in the next section, which states that the sum of two $O(n)$ bit integers can be computed by constant-depth AND-OR circuits with $O(n \log^* n)$ gates. $\square$

5.3 Trade-offs for Comparison and Addition

In this section, we present depth/size trade-off results for the functions Comparison and Addition. We first show that these two functions cannot be

computed by depth-2 polynomial-size AND-OR circuits. We know, however, from Chapter 4 that these two functions can be computed by depth-2
polynomial-size threshold circuits. Thus, these results underscore the power
of threshold gates. We then present straightforward depth-3 AND-OR circuit realizations for Comparison and Addition using $O(n)$ and $O(n^2)$ gates,
respectively.

Theorem 5.5: Any depth-2 AND-OR circuit for Comparison must be of
exponential size.

Proof: First, consider any depth-2 circuit computing Comparison with
AND gates in the first layer and an output OR gate in the second layer.
Then, the circuit is equivalent to a Sum Of Product (SOP) form (i.e., OR
of ANDs):

$$COMP_n(X,Y) = P_1 \vee P_2 \vee \ldots \vee P_m,$$

where each product term P_i corresponds to the AND function of some subset
of the variables (possibly negated), and the number of products, m, corresponds to the number of AND gates in the first layer. There are many
different SOP forms for Comparison. We show that any of these forms must
have exponentially many product terms, and thus the corresponding depth-2
circuit must have exponential size.

For convenience, we use $COMP_n(X,Y)$ to mean both the function Comparison and any of its SOP forms (in the variables $x_1, \ldots, x_n, y_1, \ldots, y_n$ and
their negations). Moreover, let $|COMP_n|$ and $|COMP_{n-1}|$ denote the min-
imum number of product terms in any SOP form of $COMP_n(X,Y)$ and
$COMP_{n-1}(X,Y)$, respectively. The key idea is to show that $|COMP_n| \geq
2|COMP_{n-1}|$, thereby implying an exponential lower bound on $|COMP_n|$.

Using standard Boolean factorization, one can always write

$$COMP_n(X,Y) = x_n y_n P_{n-1}^1 \vee x_n \overline{y}_n P_{n-1}^2 \vee \overline{x}_n y_n P_{n-1}^3 \vee \overline{x}_n \, \overline{y}_n P_{n-1}^4$$
$$\vee x_n P_{n-1}^5 \vee \overline{x}_n P_{n-1}^6 \vee y_n P_{n-1}^7 \vee \overline{y_n} P_{n-1}^8 \vee P_{n-1}^9,$$

where the P_{n-1}^j's are SOP forms that are independent of the variables x_n
and y_n.

Now, if $x_n = 0$ and $y_n = 1$, then $COMP_n(X,Y) = 0$, irrespective of the
assignments to the other variables, $x_{n-1}, \ldots, x_1$, and $y_{n-1}, \ldots, y_1$. Hence,
substituting $x_n = 0$ and $y_n = 1$ in the above expression for $COMP_n(X,Y)$,

we have $P_{n-1}^3 \equiv P_{n-1}^6 \equiv P_{n-1}^7 \equiv P_{n-1}^9 \equiv 0$. Thus, the expression for $COMP_n(X, Y)$ simplifies to

$$COMP_n(X, Y) = x_n y_n P_{n-1}^1 \vee x_n P_{n-1}^5 \vee \overline{x_n}\ \overline{y_n} P_{n-1}^4 \vee \overline{y_n} P_{n-1}^8 \vee x_n \overline{y_n} P_{n-1}^2.$$

Next, we observe that if $x_n = 1$ and $y_n = 1$, or $x_n = 0$ and $y_n = 0$, then $COMP_n(X, Y) \equiv COMP_{n-1}(X, Y)$. Hence, substituting these values of x_n and y_n in the above expression for $COMP_n(X, Y)$, we get

$$P_{n-1}^1 \vee P_{n-1}^5 \equiv COMP_{n-1}(X, Y)$$

and

$$P_{n-1}^4 \vee P_{n-1}^8 \equiv COMP_{n-1}(X, Y).$$

It follows from the above expression of $COMP_n(X, Y)$ that the number of product terms in $COMP_n\ (X, Y)$ is at least twice the number of product terms in $COMP_{n-1}(X, Y)$, i.e.,

$$|COMP_n| \geq 2|COMP_{n-1}|.$$

Thus, $COMP_n(X, Y)$ must have an exponential number of product terms in any SOP form, and the corresponding depth-2 circuit must have an exponential size.

A similar dual argument applies to any depth-2 circuit with OR gates in the first layer and an output AND gate in the second layer, and easily shows that there must be exponentially many OR gates for the function Comparison. □

As in the case of Comparison, Addition also cannot be computed by any depth-2 polynomial-size AND-OR circuit.

Theorem 5.6:   Any depth-2 AND-OR circuit for Addition must be of exponential size.

Proof: We use the same notations as in the proof of Theorem 5.5 and show that any depth-2 AND-OR circuit computing the nth sum bit s_n must have an exponential size. Since $s_n = c_n \oplus x_n \oplus y_n$, and $s_n = c_n$ for $x_n = y_n = 0$, it suffices to derive an exponential lower bound on the size of any circuit for c_n. The proof can be carried out almost exactly as in the case of Comparison (simply by renaming the variables). First, derive a recursion in the Sum Of

Product or Product of Sum forms for c_n in terms of c_{n-1}. Then show that the number of terms $|c_n| \geq 2|c_{n-1}|$ and, hence, show that $|c_n|$ must grow exponentially in n. $\qquad\Box$

Although both Comparison and Addition require any computing depth-2 AND-OR circuits to have exponential size, they can be computed by small depth-3 AND-OR circuits.

Theorem 5.7:   Comparison can be computed by a depth-3 AND-OR circuit with $3n$ gates.

Proof: We first write a Boolean expression for $COMP_n(X, Y)$ as a recursion on n.

$$COMP_1(X, Y) = x_1 \vee \overline{y}_1 = \begin{cases} 1 & \text{if } x_1 \geq y_1, \\ 0 & \text{otherwise.} \end{cases}$$

$$
\begin{aligned}
COMP_n(X, Y) &= (x_n \wedge \overline{y}_n) \bigvee [(x_n \vee \overline{y}_n) \wedge COMP_{n-1}(X, Y) \\
&= \begin{cases} 1 & \text{if } x_n > y_n \text{ or } (x_n \geq y_n \text{ and } \sum_{i=1}^{n-1} 2^i x_i \geq \sum_{i=1}^{n-1} 2^i y_i), \\ 0 & \text{otherwise.} \end{cases}
\end{aligned}
$$

Define the following Boolean expressions:

$$B_n = x_n \wedge \overline{y}_n \, ,$$

for $k = 2, \ldots, n-1$,

$$B_k = (x_k \vee \overline{y}_k) \bigwedge_{j=k+1}^{n} (x_j \wedge \overline{y}_j) \, ,$$

and

$$B_1 = \bigwedge_{j=1}^{n} (x_j \wedge \overline{y}_j).$$

Using the recursive Boolean expressions of $COMP_n(X, Y)$, it is easy to show by induction on n that $COMP_n(X, Y) = \bigvee_{j=1}^{n} B_j$.

The first layer of the circuit for Comparison has $2n - 1$ gates computing $(x_j \wedge \overline{y}_j)$ and $(x_j \vee \overline{y}_j)$. With these computed values as inputs, the second layer has n gates, each computing the B_j's. Finally, the output gate computes the OR ($\bigvee$) of all the B_j's. The total number of gates is $3n$ as claimed. □

Note that we have not made use of the full power of threshold gates in the preceding depth-3 threshold circuit for Comparison. We simply use the threshold gates to simulate the AND ($\wedge$), OR ($\vee$) operations.

In Section 4.2, we showed that each bit of the sum in the function Addition is computable by a depth-2 polynomial-size threshold circuit. The gate count can be significantly reduced to $O(n)$ if we increase the depth by 1. Since there are $n+1$ bits in the final sum, the total number of gates in the circuit computing Addition is $O(n^2)$.

Theorem 5.8:   Addition can be computed by depth-3 AND-OR circuits with $O(n^2)$ gates.

Proof: Let $X = (x_{n-1}, \ldots, x_0)$, $Y = (y_{n-1}, \ldots, y_0) \in \{1,0\}^n$ be the two n-bit numbers with x_0 and y_0 being the least significant bits, and $S = (s_n, \ldots, s_0)$ be the $(n+1)$-bit sum. Let c_k be the kth carry bit, i.e., c_k is the most significant bit of the sum $\sum_{i=0}^{k-1} 2^i(x_i + y_i)$ with $c_0 = 0$. The main idea is to compute the carry bits c_k's and their negations $\overline{c}_k$'s in parallel using depth-2 circuits. Then, the sum bits s_k can be computed as follows:

$$
\begin{aligned}
s_k &= c_k \oplus x_k \oplus y_k \\
    &= (\overline{c}_k x_k \overline{y}_k) \vee (\overline{c}_k \, \overline{x}_k y_k) \vee (c_k x_k y_k) \vee (c_k \overline{x}_k \, \overline{y}_k).
\end{aligned}
$$

The circuit computing the carry bit c_k is similar to the circuit computing Comparison. Note that $c_k = 1$ if and only if there is a $j \le k-1$ such that $x_j = y_j = 1$, but for all i, $j < i \le k-1$, we do not have $x_i = y_i = 0$. Thus,

$$
\begin{aligned}
c_k = \; & x_{k-1}y_{k-1} \bigvee [(x_{k-1} \vee y_{k-1})x_{k-2}y_{k-2}] \bigvee [(x_{k-1} \vee y_{k-1}) \\
& (x_{k-2} \vee y_{k-2})x_{k-3}y_{k-3}] \bigvee \cdots \bigvee [(x_{k-1} \vee y_{k-1})(x_{k-2} \vee y_{k-2}) \\
& (x_{k-3} \vee y_{k-3}) \cdots x_0 y_0],
\end{aligned}
$$

$$
\begin{aligned}
\overline{c}_k = \; & \overline{x}_{k-1} \, \overline{y}_{k-1} \bigvee [(\overline{x}_{k-1} \vee \overline{y}_{k-1})\overline{x}_{k-2} \, \overline{y}_{k-2}] \bigvee [(\overline{x}_{k-1} \vee \overline{y}_{k-1}) \\
& (\overline{x}_{k-2} \vee \overline{y}_{k-2})\overline{x}_{k-3} \, \overline{y}_{k-3}] \bigvee \cdots \bigvee [(\overline{x}_{k-1} \vee \overline{y}_{k-1})(\overline{x}_{k-2} \vee \overline{y}_{k-2}) \\
& (\overline{x}_{k-3} \vee \overline{y}_{k-3}) \cdots (\overline{x}_0 \vee \overline{y}_0)].
\end{aligned}
$$

As in the case of the circuit for Comparison, define Boolean expressions

$$B_k^k = x_k y_k ; \quad \tilde{B}_k^k = \overline{x}_k\,\overline{y}_k ,$$

for $j = 1, \ldots, k-1$,

$$B_j^k = (x_j y_j) \bigwedge_{i=j+1}^{k} (x_i \vee y_i) ,$$

$$\tilde{B}_j^k = (\overline{x}_j\,\overline{y}_j) \bigwedge_{i=j+1}^{k} (\overline{x}_i \vee \overline{y}_i) ,$$

and

$$\tilde{B}_1^k = \bigvee_{i=1}^{n} (\overline{x}_i \vee \overline{y}_i).$$

The first layer of the circuit consists of $2n$ gates computing the $(x_i \vee y_i)$ and $(\overline{x}_i \vee \overline{y}_i)$ for $i = 1, \ldots, n$. With these computed values, the second layer, which consists of $O(n^2)$ gates, computes for each $k = 1, \ldots, n$, $1 \leq j \leq k$, $(\tilde{B}_j^k x_k \overline{y}_k)$, $(\tilde{B}_j^k\,\overline{x}_k y_k)$, $(B_j^k x_k y_k)$, and $(B_j^k \overline{x}_k\,\overline{y}_k)$. The third layer has $n+1$ output gates and computes each sum bit s_k that is the OR of all these values:

$$s_k = \bigvee_{j=1}^{k-1} (\tilde{B}_j^{k-1} x_k \overline{y}_k) \vee (\tilde{B}_j^{k-1}\,\overline{x}_k y_k) \vee (B_j^{k-1} x_k y_k) \vee (B_j^{k-1} \overline{x}_k\,\overline{y}_k) .$$

Clearly, the gate count is dominated by the number of gates in the second layer, which is $O(n^2)$. □

Remark 5.2: In Chapter 10, we show that the number of gates in a threshold circuit with polynomially bounded integer weights for computing Comparison is $\Omega(n/\log n)$. We also present depth-3 threshold circuits of optimal size $O(n/\log n)$ for Comparison. Thus, depth-3 threshold circuits for Comparison are asymptotically optimal with respect to node complexity, and increasing the depth of threshold circuits will not lead to any significant decrease in the number of gates. □

5.4 Addition and Parallel Prefix Computation

In this section, we derive depth/size trade-offs results for Addition. The construction of the circuits is based on ideas from parallel prefix computation. These trade-off results apply to AND-OR circuits, as well as to threshold circuits.

5.4.1   Parallel Prefix Circuits

Consider a set X of elements with an associative binary operation. (X is called a *semigroup* in the mathematics literature.) We denote the binary operation simply by juxtaposition of the elements in X. For example, the AND function of the variables $x_1, x_2, \ldots, x_k$ can be considered as computing the product $x_1 x_2 \cdots x_k$ in the set $X = \{1, 0\}$, where the binary associative operation is defined as $00 = 01 = 10 = 0$ and $11 = 1$. Another example is the Parity function, where $X = \{1, 0\}$ and the binary operation (usually denoted by $\oplus$ instead of juxtaposition) is defined as $00 = 11 = 0$ and $01 = 10 = 1$.

Suppose we have a set of functional gates such that each with inputs $x_1, x_2, \ldots, x_k$ computes the product $x_1 x_2 \cdots x_k$ (under a given associative binary operation), and we have a combinational circuit that comprises these functional gates. For the moment, we do not impose any constraint on the maximum number of inputs each gate can accommodate. Given n inputs $x_1, x_2, \ldots, x_n$, we are interested in a small-depth circuit that computes every prefix product $x_1, x_1 x_2, \ldots, x_1 x_2 \cdots x_n$. A straightforward realization yields a one-layer circuit with n nodes and $(1 + 2 + \cdots + n) = O(n^2)$ edges. The following lemma provides a much-improved construction and significantly reduces the number of edges in the circuit at the expense of a small increase in the depth. For simplicity, we assume that $n = 2^m$. The details needed to handle the general case, where n is not a power of 2, is straightforward and is omitted.

Lemma 5.2:   The prefixes $x_1, x_1 x_2, \ldots, x_1 x_2 \cdots x_n$ can be computed by a depth-2 circuit of node complexity $O(n)$ and edge complexity $O(n \log n)$.

Proof: Define an *interval* I of indices $\{0, \ldots, n - 1\}$ to be an ordered subset of consecutive indices, and represent I by concatenating its indices in an increasing order, e.g., $I = (2, 3, 4, 5)$. A *principal interval* I_j^k is an interval such that for some $k = 0, \ldots, m$ and $j = 1, \ldots, 2^{m-k}$, $I_j^k = ((j - 1)2^k, \ldots, (j2^k - 1))$. Another way of visualizing the principal intervals is as follows: construct a complete binary tree whose leaves are from left to right the indices $\{0, \ldots, n - 1\}$; then each principal interval corresponds to an interior node v by concatenating the indices that are the leaves of the subtree rooted at v. Since there are $O(n)$ interior nodes in a complete binary tree, there are $O(n)$ principal intervals. The principal interval I_j^k

corresponds to the jth node (from left to right) on the kth level (the leaves are on level 0). The size of a principal interval I_j^k is 2^k. The proof of the lemma follows from the following crucial observation: every prefix of $0, \ldots, n-1$ is a concatenation of at most $\log n$ principal intervals.

In the first level of the circuit, compute in parallel the product $X_I = x_{i_1} x_{i_2} \cdots x_{i_p}$ for every principal interval $I = (i_1, i_2, \ldots, i_p)$. Since there are $O(n)$ principal intervals and $O(\log n)$ different sizes of principal intervals, the first level requires $O(n)$ nodes and $O(n \log n)$ edges.

For $k = 0, \ldots, (n-1)$, the prefix $0 \cdots k$ is the concatenation of at most $O(\log n)$ principal intervals, say $I_1, \ldots, I_d$. With the values $X_{I_1}, \ldots, X_{I_d}$ computed in the first level, we can compute in the second level the product $x_0 \cdots x_k$, which is $X_{I_1} \cdots X_{I_d}$. This level also requires $O(n)$ nodes and $O(n \log n)$ edges. Thus, the circuit has depth 2, $O(n)$ nodes, and $O(n \log n)$ edges.   □

Since there are principal intervals of size $\Omega(n)$, the fan-in of the above circuit is $\Omega(n)$. If the fan-in of each gate is constrained to be $\leq m$, then the minimum circuit depth that can be achieved is $\geq (\log n / \log m)$. It turns out that with such constraint on the fan-in, one can still construct a circuit of minimum depth with essentially the same size.

Example 5.1:   In Figure 5.1, we illustrate the concept of principal intervals with $n = 16$. For example, the prefix (0 1 2 3 4 5 6) is a concatenation of the principal intervals (0 1 2 3), (4 5), and (6). Similarly, the prefix (0 1 2 3 4 5 6 7 8 9 10 11 12) is a concatenation of the principal intervals (0 1 2 3 4 5 6 7), (8 9 10 11), and (12). Note that each prefix is a concatenation of at most $\log 16 = 4$ principal intervals.   □

Lemma 5.3:   If the fan-in is bounded by m, where $m > \log n$, then the prefixes x_1, $x_1 x_2$, $\ldots$, $x_1 x_2 \cdots x_n$ can be computed by a depth-$((\log n / \log m) + 1)$ circuit of $O(n \log n)$ edge complexity.

Proof: Note that each gate in the second level of the circuit in Lemma 5.2 has at most $\log n$ inputs. For every gate in the first level that has fan-in more than m, we can replace it by a depth $< (\log n / \log m)$ circuit using the same type of gates with fan-in m. The circuit has the structure of an m-ary tree: Each node in the tree corresponds to a gate that computes the product of its

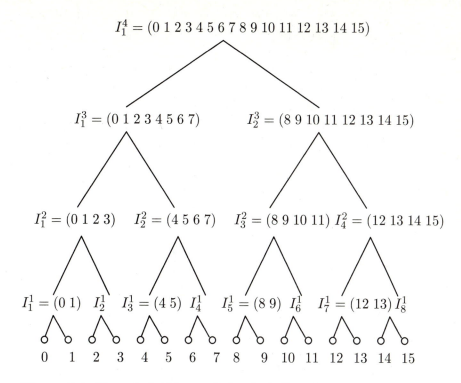

$$I_1^4 = (0\ 1\ 2\ 3\ 4\ 5\ 6\ 7\ 8\ 9\ 10\ 11\ 12\ 13\ 14\ 15)$$

$$I_1^3 = (0\ 1\ 2\ 3\ 4\ 5\ 6\ 7) \qquad I_2^3 = (8\ 9\ 10\ 11\ 12\ 13\ 14\ 15)$$

$$I_1^2 = (0\ 1\ 2\ 3) \quad I_2^2 = (4\ 5\ 6\ 7) \quad I_3^2 = (8\ 9\ 10\ 11)\ I_4^2 = (12\ 13\ 14\ 15)$$

$$I_1^1 = (0\ 1)\ \ I_2^1 \quad I_3^1 = (4\ 5)\ \ I_4^1 \quad I_5^1 = (8\ 9)\ \ I_6^1 \quad I_7^1 = (12\ 13)\ \ I_8^1$$

0   1   2   3   4   5   6   7   8   9   10   11   12   13   14   15

Figure 5.1 The principal intervals for the indices $0, \ldots, 15$, as generated using a binary tree representation.

m inputs. If d is the number of levels in the circuit, then the number of inputs to the circuit is m^d and the number of edges is $m^d + m^{d-1} + \cdots + 2 \le 2m^d$. Thus, $d \le (\log n / \log m)$ and the number of edges increases at most by a factor of two. Therefore, the new circuit has depth $\le (\log n / \log m) + 1$ and $O(n \log n)$ edges. □

Definition 5.2: For $k \ge 1$, let $\log^{(k+1)} n = \log(\log^{(k)} n)$ where $\log^{(1)} n = \log n$. Let $\log^* n = \min \{k : \log^{(k)} n \le 1\}$. □

Note that $\log^* n$ is a function that increases very slowly with n. For example, the values of $\log^* n$ at $n = 2$, 4, 16, 65536 are 1, 2, 3, 4, respectively. Applying the techniques of Lemma 5.2 and Lemma 5.3, one can compute the prefixes by a circuit of almost linear edge complexity.

Lemma 5.4: If the fan-in is bounded by m, where $m > \log n$, then the prefixes x_1, $x_1 x_2$, $\ldots$, $x_1 x_2 \cdots x_n$ can be computed by a depth-$(2(\log n / \log m) + 1)$ circuit of edge complexity $O(n \log^* n)$.

Proof: Divide the indices $\{0, \ldots, (n-1)\}$ into consecutive intervals of size $(\log n)$, divide each such interval into consecutive intervals of size $(\log \log n)$, and so forth. Then, we have a log-ary tree (compared with the binary tree in Lemma 5.2). All together we have $(\log^* n)$ different sizes of intervals.

In the first step, for each interval we compute in parallel the product of the elements with indices in that interval. Since there are $O(\log^* n)$ different sizes of intervals, this step requires $O(n \log^* n)$ edges with depth at most $(\log n / \log m)$.

Suppose an interval I has been divided into subintervals $I_1, I_2, \ldots, I_k$. Then the *subinterval prefixes* of I are $I_1, I_1 I_2, \ldots, I_1 I_2 \cdots I_k$. In the second step, all products corresponding to subinterval prefixes are computed. Observe that the subinterval prefixes can be computed with a prefix circuit whose inputs are the computed products in the first step. Then if the prefix circuit in Lemma 5.3 is used, for an interval of size s, we need $O((s / \log s) \log(s / \log s)) = O(s)$ edges and depth at most $(\log(s / \log s) / \log m)$. Summing over all intervals of different sizes, this step requires $O(n \log^* n)$ edges and depth at most $(\log n / \log m)$.

Since each prefix is a concatenation of at most $(\log^* n)$ subinterval prefixes, the final step uses $O(n \log^* n)$ edges and one level of gates to compute all the prefixes.   □

Example 5.2: Figure 5.2 illustrates the structure of a log-ary tree for $n = 16$. For example, consider the interval $I = (4\ 5\ 6\ 7)$ of size $\log n = 4$. The subintervals of I are of size $\log \log n = 2$ and the subinterval prefixes of I are $(4\ 5)$ and $(4\ 5)(6\ 7)$. Also, the prefix $(0\ 1\ 2\ 3\ 4\ 5\ 6\ 7\ 8\ 9)$ is a concatenation of the subinterval prefixes $(0\ 1\ 2\ 3)(4\ 5\ 6\ 7)$ and $(8\ 9)$.   □

5.4.2 Circuits for Addition

The complexity of an adder lies in the computation of the carry bits. It will be seen that the computation of the carry bits is equivalent to the prefix computation in a semigroup.

Consider a set Z of elements S, P, and R with associative binary operation defined by $zS = S$, $zP = z$, and $zR = R$. In the case of adding two

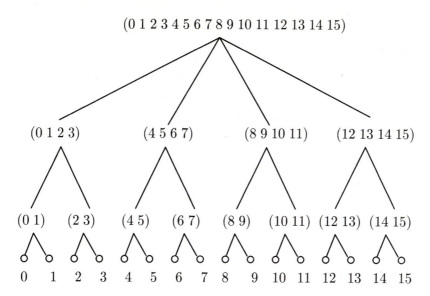

Figure 5.2 A log-ary tree for the indices $0, \ldots, 15$.

binary integers, $X = (x_{n-1}, \ldots, x_0)$ and $Y = (y_{n-1}, \ldots, y_0)$, we associate each pair of the input bits $z_i = (x_i, y_i)$ with an element in Z: z_i is associated with S if $x_i \wedge y_i = 1$ (S is the 'set carry'). Similarly, z_i is P if $x_i \vee y_i = 1$ but $x_i \wedge y_i \neq 1$ ('propagate carry') or R if $x_i \vee y_i = 0$ (reset carry). Then, computing the kth carry bit c_k of the sum is equivalent to computing the product $z_0 \, z_1 \, \cdots \, z_{k-1}$ defined by the above associative binary operation in Z. c_k is 1 if the final product has the value of S, and is 0 otherwise. Thus, the computation of the carry bits c_k for $k = 0, 1, 2, \ldots, (n - 1)$ is the same as a prefix computation.

Based on the algorithm of parallel prefix computation presented in Lemma 5.2 and Lemma 5.3, we can construct an efficient circuit that computes the carries in parallel and, therefore, the sum of two integers efficiently.

Theorem 5.9: The sum of two n-bit integers can be computed by a depth-$O(\log n / \log m)$ AND-OR circuit of edge complexity $O(n (\log^* n)^2)$ and with fan-in bounded by m.

Proof: We divide the construction into several steps. Each step takes as inputs the computed values from preceding steps. As before, denote the input integers by $X = (x_{n-1}, \ldots, x_0)$ and $Y = (y_{n-1}, \ldots, y_0)$.

Step 1: For $i = 0, 1, \ldots, (n-1)$, compute $z_i = (x_i \vee y_i)$ with one level of OR gates.

Step 2: For all principal intervals I of size $\log n$, $\log \log n, \ldots$, 1, compute $f_p(I) = \bigwedge z_i = \bigwedge_{i \in I}(x_i \vee y_i)$ using a depth-$(2 \log n / \log m + 1)$ circuit with $O(n \log^* n)$ edges as done in Lemma 5.4. f_p is to be thought of as 'propagate carry if it exists'.

Step 3: For each principal interval $I = i_1 \cdots i_k$, compute the following:

$$f_s(I) = x_{i_k} y_{i_k} \vee \bigvee_{j=1}^{k-1} (x_{i_j} y_{i_j} \wedge \bigwedge_{l=j+1}^{k} z_{i_l}).$$

We can easily modify the circuits of Lemma 5.4 with $O(k \log^* k)$ edges and depth at most $(2 \log n / \log m + 1)$ to compute the suffixes of $\bigwedge_{l=j+1}^{k} z_{i_l}$. f_s is to be thought of as 'set carry.' Now summing over all intervals gives a circuit of edge complexity $O(n(\log^* n)^2)$.

Step 4: For $k = 1, \ldots, (n+1)$, the prefix $1 \cdots k$ is the concatenation of at most $\log^* n$ intervals $I_1, \ldots, I_l$. The kth carry bit c_k can be computed as

$$c_k = f_s(I_l) \vee \bigvee_{i=1}^{l-1} [f_s(I_i) \wedge \bigwedge_{j=i+1}^{l} f_p(I_j)].$$

Using the values precomputed at Steps 2 and 3, this step uses a circuit of edge complexity $O(n(\log^* n)^2)$ and depth at most $(2 \log n / \log m + 1)$.

Clearly, the circuit computing the carries has size $O(n(\log^* n)^2)$ and depth at most $4(\log n / \log m + 1)$. Since each bit of the sum is simply $s_k = (\bar{c}_k x_k \bar{y}_k) \vee (\bar{c}_k \ \bar{x}_k y_k) \vee (c_k x_k y_k) \vee (c_k \bar{x}_k \ \bar{y}_k)$, it is easy to modify the circuit from step 4 to obtain a circuit for s_k that has fan-in bounded by m and the same depth as the circuit for c_k with additional $O(n)$ edges. Hence, the circuit for addition has size $O(n(\log^* n)^2)$ and depth $O(\log n / \log m)$. □

5.5   Trade-offs between Edge Complexity and Circuit Depth for Symmetric Functions and Multiplication

In this section, we examine the trade-offs between the number of edges and the circuit depth for symmetric functions when the fan-in is restricted. First, recall from Section 3.2 that every n-variable symmetric function can be computed by a depth-2 threshold circuit with $O(n)$ gates and $O(n^2)$ edges. However, such a circuit has a maximum fan-in of n. If the fan-in is constrained to be not more than $m << n$, then we cannot directly apply the preceding construction. Instead, we construct a small constant-depth circuit to be used as a subcircuit in our designs. By doing so, we reduce both the fan-in requirement and the size of the circuit.

The proof of the following lemma is left as an exercise.

Lemma 5.5:   For every $d > 1$, $PAR_n(X)$ can be computed by a depth-$O(d)$ threshold circuit of edge complexity $O(n^{1+1/(2^d-1)})$, node complexity $O(n^{1-1/(2^d-1)})$, and with fan-in bounded by n.

Proof: Exercise 5.5   □

Theorem 5.10:   For every fixed integer $d > 0$, $PAR_n(X)$ can be computed by a depth-$O(d \log n / \log m)$ threshold circuit of edge complexity $O(nm^{1/(2^d-1)})$, node complexity $O(nm^{-1/(2^d-1)})$, and with fan-in bounded by m.

Proof: We modify the construction of the circuit stated in Lemma 5.5. The threshold circuit has the structure of an m-ary tree: Each node in the tree corresponds to a subcircuit of $O(m^{1+1/(2^d-1)})$ edges and depth $O(d)$ that computes the Parity of m inputs, which are the outputs from the nodes in the next lower level. At the bottom level, there are n/m nodes, each computing the Parity of m inputs (the inputs are the 'leaves' of the m-ary tree). The single node at the top level of the tree outputs the Parity of the n inputs to the circuit. If l is the number of levels of the m-ary tree, then clearly the number of leaves is m^l and, hence, $l = O(\log n / \log m)$. There are $O(n/m)$ nodes in the tree and each node uses a depth-$O(d)$ threshold circuit of edge complexity $O(m^{1+1/(2^d-1)})$ and gate complexity $O(m^{1-1/(2^d-1)})$. Thus, the total edge complexity, node complexity, and circuit depth are $O(nm^{1/(2^d-1)})$, $O(nm^{-1/(2^d-1)})$, and $O(d \log n / \log m)$, respectively.   □

It is clear that if the number of gates in the threshold circuit corresponding to each node in the above circuit can be reduced, then the overall gate count will also be reduced. Depending on the technology of implementing threshold gates, sometimes we might prefer a threshold circuit with fewer threshold gates at the expense of higher edge complexity. The result in Theorem 5.10 should be compared with Theorem 5.3, where we present a threshold circuit with fewer gates but with more edges.

5.5.1   Computing the Sum of n Bits

The design of the Parity circuit shown earlier relies heavily on the periodic structure of the Parity function. For a general symmetric function, the Parity circuit design based on an m-ary tree structure does not seem to apply.

Here we constrain the fan-in of our circuit to be m and construct a threshold circuit for a general symmetric function that is efficient both in circuit size and depth. The construction consists of two parts. In the first part, we design a threshold circuit that would output the $\log n$-bit sum of its n inputs. The construction of the second part is simple. Since a symmetric function depends only on the sum of its inputs, which has been encoded into $\log n$ bits from the first part, we simply construct a look-up table for the value of the function, which costs at most $O(n \log n)$ edges.

It follows from the proof of Theorem 3.1 that one can construct a depth-2 threshold circuit with $O(m^2)$ edges to compute the $O(\log m)$-bit sum of the m inputs $(x_1, x_2, \ldots, x_m)$. Actually, the size of the circuit can be significantly improved, as stated in Lemma 5.6, which is a special case of a result in [11]. (We shall not give the proof here and refer the readers to the reference [11].) Since the sum of m bits can be expressed as a $\lceil \log m \rceil$-bit integer if m is not a power of 2, or as a $(\log m + 1)$-bit integer otherwise, for notational convenience, we use the number $\log m$ in both cases (e.g., Lemma 5.6 and Lemma 5.8).

Lemma 5.6:   Given m inputs $(x_1, x_2, \ldots, x_m) \in \{0,1\}^m$, the $\log m$-bit sum $\sum_{i=1}^{m} x_i$ can be computed by a depth-$7d$ threshold circuit of edge complexity $O(2^{-d/2} m^{1+1/(2^d-1)} \sqrt{\log m})$ and with fan-in bounded by m, for every fixed integer $d > 0$.   □

A major part of our design for symmetric functions and multiplication is a threshold circuit with fan-in bounded by m that efficiently computes the sum of many integers. We employ the block-save technique to reduce the sum of m N-bit integers to the sum of $\log m$ $(N + \log m)$-bit integers using a small-depth threshold circuit.

Lemma 5.7:   Given m N-bit integers, $x_i = 2^{N-1}x_{i_{N-1}} + \ldots + 2x_{i_1} + x_{i_0}$, $i = 1, \ldots, m$, the sum $\sum_{i=1}^{m} x_i$ can be reduced to a sum of $\log m$ integers of size at most $(N + \log m)$-bits, using a depth-$7d$ threshold circuit of edge complexity $O(2^{-d/2}Nm^{1+1/(2^d-1)}\sqrt{\log m})$ and with fan-in bounded by m.

Proof: Compute in parallel the sum of the jth bits in the integers using the circuits in Lemma 5.6. In other words, for each $j = 0, \ldots, N-1$, compute $s_j = \sum_{i=1}^{m} x_{i_j}$. Then,

$$\sum_{i=1}^{m} x_i = \sum_{j=0}^{N-1} s_j 2^j.$$

From Lemma 5.6, for each s_j, we need a depth-$7d$ threshold circuit of $O(2^{-d/2}m^{1+1/(2^d-1)}\sqrt{\log m})$ edges. Summing over all s_j's, the total number of edges of the circuit is $O(2^{-d/2}Nm^{1+1/(2^d-1)}\sqrt{\log m})$.

Since there are N s_j's, it seems that we have reduced the original sum to a sum of N integers, instead of a sum of $\log m$ integers. But notice that each s_j can be represented by only $\log m$-bits. Thus, there is no overlapping in the binary representation between the

$$s_j \cdot 2^j \quad \text{and} \quad s_{j+\log m} \cdot 2^{j+\log m} = 2^{\log m}(s_{j+\log m} \cdot 2^j)$$

for every j. If we let

$$\begin{aligned}
\tilde{s}_j &= s_j \cdot 2^j + s_{j+\log m} \cdot 2^{j+\log m} + s_{j+2\log m} \cdot 2^{j+2\log m} + \cdots \\
&= \sum_{i=0}^{(N-1)/\log m} s_{j+i\log m} \cdot 2^{j+i\log m},
\end{aligned}$$

then the binary representation of $\tilde{s}_j$ is simply a concatenation of the bits in $s_j, s_{j+\log m}, s_{j+2\log m}$, and so forth. Now note that

$$\sum_{i=1}^{m} x_i = \sum_{j=0}^{N-1} s_j 2^j = \sum_{j=0}^{\log m-1} \tilde{s}_j.$$

Each $\tilde{s}_j$ can be represented by at most $(N + \log m)$ bits. As a result, we have reduced the original sum of N integers to a sum of $\log m$ integers.   □

One can construct a threshold circuit to compute the sum of n inputs using the small-depth circuits in Lemma 5.6 and Lemma 5.7 as subcircuits.

Lemma 5.8:   Given n inputs $(x_1, x_2, \ldots, x_n) \in \{0,1\}^n$, the $O(\log n)$-bit sum $\sum_{i=1}^{n} x_i$ can be computed by a depth-$O(d \log n / \log m)$ threshold circuit of edge complexity $O(2^{-d/2} n m^{1/(2^d-1)} \sqrt{\log m})$ and with fan-in bounded by m, for every fixed integer $d > 0$.

Proof: For convenience, let $f_d(m) = 2^{-d/2} m^{1+1/(2^d-1)} \sqrt{\log m}$. Divide the inputs into n/m groups, each having m inputs. Compute the $(\log m)$-bit sum of each group in parallel using the circuits in Lemma 5.6. This step requires $7d$ levels of threshold gates and all together $O(f_d(m) \cdot n/m) = O(2^{-d/2} n m^{1/(2^d-1)} \sqrt{\log m})$ edges.

After the first step, we obtain n/m integers of size $(\log m)$-bits: $x_i = 2^{\log m-1} x_{i,\log m-1} + \ldots + 2 x_{i,1} + x_{i,0}$, $i = 1, \ldots, n/m$. Further, divide these (n/m) integers into n/m^2 groups, each having m integers of size $(\log m)$ bits. By Lemma 5.7, the sum in each group can be reduced to a sum of $(\log m)$ integers each of $(2 \log m)$ bits. This step requires $7d$ levels of threshold gates and $O(f_d(m) \log m) = O(2^{-d/2} m^{1+1/(2^d-1)} \log^{3/2} m)$ edges for each group. If we perform the reduction in parallel for all (n/m^2) groups of integers computed from the first step, then $O(n/m^2 \cdot 2^{-d/2} m^{1+1/(2^d-1)} \log^{3/2} m) = O(n \log^{3/2} m)$ edges are needed. Since each group of m integers is reduced to a group of $(\log m)$ integers, we have $(n \log m/m^2)$ integers each of $2(\log m)$ bits after the second step.

This reduction procedure is iterated until the sum is reduced to the sum of $< m$ integers each of $O(\log n)$ bits. At every step, the integers are divided into groups of m integers and the reduction procedure is applied to every group. By Lemma 5.7, the sum of m integers of size $(k \log m)$-bits in each group can be reduced to the sum of $\log m$ integers of size $(k + 1) \log m$-bits, using a depth-$7d$ threshold circuit of edge complexity $O(k f_d(m) \log m) = O(2^{-d/2} k m^{1+1/(2^d-1)} \log^{3/2} m)$. In general, after the kth step $(k > 1)$, we have $O(n(\log m)^{k-1}/m^k)$ remaining integers of size $(k \log m)$ bits, and the edge complexity in the kth step is $O(f_d(m) \cdot (k - 1) \log m \cdot n(\log m)^{k-2}/m^k) = O(n f_d(m) \cdot k \cdot (\log m)^{k-1}/m^k)$. The reduction procedure is iterated until $O(n(\log m)^{k-1}/m^k)$ drops below m, say $\tilde{m}$, or when $k = O(\log n / \log m)$. Then, apply Lemma 5.7 again with m replaced by $\tilde{m}$ and N replaced by $O(\log n)$, so that there are $< \log m$ remaining integers each of $O(\log n)$ bits. This would require a depth-$7d$ threshold circuit of edge complexity $O(f_d(m) \log n)$. Since each step uses $7d$ levels of threshold gates, there are $O(d \log n / \log m)$ levels and the $O(n/m \cdot f_d(m)) = O(2^{-d/2}nm^{1/(2^d-1)}\sqrt{\log m})$ edge complexity in the first step dominates the circuit size after the sequence of reductions.

It remains to be shown how one can efficiently compute the sum of $\log m$ $O(\log n)$-bit integers: $z_1 , \dots, z_{\log m}$. Partition each $O(\log n)$-bit integer z_i into $O(\log n/\log\log m)$ consecutive blocks $\tilde{s}_{i,j}$ with each block having $(\log\log m)$ bits, for $i = 1, \dots, \log m$, and $j = 0, \dots, O(\log n/\log\log m)$. Note that each block $\tilde{s}_{i,j}$ represents an integer $\leq 2^{\log\log m} - 1 = \log m - 1$. Thus, the sum of all j^{th} blocks, $\tilde{s}_j = \sum_{i=1}^{\log m} \tilde{s}_{i,j}$, can be formed as a sum of $(\log m)(\log m - 1) = O(\log^2 m)$ 1-bit numbers after each has been encoded, i.e., converted to unary notation. By Lemma 5.6, the sum of all j^{th} blocks $\sum_{i=1}^{\log m} \tilde{s}_{i,j}$ can be computed by a depth-$7d$ threshold circuit of edge complexity $O(f_d(\log^2 m))$ and fan-in bounded by $O(\log^2 m)$. Furthermore, each of these block sums $\tilde{s}_j = \sum_{i=1}^{\log m} \tilde{s}_{i,j}$ can be represented in $(2\log m)$ bits. Thus, there is no overlapping in the binary representation of $\tilde{s}_j 2^j$ and $\tilde{s}_{j+2} 2^{(j+2)\log m}$. Using similar arguments as in Lemma 5.7, we let

$$\tilde{s}_{even} = \sum_{j \text{ even}} \tilde{s}_j \cdot 2^{j \log m}$$

and

$$\tilde{s}_{odd} = \sum_{j \text{ odd}} \tilde{s}_j \cdot 2^{j \log m} ,$$

then the binary representation of $\tilde{s}_{even}$ is simply a concatenation of the bits in $\tilde{s}_0$, $\tilde{s}_2$, $\tilde{s}_4$, $\dots$ and so forth. Similarly, a concatenation of the bits in $\tilde{s}_1$, $\tilde{s}_3$, $\tilde{s}_5$, $\dots$ yields the binary representation of $\tilde{s}_{odd}$. Moreover, the original sum $\sum_{i=1}^{n} x_i = \tilde{s}_{even} + \tilde{s}_{odd}$. Since there are $O(\log n/\log\log m)$ $\tilde{s}_j$'s and each $\tilde{s}_j$ can be computed in parallel by a depth-$7d$ threshold circuit of edge complexity $O(f_d(\log^2 m))$, we only require a depth-$7d$ threshold circuit of edge complexity $O(f_d(\log^2 m) \log n/\log\log m)$ to reduce the sum to the two $O(\log n)$-bit integers $\tilde{s}_{odd}$ and $\tilde{s}_{even}$.

To compute the sum of the remaining two $O(\log n)$-bit integers, it only requires another constant-depth threshold circuit of almost logarithmic edge complexity by Theorem 5.9. Hence, the total edge complexity is dominated by $O(2^{-d/2}nm^{1/(2^d-1)}\sqrt{\log m})$ in the first step and the circuit depth is $O(d\log n/\log m)$. □

5.5.2   Symmetric Functions and Multiplication

Here we make use of the results developed in the preceding section to design efficient threshold circuits for computing symmetric functions and the product of two numbers.

Lemma 5.9: Let $f(x_1, x_2, \ldots, x_n)$ be a general symmetric function. If the $\log n$-bit sum $\sum_{i=1}^{n} x_i$ is given, then the function f can be computed by a depth-$O(\log n / \log m)$ AND-OR circuit of edge complexity $O(n \log n)$ and with fan-in bounded by m.

Proof: Since a symmetric function depends only on the sum of its inputs $\sum_{i=1}^{n} x_i = \sum_{j=0}^{\log n - 1} s_j 2^j$, which is given, we can treat f as a function of $\log n$ inputs $s_0, s_1, \ldots, s_{\log n - 1}$. Express f in a Sum Of Products form. Then, the number of product terms is at most $O(2^{\log n}) = O(n)$. The first level of the circuit consists of at most n AND gates with fan-in $\leq \log n$, computing all the product terms, if $m \geq \log n$. If $m < \log n$, we can replace each AND gate by an m-ary tree of AND gates with fan-in m. This requires at most $O(\log \log n / \log m)$ depth and $O(n \log n)$ edges. To compute the OR of these product terms, we can use an m-ary tree of OR gates with fan-in m. This costs at most $2n$ edges and $O(\log n / \log m)$ depth. Thus, the overall edge complexity is $O(n \log n)$ and the circuit depth is $O(\log n / \log m)$.   □

As a consequence of Lemma 5.8 and Lemma 5.9, we have the following result:

Theorem 5.11: Any n-variable symmetric function can be computed by a depth-$O(d \log n / \log m)$ threshold circuit of edge complexity $O(2^{-d/2} n m^{1/(2^d - 1)} \sqrt{\log m})$ or $O(n \log n)$ if $m^{1/(2^d - 1)} \sqrt{\log m} = o(\log n)$ and with fan-in bounded by m, for every fixed integer $d > 0$.

Proof: Let $f(x_1, x_2, \ldots, x_n)$ be a symmetric function. By Lemma 5.8, the $\log n$-bit sum of the inputs $\sum_{i=1}^{n} x_i$ can be computed by a depth-$O(d \log n / \log m)$ threshold circuit of edge complexity $O(2^{-d/2} n m^{1/(2^d - 1)} \sqrt{\log m})$ and with fan-in bounded by m. It follows from Lemma 5.9 that another depth-$O(\log n / \log m)$ circuit of edge complexity $O(n \log n)$ and with fan-in $\leq m$ can be used to compute the final value.   □

Theorem 5.12: The product of two n-bit integers can be computed using a depth-$O(d \log n / \log m)$ threshold circuit of edge complexity

$O(2^{-d/2}n^2m^{1/(2^d-1)}\sqrt{\log m})$ and with fan-in bounded by m, for every fixed integer $d > 0$.

Proof: Again for convenience, let $f_d(m) = 2^{-d/2}m^{1+1/(2^d-1)}\sqrt{\log m}$. Let the two input binary integers be $x = x_{n-1}x_{n-2}\ldots x_0$, $y = y_{n-1}y_{n-2}\ldots y_0$. The first level of our circuit computes the n binary integers

$$z_i = z_{i_{2n-1}}z_{i_{2n-2}}\cdots z_{i_0}$$

each of size $2n$ bits, for $i = 0, \ldots, n-1$, where

$$z_i = \underbrace{0\ldots0}_{n-i}(x_{n-1}\wedge y_i)(x_{n-2}\wedge y_i)\ldots(x_0\wedge y_i)\underbrace{0\ldots0}_{i}.$$

It is easy to see that the product of x and y is simply the sum of the z_i's. This step requires one level of AND/OR gates with fan-in 2 and $O(n^2)$ edges.

The problem is reduced to the computation of a sum of n integers z_i of size $2n$ bits. We proceed as in Lemma 5.8. Divide the n integers into n/m groups, each having m integers. Apply the reduction procedure of Lemma 5.7 (with $N = 2n$ in this step) in parallel to each group of integers. After the reduction, there are $(\log m)$ integers remaining in each group. The reduction for each group uses a depth-$7d$ threshold circuit of edge complexity $O(nf_d(m)) = O(2^{-d/2}nm^{1+1/(2^d-1)}\sqrt{\log m})$ by Lemma 5.7. Since there are n/m groups, there will be $(n\log m/m)$ remaining integers, and the edge complexity in this step is $O(nf_d(m) \cdot n/m) = O(2^{-d/2}n^2m^{1/(2^d-1)}\sqrt{\log m})$.

Now apply the reduction procedure iteratively until the multiple sum is reduced into a sum of $< m$ integers each with $O(n)$-bits. A simple induction shows that in general at the end of the kth step (for $k > 1$), we will have $O(n(\log m)^k/m^k)$ integers of size at most $(2n + k\log m)$ bits, and the edge complexity required at the kth step is $O((2n + k\log m) \cdot n(\log m)^{k-1}/m^k \cdot f_d(m))$. So $k = O(\log n/\log m)$. Another application of Lemma 5.7 with N replaced by $O(n)$ reduces the sum to a sum of $< \log m$ integers each of $O(n)$ bits, using a depth-$7d$ threshold circuit of edge complexity $O(nf_d(m))$. A similar block-partitioning argument as in Lemma 5.8 further reduces this sum of $< \log m$ integers to a sum of two $O(n)$-bit integers with a depth-$7d$ threshold circuit of edge complexity $O(nf_d(\log^2 m))$. At the final step, the sum of the two integers can be computed using the circuit in Theorem 5.9, which costs almost linear number of edges and depth $O(\log n/\log m)$.

Thus, the overall edge complexity in the circuit is dominated by the first step, where $O(2^{-d/2}n^2 m^{1/(2^d-1)}\sqrt{\log m})$ edges are introduced. Since each reduction requires $7d$ levels of threshold gates, the overall circuit depth is $O(d \log n / \log m)$. $\square$

Exercises

5.1:   [*Bounded vs. unbounded fan-in*]
Let $\{f_n\}$ be a family of Boolean functions of n variables that can be computed by depth-d_n circuits comprising s_n bounded fan-in AND/OR gates. Show that for any $c > 0$, $\{f_n\}$ can be computed by circuits of depth $O(\dfrac{d_n}{c \log \log n})$ and $O(s_n 2^{(\log n)^c})$ unbounded fan-in AND/OR gates.

5.2:   [*Computing Parity with a minimum number of threshold gates*]
Give a construction of a threshold circuit with $\lfloor \log n \rfloor + 1$ gates and $\lfloor \log n \rfloor + 1$ depth that computes the Parity of n variables.

5.3:   [*Computing symmetric functions with a minimum number of threshold gates*]
Using the fact that any Boolean function of n variables can be computed by a constant-depth threshold circuit with $O(2^{n/2}/\sqrt{n})$ gates (see Bibliographic Notes in Chapter 3) and Lemma 5.1, show that any symmetric function of n variables can be computed by a constant-depth threshold circuit with $O(\sqrt{n/\log n})$ gates. This upper bound is tight by Corollary 3.3.

5.4:   [*Edge-efficient depth-4 threshold circuit for Parity*]
Partition n variables into b groups of n/b variables. By choosing an optimal value of b, show that the Parity of n variables can be computed by depth-4 threshold circuits of $n^{4/3}$ edges.

5.5:   [*Edge-efficient depth-d threshold circuit for Parity*]
For every $d > 1$, give a construction of a threshold circuit for the n-variable Parity function $PAR_n(X)$ with $O(n^{1-1/(2^d-1)})$ gates, $O(n^{1+1/(2^d-1)})$ edges, depth $O(d)$, and fan-in bounded by n.

5.6:   [*Sum of two n-bit integers vs. sum of n 1-bit integers*]
Show that Parity cannot be computed by constant-depth polynomial-size

circuits using only AND/OR gates and gates that compute the sum of 2 n-bit integers. However, show that all arithmetic functions considered in this chapter can be computed by constant-depth polynomial-size circuits using only AND/OR and gates that compute the sum of n 1-bit numbers.

5.7:   [*Computing the sum of* $\log n$ *n-bit integers*]
Show that the sum of $\log n$ n-bit integers can be computed by AC^0 circuits. Give an upper bound on the depth of the corresponding AND-OR circuits.

5.8:   [*Maximum vs. Sorting*]
Recall that the problem Maximum is to find the maximum of a given list of n n-bit integers. Construct AC^0 circuits for Maximum. Hence, conclude that Sorting cannot be computed by constant-depth polynomial-size circuits using only AND/OR gates and gates that compute Maximum.

5.9:   [*Unary to Binary*]
The *unary* representation of an integer k in $\{0, \ldots, n\}$ has the form $1 \cdots 1 0 \cdots 0$ where k 1's are followed by $(n - k)$ 0's. Show that converting the unary representation of an integer to its binary representation is in AC^0.

5.10:   [*Computing Parity with sorters*]
Show how to compute Parity with constant-depth polynomial-size circuits of AND/OR gates and sorters that can sort n 1-bit numbers.

5.11:   [*Restricted Fan-in Sorters*]
Design a sorting circuit that sorts n n-bit integers using AND/OR gates and circuits that sort m 1-bit integers, where $m < n$.

Bibliographic Notes

Efficient design of parallel adders has been known for more than three decades. In 1963, Ofman [91] designed a parallel adder that has circuit depth $O(\log n)$ and gate count $O(n)$. Ofman's design later on has been generalized by Khrapchenko [65], Ladner and Fischer [68], Fich [35], and Snir [133].

The first conceptual breakthrough in the design of a fast multiplier was provided by Ofman [91] and Wallace [139]. Their circuit design is based on a fast algorithm for computing the sum of multiple binary integers and is now called the carry-save adder or the Ofman-Wallace tree. This multiplier has propagation delay $O(\log n)$. In 1971, Schonhage and Strassen [118] designed a multiplier based on the Fast Fourier Transform algorithm [30] with $O(\log n)$ propagation delay and $O(n \log n \log \log n)$ gates, which remains the best-known result today. In [97, 98], the ideas of Ofman and Wallace are exploited to yield (fan-in two) multiplication circuits of depth $4.57 \log n$. See [79] for an efficient VLSI implementation of a multiplier.

While the design of circuits mentioned in the preceding paragraph is based on a bounded fan-in circuit model, we examine unbounded fan-in circuits in this chapter. The construction based on the carry-look-ahead technique of the depth-3 AND-OR circuit with $O(n^3)$ edges for Addition (Theorem 5.8) is implicit in the work of Weinberger and Smith [142]. The fact that any depth-2 AND-OR circuit for Addition requires exponential size is well known in the literature. However, we are unable to find a reference that gives the formal proof of this fact as presented here.

The construction of the constant-depth AND-OR circuits for Addition with $O(n(\log^* n)^2)$ edges (Theorem 5.9) is taken from Chandra, Fortune, and Lipton [25]. Similar ideas based on parallel prefix computation in the design of adders have been used in Ofman [91]. A stronger result than the one stated here is proved in [25]. It is shown [25] that the edge complexity of the AND-OR circuits for Addition can be reduced to $O(nf^{-1}(n))$ for any primitive recursive function f, while keeping the depth to be a constant independent of n. Furthermore, the edge complexity can be reduced to linear if the depth is allowed to increase at the rate of the inverse Ackermann function. However, some of the gates in the circuit suffer from a large fan-in of $\Omega(n)$. When there are constraints on the fan-in, the construction in [25] can be modified to provide a trade-off between the fan-in and the depth of the circuit, while the edge complexity of the circuit remains essentially unchanged. More precisely, the sum of two n-bit integers can be computed using an AND-OR circuit of edge complexity $O(nf^{-1}(n))$, depth $O(\log n / \log m)$, and fan-in bounded by m for any increasing primitive recursive function $f(n)$.

On the combinational complexity of a general symmetric function, it is shown in Muller and Preparata [83] that any n-variable symmetric function can be constructed using a bounded fan-in circuit of size $O(n)$ and depth $O(\log n)$. For bounded fan-in circuits, this construction is (within a constant factor) optimal both in circuit size and depth. For unbounded fan-in AND-OR circuits, a technique in [26] shows that the depth of circuitd computing symmetric functions can be improved to $O(\log n / \log \log n)$ by having arbitrarily large fan-in gates, an improvement which is shown to be the best possible (asymptotically) in Håstad [50].

It is shown in Siu, Roychowdhury, and Kailath [126] that Parity can be computed by depth-$(d+1)$ threshold circuits of $O(dn^{1/d})$ gates. Paturi and Saks [100] show that Parity can be computed by depth-d threshold circuits with edge complexity $O(n^{1+1/\phi^d})$ (but with $\Theta(n)$ gates), where $\phi = (1 + \sqrt{5})/2$ is the Fibonacci constant. Lemmas 5.5 and 5.8 are from Beame, Brisson, and Ladner [11]. Some of the results discussed in this chapter are also presented in Siu, Roychowdhury, and Kailath [130].

Chapter 6

Computing with Small Weights

6.1 Introduction

In this chapter, we study issues in the trade-offs between the weights and the depth in threshold circuits. We show that exponential weights in polynomial-size threshold circuits can be removed at the expense of an increase in the circuit depth by one, while keeping the circuit size to be polynomially bounded.

We first derive a weaker result stating that any depth-d polynomial-size threshold circuit with arbitrary weights can be simulated by a depth-$(2d+1)$ polynomial-size threshold circuit with polynomially bounded integer weights, i.e., $LT_d \subset \widehat{LT}_{2d+1}$. This result can be established using the techniques developed in the previous chapters. Then, we develop more combinatorial techniques that allow us to show a stronger result: $LT_d \subset \widehat{LT}_{d+1}$. The results are also applied to derive depth-optimal threshold circuits for computing multiplication and division of two n-bit integers. In particular, we show that the sum of n n-bit integers can be computed by depth-2 polynomial-size threshold circuits and that both multiplication and division of two n-bit integers can be computed by depth-3 polynomial-size threshold circuits.

6.2 Necessity of Exponential Weights

In Chapter 2, we show that there are $2^{\Omega(n^2)}$ n-variable different linear threshold functions $\mathrm{sgn}(w_0 + \sum_{i=1}^n w_i x_i) : \{1, -1\}^n \to \{1, -1\}$. For each linear threshold function, assume that we always choose the integer weights with

203

the smallest possible magnitudes to realize the function. For $i = 0, \ldots, n$, let m_i be the maximum among the magnitudes of all possible integer weights w_i. Then, an upper bound on the number of linear threshold functions in n variables is $\prod_{i=0}^{n}(2m_i + 1)$. It follows that $\sum_{i=0}^{n} \log(2m_i + 1) = \Omega(n^2)$. Hence, there exists an i such that $\log(2m_i+1) = \Omega(n)$, or $m_i = 2^{\Omega(n)}$. This counting argument shows the necessity of exponentially large weights in realizing a general linear threshold element.

Can one give an explicit construction of a linear threshold function that requires exponentially large weights? In the following, we show that Comparison is an example of linear threshold functions that cannot be realized with small weights.

Theorem 6.1:   Comparison $\notin \widehat{LT}_1$.

Proof: Recall that the function Comparison can be realized as

$$COMP_n(X,Y) = \text{sgn}(\sum_{i=1}^{n} 2^i(x_i - y_i) + 1),$$

where $X = (x_1, \ldots, x_n)$, $Y = (y_1, \ldots, y_n) \in \{1, -1\}^n$. We write the values of the function $COMP_n(X,Y)$ in the form of a $2^n \times 2^n$ matrix in such a way that each row corresponds to the values of the function over the variables y_i's for a fixed value of X. For example, take $n = 2$, we have:

$$\begin{bmatrix} 1 & 1 & 1 & 1 \\ -1 & 1 & 1 & 1 \\ -1 & -1 & 1 & 1 \\ -1 & -1 & -1 & 1 \end{bmatrix}.$$

The rows are arranged from top to bottom in the order $X = (1,1)$, $(1,-1)$, $(-1,1)$, $(-1,-1)$. The columns are arranged from left to right in the same order for the Y's. Observe that every row in the matrix is distinct. This is true for general values of n.

Suppose we realize $COMP_n(X,Y)$ using a different set of integer weights, say $\text{sgn}(w_0 + \sum_{i=1}^{n}(w_i x_i + \tilde{w}_i y_i))$. The fact that there are 2^n distinct rows in the matrix implies that there are 2^n distinct values of $\sum_{i=1}^{n} w_i x_i$. Thus, there is at least one integer weight w_i with $|w_i| = \Omega(2^n/n)$. $\quad\square$

We next prove that if n is a power of 2, then there is an n-variable threshold function that requires weights of size around $2^{(n \log n)/2 - n}$. This almost matches the upper bound on the size of weights derived in Chapter 2.

Let us assume that n is a power of 2 and that $n = 2^m$. We adopt the $\{-1, 1\}$ notation in this section and interpret vectors in $\{-1, 1\}^n$ as functions from $\{-1, 1\}^m$ to $\{-1, 1\}$. Thus, an input $f \in \{-1, 1\}^n$ to an n-variable function $F : \{-1, 1\}^n \to \{-1, 1\}$, is itself an m-variable function: $f : \{-1, 1\}^m \to \{-1, 1\}$. In this section, we reserve capital letters for functions on n inputs. This geometric approach of treating Boolean functions as vectors is discussed in detail in Chapter 8. Given any m-variable Boolean function f, let $(f)_j$ denote the jth entry of the corresponding vector representation of the function, and let $f(v)$, $v \in \{-1, 1\}^m$, denote its value as evaluated for the input v.

For $\alpha \subseteq \{1, \ldots, m\}$, let φ_α be the Parity function of the corresponding variables, i.e., $\varphi_\alpha(v) = \prod_{i \in \alpha} v_i$, where we let $\varphi_\emptyset$ be the function that is identically 1. Properties of the set of Parity functions are discussed in Chapter 8, and the orthogonality property used in this section can be stated as follows:

$$(\varphi_{\alpha_i}, \varphi_{\alpha_j}) = \varphi_{\alpha_i}^T \varphi_{\alpha_j} = \sum_{v \in \{-1,1\}^m} \varphi_{\alpha_i}(v)\varphi_{\alpha_j}(v) = \begin{cases} n & \text{if } i = j, \\ 0 & \text{if } i \neq j. \end{cases}$$

Also, it follows from the spectral representation of a function $f : \{-1, 1\}^m \to \{-1, 1\}$ that it can be represented as

$$f(v) = \sum_{\alpha \subseteq \{1, \ldots, m\}} a_\alpha \varphi_\alpha(v),$$

where $a_\alpha = (f, \varphi_\alpha)/n$ are the spectral coefficients.

Choose an ordering of $\alpha_0, \alpha_1, \ldots, \alpha_{n-1}$ of the sets such that:

1. $|\alpha_i| \leq |\alpha_j|$ for $i \leq j$.

2. $|\alpha_i \Delta \alpha_{i+1}| \leq 2$ for all i, where $|\alpha_i \Delta \alpha_{i+1}|$ is the symmetric difference of the two sets α_i and α_{i+1}.

This implies that $\alpha_0 = \emptyset$ and $\alpha_1, \alpha_2, \ldots, \alpha_m$ are the singletons, and that $|\alpha_i \Delta \alpha_{i+1}| = 1$ when $|\alpha_{i+1}| = |\alpha_i| + 1$, while $|\alpha_i \Delta \alpha_{i+1}| = 2$ otherwise.

Lemma 6.1:   There is an ordering that satisfies conditions 1 and 2 above.

Proof: Exercise 6.1.   □

Let the inner product of the functions f and g be denoted as

$$(f,g) = f^T g = \sum_{v \in \{-1,1\}^m} f(v)g(v) \ .$$

Define $F(f) : \{-1,1\}^n \to \{-1,1\}$ as $\mathrm{sgn}((f, \varphi_{\alpha_i}))$ where i is the largest index such that $(f, \varphi_{\alpha_i}) \neq 0$. This function is a threshold function because

$$F(f) = \mathrm{sgn}\left(\sum_{i=0}^{n-1} (n+1)^i (f, \varphi_{\alpha_i}) \right)$$

and (f, φ_{α_i}) is a linear function in the values $(f)_j$. This expression is correct since $|(f, \varphi_{\alpha_i})| \leq n$ for all i.

We want to prove that if

$$F(f) = \mathrm{sgn}\left(\sum_{j=0}^{n-1} w_j (f)_j - t \right), \tag{6.1}$$

then one of the w_j's is large. Lemma 6.2 shows that one can remove the threshold t from the above expression.

Lemma 6.2:   In Eqn. (6.1), the value of t can be taken to be 0.

Proof: Note that $F(f) = -F(-f)$. Hence, $|\sum_{j=0}^{n-1} w_j (f)_j| > |t|$ for any f, and we can set $t = 0$ without changing the function.   □

It is easier to work with expressions of the form

$$\mathrm{sgn}\left(\sum_{i=0}^{n-1} w_i' (f, \varphi_{\alpha_i}) \right), \tag{6.2}$$

and the following lemma shows how the weights in Eqns. (6.1) and (6.2) can be related.

Lemma 6.3:   For all $f \in \{1, -1\}^n$,

$$\sum_{j=0}^{n-1} w_j (f)_j = \sum_{i=0}^{n-1} w_i' (f, \varphi_{\alpha_i})$$

if and only if

$$w_i' = \frac{1}{n} \sum_{j=0}^{n-1} w_j (\varphi_{\alpha_i})_j$$

and

$$w_j = \sum_{i=0}^{n-1} w_i' (\varphi_{\alpha_i})_j.$$

Proof: The second statement follows from rearranging the terms. To prove the first statement, note that by the spectral representation,

$$(f)_j = \frac{1}{n} \sum_{i=0}^{n-1} (\varphi_{\alpha_i})_j (f, \varphi_{\alpha_i}).$$

This implies

$$\sum_{j=0}^{n-1} w_j (f)_j = \sum_{j=0}^{n-1} w_j \frac{1}{n} \sum_{i=0}^{n-1} (\varphi_{\alpha_i})_j (f, \varphi_{\alpha_i}) = \frac{1}{n} \sum_{i=0}^{n-1} (f, \varphi_{\alpha_i}) \sum_{j=0}^{n-1} w_j (\varphi_{\alpha_i})_j,$$

and the lemma follows.   □

We first establish that if $F(f) = \text{sgn}(\sum_{i=0}^{n-1} w_i (f, \varphi_{\alpha_i}))$, then some of the weights have to be large (see Theorem 6.2). We then apply the above lemma to show that large weights in Eqn. (6.2) would imply large weights in Eqn. (6.1) as well (see Theorem 6.3).

By choosing a suitable sequence of test functions f, we shall prove that w_i's have to grow exponentially (see Lemmas 6.5, 6.6, and 6.7). Since the ordering of the α_i is not explicit, we shall sometimes use the notation w_α, which should be read as w_i, where i is chosen such that $\alpha_i = \alpha$. Lemma 6.4 makes an easy but useful observation.

Lemma 6.4:   For any i, $w_i > 0$.

Proof: This follows from setting $f = \varphi_{\alpha_i}$ and noting that $F(f) = 1$ for such f.   □

Lemma 6.5:   Suppose $|\alpha_{i+1}| = |\alpha_i| = k$, where $2 \leq k \leq n-1$ and that $\alpha_i \Delta \alpha_{i+1} = \{a, b\}$. Let $v \in \{-1, 1\}^m$ be any point with $v_a = v_b$, then

$$w_{i+1} > (2^{k-1} - 1) w_i - \sum \varphi_{\alpha_i}(v) \varphi_\alpha(v) w_\alpha ,$$

where the sum extends over all α such that $\alpha \subset \alpha_i \bigcup \alpha_{i+1}$, α contains exactly one of a and b, and α is not equal to α_i or α_{i+1}.

Proof: Let us assume that $\alpha_i = \{1, 2, \ldots, k\}$ and $\alpha_{i+1} = \{1, 2, \ldots, k - 1, k + 1\}$. Let v^1 be a vector of length $k + 1$ that satisfies $v_j^1 = v_j$ for $1 \leq j \leq k + 1$. Furthermore, let v^2 be a similar vector such that $v_j^2 = v_j$ for $j < k$ and $v_j^2 = -v_j$ for $j = k$ and $j = k + 1$. Define the following function on the first $k + 1$ variables:

$$f(u) = \begin{cases} \varphi_{\alpha_i}(u) & \text{if } u = v^1 \text{ or } u = v^2, \\ -\varphi_{\alpha_i}(u) & \text{otherwise.} \end{cases}$$

We extend f to a (degenerate) function of m variables. First, note that $(f, \varphi_\alpha) = 0$ for any α that contains an element larger than $k + 1$. Then, $(f, \varphi_{\alpha_i}) = 2^{m-k-1}(4 - 2^{k+1})$, while for other $\alpha \subseteq \{1, 2 \ldots k + 1\}$, we have

$$\begin{aligned} (f, \varphi_\alpha) &= (-\varphi_{\alpha_i}, \varphi_\alpha) + 2^{m-k}\varphi_{\alpha_i}(v^1)\varphi_\alpha(v^1) + 2^{m-k}\varphi_{\alpha_i}(v^2)\varphi_\alpha(v^2) \\ &\quad - 2^{m-k}\varphi_{\alpha_i}(v^1)\varphi_\alpha(v^1)(1 + \varphi_{\alpha_i}(v^1 v^2)\varphi_\alpha(v^1 v^2)), \end{aligned}$$

where $v^1 v^2$ is the pointwise product of v^1 and v^2. Since this vector is 1 except for coordinates k and $k + 1$, we get a nonzero inner product if and only if $\alpha \subset \{1, 2, \ldots k + 1\}$ and α contains exactly one of the elements k and $k + 1$. The only two sets of size k with these properties are α_{i+1} and α_i, while all other sets with these properties have cardinality at most $k - 1$. Note that by property 1 of our ordering, all these sets appear before α_i, and this implies that $F(f) = \text{sgn}((\varphi_{\alpha_{i+1}}, f)) = 1$. Writing this statement as an inequality of the weights yields the inequality of the lemma. $\square$

Lemma 6.6: Suppose $|\alpha_{i+1}| = 1 + |\alpha_i| = k$, where $2 \leq k \leq n - 1$ and $\alpha_{i+1} = \alpha_i \bigcup \{a\}$. Then, for any vector v with $v_a = 1$, we have

$$w_{i+1} > (2^{k-1} - 1)w_i - \sum_{\alpha \subset \alpha_{i+1}, \alpha \neq \alpha_i} \varphi_{\alpha_i}(v)\varphi_\alpha(v)w_\alpha .$$

Proof: Let us assume that $\alpha_i = \{1, 2, \ldots, k - 1\}$ and $\alpha_{i+1} = \{1, 2, \ldots, k\}$, and let v^1 be the vector of length k that satisfies $v_j^1 = v_j$ for $1 \leq j \leq k$. Define the following function on the first k variables:

$$f(u) = \begin{cases} \varphi_{\alpha_i}(u) & \text{if } u = v^1, \\ -\varphi_{\alpha_i}(u) & \text{otherwise.} \end{cases}$$

Extend f to a (degenerate) function of m variables. Again $(f, \varphi_\alpha) = 0$ for any α that contains an element larger than k. Clearly, $(f, \varphi_{\alpha_i}) = 2^{m-k}(2 -$

2^k), while for other $\alpha \subset \{1, 2 \ldots k\}$, we have

$$(f, \varphi_\alpha) = (-\varphi_{\alpha_i}, \varphi_\alpha) + 2^{m+1-k}\varphi_{\alpha_i}(v^1)\varphi_\alpha(v^1) = 2^{m+1-k}\varphi_{\alpha_i}(v^1)\varphi_\alpha(v^1).$$

$F(f) = \text{sgn}((\varphi_{\alpha_{i+1}}, f)) = 1$, and writing out the corresponding inequality for the weights gives the lemma. $\square$

Using Lemmas 6.5 and 6.6, we can now prove the following main lemma in this section.

Lemma 6.7:   For each i such that $|\alpha_{i+1}| \geq 2$, we have $w_{i+1} > (2^{|\alpha_{i+1}|-1} - 1)w_i$. Furthermore, if $\alpha = \{a, b\}$, then

$$w_\alpha > w_{\{a\}} + w_{\{b\}} + w_0.$$

Proof: We establish the lemma by induction over i and we need to handle the cases, where $|\alpha_{i+1}|$ is small, separately.

Let us first establish the lower bounds for w_α when $|\alpha| = 2$. Suppose $a = 1$ and $b = 2$. Then, apply the proof of Lemma 6.6 with $\alpha_{i+1} = \alpha$, $\alpha_i = \{1\}$, and $v^1 = (-1, 1)$. This shows that

$$w_\alpha > w_{\{1\}} + w_{\{2\}} + w_0.$$

Let us now establish the lemma when $|\alpha_{i+1}| = |\alpha_i| = 2$. Suppose $\alpha_{i+1} = \{1, 3\}$ and $\alpha_i = \{1, 2\}$. Then apply Lemma 6.5 with $v^1 = (-1, 1, 1)$. This gives

$$w_{i+1} > w_i + w_{\{2\}} + w_{\{3\}}.$$

Next, suppose $|\alpha_{i+1}| = 3$ and $|\alpha_i| = 2$. We might assume that $\alpha_{i+1} = \{1, 2, 3\}$ and $\alpha_i = \{1, 2\}$. Apply Lemma 6.6 with $v^1 = (-1, -1, 1)$. This gives

$$w_{i+1} > 3w_i + w_{\{1,3\}} + w_{\{2,3\}} + w_{\{1\}} + w_{\{2\}} - w_{\{3\}} - w_0 > 3w_i \ ,$$

where the last inequality follows from the already established bounds.

Now consider the case $|\alpha_{i+1}| = |\alpha_i| = 3$, where we assume that $\alpha_{i+1} = \{1, 2, 4\}$ and $\alpha_i = \{1, 2, 3\}$. Apply Lemma 6.5 with $v^1 = (-1, -1, 1, 1)$. This gives

$$w_{i+1} > 3w_i + w_{\{1,3\}} + w_{\{2,3\}} + w_{\{1,4\}} + w_{\{2,4\}} - w_{\{3\}} - w_{\{4\}} > 3w_i.$$

Next consider the case when $|\alpha_{i+1}| = k$ and $|\alpha_i| = k-1$ where $k \geq 4$. We can assume that $\alpha_{i+1} = \{1, 2, \ldots k\}$ and $\alpha_i = \{1, \ldots, k-1\}$. Suppose that α_j is the proper subset of α_{i+1} other than α_i that has the highest index. Choose a vector v with $v_k = 1$ such that $\varphi_{\alpha_i}(v)\varphi_{\alpha_j}(v) = -1$. Apply Lemma 6.6 with this v. We have

$$w_{i+1} > (2^{k-1} - 1)w_i + w_j - \sum_{\alpha \subset \alpha_{i+1}, \alpha \neq \alpha_j, \alpha_i} \varphi_{\alpha_i}(v)\varphi_\alpha(v)w_\alpha .$$

Thus, the lemma will follow from establishing

$$\sum_{\alpha \subset \alpha_{i+1}, \alpha \neq \alpha_j, \alpha_i} |w_\alpha| \leq w_j.$$

We divide the sum into those α of size at least 3 and those of size at most 2. To bound the first sum, we note that since $w_{l+1} > 3w_l$ when $|\alpha_{l+1}| \geq 3$ and $l < j$, the sum over all sets of size at least 3 and index less than j is bounded by $w_j/2$.

To bound the second sum, we observe that there are at most $k(k-1)/2 + k + 1$ terms and each is bounded by the maximal weight of a set of size 2. Now there are at least $k - 2$ sets of size 3 before α_j in the enumeration (this bound is tight for $k = 4$ but very weak otherwise), which implies by induction that w_j is at least 3^{k-1} times the maximal weight of any set of size 2. The inequality $3^{k-1} > k(k - 1) + 2k + 2$ valid for $k \geq 4$ concludes this case.

Finally, suppose $|\alpha_{i+1}| = |\alpha_i| = k$ and $k \geq 4$. We assume that $\alpha_{i+1} = \{1, 2, \ldots, k - 1, k + 1\}$ and $\alpha_i = \{1, \ldots, k\}$. Suppose that α_j is the set of the highest index that appears in the sum of Lemma 6.5. Choose a vector v with $v_k = v_{k+1} = 1$ such that $\varphi_{\alpha_i}(v)\varphi_{\alpha_j}(v) = -1$. Now apply Lemma 6.5 with this v. The analysis is similar to the last case. This finishes the proof of the lemma.   □

The main theorem follows as a direct consequence of Lemma 6.7.

Theorem 6.2:   If n is a power of 2 and

$$F(f) = \text{sgn}\left(\sum_{i=0}^{n-1} w_i(f, \varphi_{\alpha_i})\right),$$

where w_i are integers, then $w_{n-1} \geq e^{-4n^\beta} 2^{(n \log n)/2 - n}$, where $\beta = \log(3/2)$.

Proof: Let α_{i_0} be the last set of size 2 in our ordering. Then

$$w_{n-1} \geq \prod_{|\alpha_i| > 2} (2^{|\alpha_i|-1} - 1)w_{i_0} > 2^{\sum_{i=1}^{n-1} |\alpha_i|-1} \prod_{|\alpha_i| \geq 2} (1 - 2^{1-|\alpha_i|}),$$

since $w_{i_0} > 1$ and the two factors introduced when $|\alpha_i| = 2$ cancel each other.

The first factor equals $2^{\frac{nm}{2}+1-n}$. This follows since the average size of a subset of $\{1, \ldots, m\}$ is $m/2$ and that there are n such sets. The extra $(1-n)$ term in the exponent is the result of -1 inside the summation. To estimate the second factor, we observe that if $0 < x < \frac{1}{2}$, then $(1 - x) > e^{-2x}$ and, hence,

$$\prod_{|\alpha_i| \geq 2} (1 - 2^{1-|\alpha_i|}) > \exp(-\sum_{k=2}^{n} \binom{m}{k} 2^{2-k}) \geq \exp(-4(1 + \frac{1}{2})^m) = \exp(-4n^\beta),$$

and the theorem follows. $\square$

The following theorem uses Lemma 6.3 to show the necessity of exponential weights in Eqn. (6.1).

Theorem 6.3: Assume that n is a power of 2 and

$$F(f) = \text{sgn}\left(\sum_{j=0}^{n-1} w_j(f)_j\right)$$

where w_j are integers, then for some j, we have $|w_j| \geq \frac{1}{n} e^{-4n^\beta} 2^{(n \log n)/2 - n}$, where $\beta = \log(3/2)$.

Proof: If we have an expansion of the kind given in the theorem, then by Lemma 6.3, setting

$$w_i' = \frac{1}{n} \sum_{j=0}^{n-1} w_j(\varphi_{\alpha_i})_j,$$

we can convert it to an expansion of the form considered in Theorem 6.2. The w_i' might not be integers, but nw_i' are integers for all i. Multiplying every weight by the same integer does not change the function and hence, by Theorem 6.2, $nw_{n-1} \geq e^{-4n^\beta} 2^{(n \log n)/2 - n}$. This is equivalent to saying that

$$\sum_{j=0}^{n-1} w_j(\varphi_{\alpha_i})_j \geq e^{-4n^\beta} 2^{(n \log n)/2 - n}$$

and the theorem follows.   □

If we use the full strength of Lemma 6.7, then we can actually strengthen Theorem 6.3 to apply to all weights in Eqn. (6.1).

Theorem 6.4:   Assume that n is a power of 2 and

$$F(f) = \text{sgn}\left(\sum_{j=0}^{n-1} w_j(f)_j\right),$$

where w_j are integers. Then, for $n \geq 8$ and all j, we have $|w_j| \geq \frac{1}{2n}e^{-4n^\beta}2^{(n\log n)/2-n}$, where $\beta = \log(3/2)$.

Proof: Use Lemma 6.3 to get a corresponding expansion

$$\sum_{i=0}^{n-1} w_i'(f, \varphi_{\alpha_i}),$$

where

$$w_i' = \frac{1}{n}\sum_{j=0}^{n-1} w_j(\varphi_{\alpha_i})_j.$$

We know by Lemma 6.7 that $w_{n-1}' \geq (2^{m-1} - 1)w_{n-2}' \geq (2^{m-1} - 1)(2^{m-2} - 1)w_{n-3}'\ldots$, where $n = 2^m$. By Lemma 6.3,

$$w_j = \sum_{i=0}^{n-1} w_i'(\varphi_{\alpha_i})_j,$$

and for $n \geq 8$,

$$|\sum_{i=0}^{n-2} w_i'(\varphi_{\alpha_i})_j| \leq \frac{1}{2}w_{n-1}'.$$

Since

$$w_{n-1}' \geq \frac{1}{n}e^{-4n^\beta}2^{(n\log n)/2-n},$$

the theorem follows.   □

6.3 Depth/Weight Trade-offs

In this section, we are going to show that for a general threshold circuit, we can trade off exponential weights with polynomially bounded weights at the expense of a polynomial increase in size and an additional layer, i.e., $LT_d \subset \widehat{LT}_{d+1}$. The derivation of this result requires somewhat involved combinatorial arguments. We first derive a weaker theorem that follows easily from the results established in previous chapters.

Theorem 6.5: For every integer $d \geq 1$, we have $LT_d \subset \widehat{LT}_{2d+1}$.

Proof: We first prove the result for $d = 1$. By considering the binary representation of the weights w_i, we can introduce more variables and assign some constant values to the renamed variables in such a way that any linear threshold function can be assumed to be of the following generic form:

$$f(X) = \text{sgn}(F(X)),$$

where

$$F(X) = \sum_{i=1}^{O(n \log n)} 2^i(x_{1_i} + x_{2_i} + \cdots + x_{n_i}).$$

The block-save technique developed in Section 4.3 can be applied to reduce $F(X)$ to the sum of two $O(n \log n)$-bit numbers using one layer of threshold gates. The details of the reduction will not be repeated here. Now the sign of the sum of the resultant two numbers is simply the Comparison function, which can be computed by another two layers of threshold gates. Thus, $\text{sgn}(F(X))$ is in $\widehat{LT}_3$.

The case for general d can be proved by induction on d. In fact, we shall prove a slightly stronger result: Any function in LT_d can be closely approximated by a sum of polynomially many $\widehat{LT}_{2d}$ functions (in the sense defined in Section 4.2). From the preceding argument and the fact that Comparison can be closely approximated by a sum of polynomially many $\widehat{LT}_1$ functions, the case for $d = 1$ is proved.

Let C_d be a depth-d polynomial-size threshold circuit and let

$$f_1(X), \ldots, f_m(X)$$

be the outputs from the $(d-1)^{th}$ layer, and $g(f_1(X), \ldots, f_m(X))$ be the output of C_d. By the preceding argument, g can be closely approximated by a sum of polynomially many $\widehat{LT}_2$ functions with $f_i(X)$'s as inputs to the first layer. Since $f_i(X) \in LT_{d-1}$, by the inductive hypothesis, each $f_i(X)$ can be closely approximated by a sum of polynomially many $\widehat{LT}_{2d-2}$ functions. Now replace the output gate g and each of its input functions $f_i(X)$ by the corresponding approximations. Then, we obtain a sum of polynomially many $\widehat{LT}_{2d}$ functions that approximate the output of the circuit. Therefore, by induction we have shown that any function in LT_d can be closely approximated by a sum of polynomially many $\widehat{LT}_{2d}$ functions and thus can be computed by $\widehat{LT}_{2d+1}$ circuits. □

6.3.1 Approximating Functions in LT_1

The first step in simulating a general depth-d polynomial-size threshold circuit with arbitrary weights by a depth-$(d+1)$ polynomial-size threshold circuit with small weights is to construct a good approximation for a general linear threshold element by elements with small weights.

Before we present a formal proof of the main result, let us discuss the underlying idea. Suppose we are given a general linear threshold function $f(X) = \text{sgn}(F(X)) = \text{sgn}(\sum_{i=1}^n w_i x_i)$, where $X \in \{1, -1\}^n$ and $F(X) = \sum_{i=1}^n w_i x_i$. First, we approximate $f(X)$ by another function $\tilde{f}_{(\alpha,p)}(X)$, where α and p are the parameters, and each $\tilde{f}_{(\alpha,p)}(X)$ can be written as a sum of polynomially many $\widehat{LT}_1$ functions. Moreover, α and p are polynomially bounded integers. We then show that for every input X, $\tilde{f}_{(\alpha,p)}(X) \neq f(X)$ for only a polynomially small fraction of the possible values of the parameters α and p. If we consider the sum of all the $\tilde{f}_{(\alpha,p)}(X)$, we can show that $\text{sgn}(\sum_{(\alpha,p)} \tilde{f}_{(\alpha,p)}(X)) = f(X)$.

We now construct an approximation $\tilde{f}(X)$ that will give the same output as $f(X)$ for most inputs X. Then from $\tilde{f}(X)$, we construct the functions $\tilde{f}_{(\alpha,p)}(X)$ described above.

As usual, let us assume that the weights in $f(X)$ are integers, and that $F(X) \neq 0$ for all inputs X. Let c be a fixed integer > 2. Consider a fixed input X. For each integer $\ell \geq 0$, let

$$\tilde{F}^\ell(X) = \sum_{i=1}^n \lfloor \frac{n^c w_i}{2^\ell} \rfloor x_i.$$

Note that if $2^\ell \leq |F(X)| < 2^{\ell+1}$, then $n^c \leq |\frac{n^c}{2^\ell} F(X)| < 2n^c$. Moreover, since

$$f(X) = \text{sgn}(F(X)) = \text{sgn}(\frac{n^c}{2^\ell}F(X)) = \text{sgn}(\sum_{i=1}^{n} \frac{n^c w_i}{2^\ell} x_i),$$

and

$$|\sum_{i=1}^{n} \frac{n^c w_i}{2^\ell} x_i - \tilde{F}^\ell(X)| \leq n,$$

we have

$$f(X) = \text{sgn}(\tilde{F}^\ell(X)), \tag{6.3}$$

if $2^\ell \leq |F(X)| < 2^{\ell+1}$.

In general, the weights in $\tilde{F}^\ell(X)$ can still be exponential. We now construct from $\tilde{F}^\ell(X)$ a gadget $\hat{F}^\ell(X)$ that will be useful later on. The idea is to consider $\tilde{F}^\ell(X)$ modulo some polynomially bounded prime integer $p > n^{2c}$. However, we need to modify the usual definition of the modulo operation for our purpose, as follows.

Definition 6.1: For an integer a and a prime $p \neq 2$, we define $a \tilde{\mod} p$ as the unique integer $b \in [\frac{-(p-1)}{2}, \frac{(p-1)}{2}]$ such that $a \equiv b \pmod{p}$. $\square$

For each integer $\ell \geq 0$, let

$$w_i^\ell = \lfloor \frac{n^c w_i}{2^\ell} \rfloor \tilde{\mod} p,$$

$$\hat{F}^\ell(X) = \sum_{i=1}^{n} w_i^\ell x_i,$$

and

$$\hat{f}^\ell(X) = \sum_{j=-n}^{n} \frac{1}{2}[\text{sgn}(\hat{F}^\ell(X) - j \cdot p - n^c) - \text{sgn}(\hat{F}^\ell(X) - j \cdot p - 2n^c)$$

$$+\text{sgn}(\hat{F}^\ell(X) - j \cdot p + n^c) - \text{sgn}(\hat{F}^\ell(X) - j \cdot p + 2n^c)]. \tag{6.4}$$

Remark 6.1: Observe that $\hat{f}^\ell(X)$ is simply a sum of polynomially many $\widehat{LT}_1$ functions. Note that since $p > n^{2c}$, there is no overlapping between the intervals $[j \cdot p - 2n^c, j \cdot p - n^c]$ and $[j \cdot p + n^c, j \cdot p + 2n^c]$, for each integer $j \in [-n, n]$. Moreover, the summand for each j is non-zero only when $\hat{F}^\ell(X)$ belongs to $[j \cdot p - 2n^c, j \cdot p - n^c]$ or $[j \cdot p + n^c, j \cdot p + 2n^c]$. It follows that for any ℓ, at most one j gives a non-zero contribution to the sum in $\hat{f}^\ell(X)$. $\square$

We have the following observations.

Lemma 6.8:

$$\hat{F}^\ell(X) = (\tilde{F}^\ell(X) \bmod p) + k \cdot p$$

for some integer k. Moreover, if

$$n^c \leq |\tilde{F}^\ell(X)| < 2n^c,$$

then

$$\hat{F}^\ell(X) = \tilde{F}^\ell(X) + k \cdot p$$

for some integer $k \in [-n, n]$.

Proof: The first statement of the lemma follows essentially from the linearity of the standard modulo operation. The details of the argument are left as an exercise.

Since $p > n^{2c}$ and $n^c \leq |\tilde{F}^\ell(X)| < 2n^c$, we have $\tilde{F}^\ell(X) \bmod p = \tilde{F}^\ell(X)$. Now, the second statement of the lemma follows from the fact that $|\hat{F}^\ell(X)| < np/2$. $\square$

Lemma 6.9:

$$\hat{f}^\ell(X) = \begin{cases} \mathrm{sgn}(\tilde{F}^\ell(X) \bmod p) & \text{if } |\tilde{F}^\ell(X) \bmod p| \in [n^c, 2n^c], \\ 0 & \text{otherwise.} \end{cases}$$

Proof: By Lemma 6.8, we have

$$\hat{F}^\ell(X) = (\tilde{F}^\ell(X) \bmod p) + k \cdot p$$

for some integer k. Recall from Remark 6.1 that $\hat{f}^\ell(X)$ is non-zero only when $\hat{F}^\ell(X)$ belongs to $[j \cdot p - 2n^c, j \cdot p - n^c]$ or $[j \cdot p + n^c, j \cdot p + 2n^c]$, for some integer j. It follows from the definition of $\bmod\, p$ (Definition 6.1) that $\hat{f}^\ell(X)$ is non-zero only when $\tilde{F}^\ell(X) \bmod p$ belongs to $[-2n^c, -n^c]$ or $[n^c, 2n^c]$. (See Exercise 6.3.) Moreover, if $|\tilde{F}^\ell(X) \bmod p| \in [n^c, 2n^c]$, then

$$\begin{aligned} \hat{f}^\ell(X) &= \frac{1}{2}[\mathrm{sgn}(\tilde{F}^\ell(X) \bmod p - n^c) - \mathrm{sgn}(\tilde{F}^\ell(X) \bmod p - 2n^c) \\ &\quad + \mathrm{sgn}(\tilde{F}^\ell(X) \bmod p + n^c) - \mathrm{sgn}(\tilde{F}^\ell(X) \bmod p + 2n^c)] \\ &= \mathrm{sgn}(\tilde{F}^\ell(X) \bmod p). \quad \square \end{aligned}$$

Lemma 6.10:   If for some integer ℓ, we have

$$2^\ell(1 + \frac{1}{n^{c-2}}) < |F(X)| < 2^\ell(2 - \frac{1}{n^{c-2}}),$$

then $\hat{f}^\ell(X) = f(X)$.

Proof: Recall that

$$\tilde{F}^\ell(X) = \sum_{i=1}^{n} \lfloor \frac{n^c w_i}{2^\ell} \rfloor x_i,$$

and

$$|\sum_{i=1}^{n} \frac{n^c w_i}{2^\ell} x_i - \tilde{F}^\ell(X)| = |\frac{n^c}{2^\ell} F(X) - \tilde{F}^\ell(X)| \le n.$$

By the assumption of the lemma, we have

$$n^c + n^2 < \frac{n^c}{2^\ell}|F(X)| < 2n^c - n^2.$$

It follows that

$$n^c < |\tilde{F}^\ell(X)| < 2n^c.$$

By Lemma 6.8, we have

$$\hat{F}^\ell(X) = \tilde{F}^\ell(X) + \hat{k} \cdot p$$

for some integer $\hat{k} \in [-n, n]$. Recall the definition of $\hat{f}^\ell(X)$, where

$$\hat{f}^\ell(X) = \sum_{j=-n}^{n} \frac{1}{2}[\text{sgn}(\hat{F}^\ell(X) - j \cdot p - n^c) - \text{sgn}(\hat{F}^\ell(X) - j \cdot p - 2n^c)$$

$$+\text{sgn}(\hat{F}^\ell(X) - j \cdot p + n^c) - \text{sgn}(\hat{F}^\ell(X) - j \cdot p + 2n^c)].$$

In the above expression, when $j = \hat{k}$, the summand will become

$$\frac{1}{2}[\text{sgn}(\tilde{F}^\ell(X) - n^c) - \text{sgn}(\tilde{F}^\ell(X) - 2n^c) + \text{sgn}(\tilde{F}^\ell(X) + n^c) - \text{sgn}(\tilde{F}^\ell(X) + 2n^c)]$$

$$= \text{sgn}(\tilde{F}^\ell(X)).$$

By Remark 6.1, for other values of j, the summands are all zero. Hence, from Eqn. (6.3), we have $\hat{f}^\ell(X) = \text{sgn}(\tilde{F}^\ell(X)) = f(X)$. $\square$

Lemma 6.10 gives conditions on the values of ℓ and $F(X)$ under which we have $\hat{f}^\ell(X) = f(X)$. What happens to other values of ℓ that do not satisfy the conditions? We shall see later that for most inputs X such that $|F(X)| \notin [2^\ell, 2^{\ell+1}]$, we have $\hat{f}^\ell(X) = 0$. Thus, intuitively, if we sum the $\hat{f}^\ell(X)$ for all possible values of ℓ, we should obtain a good approximation to $f(X)$. More formally, consider the following sum

$$\tilde{f}(X) = \sum_{\ell=0}^{\lceil \log(nW) \rceil} \hat{f}^\ell(X),$$

where W is an integer such that all integer weights $|w_i| \leq W$, and $|F(X)| \leq 2^{\lceil \log(nW) \rceil}$. Then, for most inputs X, we will have $\tilde{f}(X) = f(X)$.

By Corollary 2.3 in Chapter 2, W can be chosen to be $2^{O(n \log n)}$. So when $\tilde{f}(X) \neq f(X)$, $\tilde{f}(X)$ is an integer bounded by $\lceil \log(nW) \rceil = O(n \log n)$. Moreover, $\tilde{f}(X)$ is a sum of polynomially many $\widehat{LT}_1$ functions.

Now let us examine the cases when $\tilde{f}(X) \neq f(X)$. This might happen if any of the following two situations occurs:

1. If for no integer ℓ, the condition stated in Lemma 6.10 can be satisfied. Or equivalently, there is some ℓ such that $||F(x)| - 2^\ell| \leq \frac{2^\ell}{n^{c-2}}$.

2. Even if the condition in Lemma 6.10 is satisfied for some integer ℓ_o (so that $\hat{f}^{\ell_o}(X) = f(X)$), there is some $\ell \neq \ell_o$ such that $\hat{f}^\ell(X) \neq 0$.

If the above situations do not occur, then by the preceding discussions, we must have $\tilde{f}(X) = f(X)$. If we only consider $\tilde{f}(X)$ for a fixed p, it seems that there is no way of avoiding the two situations to happen for some input X. The idea is to consider a parametric family of functions $\tilde{f}_{(\alpha,p)}(X)$ derived from $\tilde{f}(X)$ by varying the prime integer p and replacing $F(X)$ by $\alpha F(X)$ for a positive integer α. Note that since $\alpha > 0$, we have $\text{sgn}(\alpha F(X)) = \text{sgn}(F(X)) = f(X)$.

The multiplier α we consider lies within the range $[1, n^{c-2}]$, and p is chosen among the first n^{2c} prime integers larger than n^{2c}. Note that by the Prime Number Theorem, $p = O(n^{2c} \log n)$.

We now show that for any fixed input X, the condition in Lemma 6.10 is satisfied for most α's within the possible range $[1, n^{c-2}]$. In other words, the first situation described above does not happen for most α's.

Lemma 6.11:   For any fixed X and for $1 - o(n^{-c+3})$ fraction of the α's in $[1, n^{c-2}]$, we have

$$2^\ell (1 + \frac{1}{n^{c-2}}) < |\alpha F(X)| < 2^\ell (2 - \frac{1}{n^{c-2}})$$

for some ℓ.

Proof: Let us say that the α's are 'bad' when $|2^{\ell} - |\alpha F(X)|| \leq \frac{2^{\ell}}{n^{c-2}}$ for some ℓ. To prove Lemma 6.11, it is equivalent to prove that only $o(n^{-c+3})$ fraction of the α's are bad.

Consider a fixed integer ℓ. If $|F(X)| \leq \frac{2^{\ell-1}}{n^{c-2}}$, then since $\alpha \leq n^{c-2}$, we have $|2^{\ell} - |\alpha F(X)|| > 2^{\ell-1} > \frac{2^{\ell}}{n^{c-2}}$. Now suppose $|F(X)| > \frac{2^{\ell-1}}{n^{c-2}}$. If α_o is the minimal α such that $|2^{\ell} - |\alpha F(X)|| \leq \frac{2^{\ell}}{n^{c-2}}$, then for all $\alpha > \alpha_o + 4$, we have $|2^{\ell} - |\alpha F(X)|| > \frac{2^{\ell}}{n^{c-2}}$. Hence, for each ℓ, at most five different α's are bad.

Since there are $O(n \log n)$ ℓ's, the number of bad α's is also bounded by $O(n \log n)$. Hence, the fraction of bad α's is $O(n \log n / n^{c-2}) = o(n^{-c+3})$. $\square$

Next we show that for any fixed input X and α, the second situation does not occur for most of the prime integers we consider. For simplicity, let us call those relevant prime integers p's such that the second situation occurs as the 'bad' p's.

Lemma 6.12: For any fixed X and α, the fraction of bad p's is $o(n^{-c+3})$.

Proof: Consider a fixed integer ℓ. Let

$$\tilde{F}_{\alpha}^{\ell}(X) = \sum_{i=1}^{n} \lfloor \alpha \frac{n^c w_i}{2^{\ell}} \rfloor x_i.$$

From the proof of Lemma 6.10, if $n^c \leq |\tilde{F}^{\ell}(X)| \leq 2n^c$, then p cannot be bad. By Lemma 6.9, p can be bad only when $|\tilde{F}_{\alpha}^{\ell}(X) \bmod p| \in [n^c, 2n^c]$ and $|\tilde{F}_{\alpha}^{\ell}(X)| > 2n^c$. Recall that p is chosen among the first n^{2c} prime integers larger than n^{2c}. Let N be the number of bad p's with respect to ℓ. Then by the pigeon-hole principle, there is some integer r where $|r| \in [n^c, 2n^c]$ such that $r = \tilde{F}_{\alpha}^{\ell}(X) \bmod p$ for $\Omega(N/2n^c)$ different p's.

Note that

$$|\tilde{F}_{\alpha}^{\ell}(X)| \leq \alpha n^c |F(X)| = 2^{O(n \log n)},$$

and the product of all bad p's is $> n^{2c(N/2n^c)} = 2^{cN \log n / n^c}$.

If the product of bad p's is larger than $|\tilde{F}_{\alpha}^{\ell}(X)|$, by Chinese Remainder Theorem, we must have $r = \tilde{F}_{\alpha}^{\ell}(X)$ and, hence, $|\tilde{F}_{\alpha}^{\ell}(X)| \leq 2n^c$, a contradiction. Thus, we must have $N = O(n^{c+1})$.

Since there are at most $O(n \log n)$ ℓ's, the fraction of bad p's is $O(Nn \log n/n^{2c}) = o(n^{-c+3})$.   □

With Lemma 6.12, we are now ready to prove the main result in this section.

Theorem 6.6:

$$LT_1 \in \widehat{LT}_2 \, .$$

Proof: Consider the function

$$g(X) = \mathrm{sgn}(\sum_{(\alpha,p)} \tilde{f}_{(\alpha,p)}(X)),$$

where the parameter $c = 5$. Let N be the total number of (α, p). Observe the following:

1. $\tilde{f}_{(\alpha,p)}(X) = f(X)$ for $1 - o(n^{-c+3}) = 1 - o(n^{-2})$ of the terms.

2. When $\tilde{f}_{(\alpha,p)}(X) \neq f(X)$, $|\tilde{f}_{(\alpha,p)}(X)| = O(n \log n)$.

It follows that

$$|f(X) - \frac{1}{N} \sum_{(\alpha,p)} \tilde{f}_{(\alpha,p)}(X)| = 1 - o(n \log n/n^2).$$

Since $f(X) = \pm 1$, we have $g(X) = f(X)$. Note that each $\tilde{f}_{(\alpha,p)}(X)$ is a sum of polynomially many $\widehat{LT}_1$ functions and the number of terms in the above sum is polynomially bounded and, hence, $f(X) = g(X) \in \widehat{LT}_2$. Since $f(X)$ is a general LT_1 function, we have proved that $LT_1 \in \widehat{LT}_2$.   □

6.3.2   Simulating LT_d in $\widehat{LT}_{d+1}$

We can generalize the result in the preceding section to general depth-d threshold circuits, for $d > 1$. In particular, we show that any depth-d polynomial-size threshold circuit with arbitrary weights can be simulated by a depth-$(d+1)$ polynomial-size threshold circuit with small weights.

Theorem 6.7:   For each integer $d > 0$, $LT_d \subset \widehat{LT}_{d+1}$.

Proof: Let C be a depth-d threshold circuit with size $\leq n^s$ and $C(X)$ be the function it computes. In other words, $C(X) \in LT_d$. For simplicity, assume without loss of generality that each threshold gate f_i in C, $i = 1, \ldots, n^s$, has exactly n^s wires leading into it (by adding dummy wires with weight zero, if necessary).

The idea is to replace each gate f_i in the circuit C by its corresponding approximation $f_{i,(\alpha,p)}$ as described in the preceding section, where we set the parameter $c = 5$. Moreover, all gates use the same pair of parameters (α,p), and the fraction of pairs that are bad for each gate is $o(n^{-3s})$. Let Q be the total number of different pairs of parameters (α,p) (which is polynomially bounded). Let $\tilde{C}_{(\alpha,p)}(X)$ be the corresponding output of such an approximation to the circuit.

Note that for any X and (α,p), we have $|\tilde{C}_{(\alpha,p)}(X)| = O(n^s \log n)$. Consider

$$\tilde{C}(X) = \sum_{(\alpha,p)} \tilde{C}_{(\alpha,p)}(X).$$

Observe that for any X,

$$
\begin{aligned}
C(X)\tilde{C}(X) \;\geq\; & |\{(\alpha,p) : \tilde{C}_{(\alpha,p)}(X) = C(X)\}| - \\
& O(n^s \log n)|\{(\alpha,p) : \tilde{C}_{(\alpha,p)}(X) \neq C(X)\}| \\
\geq\; & (1 - o(n^{-3s}))Q - \\
& O(n^s \log n)|\{(\alpha,p) : (\alpha,p) \text{ is bad for some } f_{i,(\alpha,p)}\}| \\
\geq\; & (1 - o(n^{-3s}))Q - \\
& O(n^s \log n)\sum_i |\{(\alpha,p) : (\alpha,p) \text{ is bad for } f_{i,(\alpha,p)}\}| \\
\geq\; & (1 - o(n^{-3s}))Q - O(n^s \log n) \cdot O(n^s) \cdot o(n^{-3s})Q \\
>\; & 0.
\end{aligned}
$$

It follows that $\operatorname{sgn}(\tilde{C}(X)) = C(X)$. Further, each $\tilde{C}_{(\alpha,p)}(X)$ can be realized by a sum of polynomially many $\widehat{LT}_d$ functions, thus $\operatorname{sgn}(\tilde{C}(X)) = \operatorname{sgn}(\sum_{(\alpha,p)} \tilde{C}_{(\alpha,p)}(X))$ can be realized in $\widehat{LT}_{d+1}$. Since $C(X)$ can be any circuit in LT_d, we have shown that $LT_d \subset \widehat{LT}_{d+1}$. $\square$

6.4   Optimal Depth Threshold Circuits for Multiplication and Division

Here we use the results in the preceding sections to construct threshold circuits with depth smaller than the ones presented in Chapter 4 for Multiple Sum, Multiplication, POWERING, Division, and Multiple Product. Since the proofs in this section are simple modifications of those in Chapter 4, we follow the same notations as in Chapter 4 and only indicate the necessary changes in the proofs.

We start with two key lemmas. The first lemma is a generalization of the result for symmetric functions presented as Theorem 3.2 in Section 3.2. Informally, it says that if a function is 1 when a weighted sum (possibly exponential) of its inputs lies in one of polynomially many intervals, and is 0 otherwise, then the function can be expressed as a sum of polynomially many LT_1 functions. The second lemma states that a linear threshold function of arbitrary weights can be closely approximated by a linear combination of polynomially many $\widehat{LT}_1$ functions.

Definition 6.2: $\widetilde{LT}_d$ is defined to be the class of Boolean functions computable by depth-d polynomial-size threshold circuits where the weights at the output gate are polynomially bounded integers (with no restriction on the weights of the other gates).   □

Lemma 6.13:   Let $S = \sum_{i=1}^{n} w_i x_i$ and $f(X)$ be a function such that $f = 1$ if $S \in [l_i, u_i]$ for $i = 1, \ldots, N$, and $f = 0$ otherwise, where N is polynomially bounded. Then, f can be computed as a sum of polynomially many LT_1 functions and thus, $f \in \widetilde{LT}_2$.

Proof: The proof is very similar to the proof of Theorem 3.2 in Chapter 3. For $j = 1, \ldots, N$, let

$$y_{l_j} = \text{sgn}(\sum_{i=1}^{n} w_i x_i - l_j),$$

$$y_{u_j} = \text{sgn}(u_j - \sum_{i=1}^{n} w_i x_i).$$

Then, $f(X) = \sum_{j=1}^{N}(y_{l_j} + y_{u_j}) - N$ and, hence, $f(X) = \text{sgn}\{\sum_{j=1}^{N}(y_{l_j} + y_{u_j}) - N - 1\}$.   □

The next lemma follows directly from the proof of Theorem 6.6.

Lemma 6.14:   Let $f(X) \in LT_1$. Then for any $k > 0$, there exist $t_j(X) \in \widehat{LT}_1$, $j = 1, \ldots, m$, such that

$$|f(X) - \frac{1}{N} \sum_{j=1}^{m} t_j(X)| \leq n^{-k},$$

where m and N are integers bounded by a polynomial in n.   □

As a consequence of Lemma 6.14, we have the following result, whose proof is left as an exercise.

Lemma 6.15: $\widetilde{LT}_d = \widehat{LT}_d$ for any fixed integer $d \geq 1$.

Theorem 6.8:   Multiple Sum $\in \widehat{LT}_2$.

Proof: Given n n-bit integers $z_i = \sum_{j=0}^{n-1} z_{i,j} 2^j$, $i = 1, \ldots, n$, the sum $\tilde{S} = \sum_{i=1}^{n} z_i$ can be represented as an $(n + \log n)$-bit integer, $\tilde{S} = \sum_{i=0}^{n + \log n - 1} \tilde{s}_i 2^i$. Clearly, the kth bit of $\tilde{S}$, $\tilde{s}_{k-1}$, is the same as the kth bit of the sum of the first k bits of the integers, i.e., the kth bit of $\sum_{i=1}^{n} \sum_{j=0}^{k-1} z_{i,j} 2^j$. Thus, to prove the theorem, it suffices to show that the kth bit of n k-bit integers can be computed by an $\widehat{LT}_2$ circuit, for $k = 1, \ldots, n + \log n$.

So let $S = \sum_{i=1}^{n} \sum_{j=0}^{k-1} 2^j z_{i,j} = \sum_{l=0}^{\log n + k - 1} 2^l s_l$ be the sum of n k-bit integers. Note that the kth bit of S, s_{k-1}, is 1 if $S \in I_{j,k} = [j2^{k-1}, (j+1)2^{k-1} - 1]$ for $j = 1, 3, 5, \ldots, 2^{\log n + 1} - 1$, and is 0 otherwise. Since there are only polynomially many intervals $I_{j,k}$, it follows from Lemma 6.13 the kth bit can be computed in $\widetilde{LT}_2$. Now apply Lemma 6.15 for $d = 2$, thus the kth bit can be computed in $\widehat{LT}_2$.   □

As a direct consequence of the above result, we have the following:

Theorem 6.9:   Multiplication $\in \widehat{LT}_3$.

To prove the results for Division and related problems, we need a stronger version of Theorem 6.8.

Lemma 6.16:   Let s_i be any of the outputs in Multiple Sum. Then for any $\hat{k} > 0$, there exist $\hat{t}_j(X) \in \widehat{LT}_1$ such that

$$|s_i - \frac{1}{\hat{N}} \sum_{j=1}^{\hat{m}} t_j(X)| \leq n^{-\hat{k}},$$

where $\hat{m}$ and $\hat{N}$ are integers bounded by a polynomial in n.

Proof: From Lemma 6.13 and from the proof of Theorem 6.8, we see that each s_i can be expressed as a sum of polynomially many LT_1 functions. Now apply Lemma 6.14 to approximate each of the LT_1 functions by a sum of polynomially many $\widehat{LT}_1$ functions with a sufficiently small error, the result follows.   □

To avoid cumbersome explanations in the following discussions, we say informally that every LT_1 function and each output bit in Multiple Sum can be closely approximated by a sum of polynomially many $\widehat{LT}_1$ functions, in the sense of Lemma 6.14 and Lemma 6.16.

Theorem 6.10:   POWERING $\in \widehat{LT}_3$.

Proof: Since the proof follows similar steps as in the proof of Theorem 4.10, we only indicate the modifications in order to reduce the circuit depth.

In Step 2 of the proof of Theorem 4.10, the m_i's are fixed integers (possibly exponentially large) and therefore Step 2 is in fact Multiple Sum. We can compute the values r_i in Step 1 as a sum of polynomially many $\widehat{LT}_1$ functions. By Lemma 6.14 and Lemma 6.16, each z_{jk} and each $EQ_k(Z)$ can be closely approximated by a sum of polynomially many $\widehat{LT}_1$ functions with variables r_i. Thus, $EQ_k(Z)$ and z_{jk} can be closely approximated by a sum of the outputs from polynomially many depth-2 threshold circuits whose inputs are the variables X. Then, by Lemma 4.7, $(EQ_k(Z) \wedge z_{jk})$ can also be closely approximated by a sum of the outputs from polynomially many $\widehat{LT}_2$ circuits. Hence, each of the outputs $\bigvee_{0 \leq k \leq n^4} (EQ_k(Z) \wedge z_{jk})$ can be computed by a $\widehat{LT}_3$ circuit.   □

Remark 6.2:   In the proof of Theorem 4.10, each $EQ_k(Z)$ and z_{jk} is closely approximated by a sum of outputs from polynomially many $\widehat{LT}_3$ circuits. Lemma 6.14 and Lemma 6.16 enable us to save one level of threshold gates in computing them.   □

Theorem 6.11:   Multiple Product $\in \widehat{LT}_4$.

Sketch of Proof: Let $Z = \prod_{j=1}^{n} z_j$, where each z_i is an n-bit integer. The proof for Theorem 6.11 is very similar to that of Theorem 4.10. We can compute Z using the same three steps as in Theorem 4.10. The only difference is that now each $r_i = Z \bmod p_i$ is computed by a sum of the outputs from polynomially many $\widehat{LT}_2$ circuits.   □

Theorem 6.12: $\mathrm{DIV}_k(X/Y) \in \widehat{LT}_3$.

Proof: Applying the same argument as in Theorem 6.10, one can show that each $(I_j \wedge EQ_k(Z_j) \wedge z_{j,k,i})$ can be closely approximated by a sum of outputs from polynomially many $(\widehat{LT}_2)$ circuits. Hence, the final result can be computed by a $\widehat{LT}_3$ circuit. □

Remark 6.3: In the proof of Theorem 4.12, each $(I_j \wedge EQ_k(Z_j) \wedge z_{j,k,i})$ is closely approximated by a sum of the outputs from polynomially many $(\widehat{LT}_3)$ circuits. Here again Lemma 6.14 and Lemma 6.16 enable us to save one level of threshold gates in computing them. □

Exercises

6.1: [*An ordering of the subsets of* $\{1, \ldots, m\}$]
Show that there exists an ordering $\alpha_0, \alpha_1, \ldots, \alpha_{n-1}$ of all the subsets of $\{1, \ldots, m\}$ $(n = 2^m)$ such that: (1) $|\alpha_i| \leq |\alpha_j|$ for $i \leq j$, and (2) $|\alpha_i \Delta \alpha_{i+1}| \leq 2$ for all i, where $|\alpha_i \Delta \alpha_{i+1}|$ is the symmetric difference of the two sets α_i and α_{i+1}.

6.2: Let $x_i \in \{1, -1\}$, $S = \sum_{i=1}^{n} w_i x_i$, and $\tilde{S} = \sum_{i=1}^{n} (w_i \tilde{\bmod} p) x_i$, for prime $p \neq 2$ and integer weights w_i's. Show that $S = \tilde{S} + k \cdot p$ for some integer k. This justifies the first statement of Lemma 6.8.

6.3: [$\tilde{\bmod} p$ *vs.* $\bmod p$]
Suppose we define $a \bmod p$ as the unique integer $b \in [0, p-1]$ such that $a \equiv b \pmod{p}$. Show that $\hat{f}^\ell(X)$ (Eqn. (6.4)) could be non-zero even if $\tilde{F}^\ell(X) \bmod p$ does not belong to $[n^c, 2n^c]$. This is the motivation for changing the usual definition of mod to Definition 6.1.

6.4: [*Threshold circuits with polynomially bounded weights in the output layer*]
Show that $\widetilde{LT}_d = \widehat{LT}_d$ for any fixed integer $d \geq 1$.

6.5: [*Computing rounded quotient in depth-3 threshold circuits*]
Show that the rounded quotient (see Exercise 4.11) is in $\widehat{LT}_3$.

6.6: $[LT_1 \notin \widehat{PT}_1]$

Let $\widehat{PT}_1$ be the class of Boolean functions that can be expressed as a sign of sparse polynomials (in $\{1, -1\}$ notation) with polynomially bounded integer coefficients. It can be shown [40] that the Boolean function

$$\mathrm{sgn}\Big(1 + 2 \sum_{i=0}^{n-1} \sum_{j=0}^{2n-1} 2^i y_j (x_{i,2j} + x_{i,2j+1})\Big)$$

is not in $\widehat{LT}_2$. Use this result to construct an LT_1 function that is not in $\widehat{PT}_1$.

Bibliographic Notes

It is implicitly shown in Chandra, Stockmeyer, and Vishkin [26] that a depth-d polynomial-size threshold circuit with arbitrary weights can be simulated by a depth-αd polynomial-size threshold circuit with small weights, where α is a fixed integer. This result follows from the fact that the sum of n n-bit integers (Multiple Sum) can be computed by a constant-depth polynomial-size threshold circuit with small weights, which is shown in [26]. However, the issue of minimizing the constant factor α is not addressed in [26].

The first attempt to minimize the constant factor α was made by Siu and Bruck [123]. It is shown in [123] that any LT_1 function can be simulated in $\widehat{LT}_3$ and for $d > 1$, $LT_d \subset \widehat{LT}_{2d+1}$. The idea in [123] is essentially to construct of a small-depth threshold circuit for computing the sum of n n-bit integers. This result is stated as Theorem 6.5 in Section 6.3. It is conjectured in [123] that $LT_d \subset \widehat{LT}_{d+c}$ for some fixed integer c. The observation that Comparison requires exponentially large weights can be also found in [123].

The conjecture in [123] is proved in Goldmann, Håstad, and Razborov [40]. Among other results, it is shown in [40] that in fact for every integer $d > 0$, $LT_d \subset \widehat{LT}_{d+1}$. However, since the results in [40] are based on probabilistic arguments, only the existence of the simulating small-weight circuits is shown.

The result in [48] indicates that any polynomial-size threshold circuit (with small weights) computing Multiplication must have depth at least 3, and the question of whether depth-3 threshold circuits suffice is left unresolved in [48]. This question is answered affirmatively in Siu and Roychowdhury [125]. Using the results in [40], optimal-depth threshold circuits for the problems Multiple Sum, Multiplication, and Division are shown in [125]. These results are presented in Section 6.4.

Although the size of the simulating circuit in [123] increases as a polynomial in the size of the original circuit, the exponent in the polynomial factor increases rapidly with the depth d. Using a similar intuition as in [40], Goldmann and Karpinski [41] further improve the results in [40] in two respects. First, an explicit construction of the simulating circuits is given in [41]. Second, the increase in the size of the simulating circuits can be made independently of the depth d of the original circuit. Consequently, the threshold circuits presented in Section 6.4 all have explicit constructions. Our presentation in Section 6.3 follows closely the one in [41].

A construction of linear threshold functions that requires exponentially large weights can be found in [85]. However, the magnitude of the weights increases only exponentially (i.e., $O(2^{cn})$ for some $c > 0$) and does not match the best-known upper bounds ($2^{O(n \log n)}$). Håstad [51] provides a construction where the magnitude of the weights almost matches the best-known upper bounds, and the corresponding results are presented in Section 6.2.

Chapter 7

Rational Approximation and Optimal-Size Circuits

7.1   Introduction

In this chapter, we introduce the techniques of rational approximation and derive lower bound results on the number of neural elements required for the computation of certain functions using feedforward networks. The techniques and results in this chapter apply to both threshold circuits and γ networks. As noted in the first chapter, one of the challenges in complexity theory is to establish nontrivial lower bounds on the size of circuits computing specific families of functions. For example, in Chapter 3 we proved a lower bound result showing that the size of a threshold circuit (with no restriction on the weights and the depth) computing a general n-variable symmetric function is $\Omega(\sqrt{n/\log n})$. We also showed that this lower bound result is almost tight by developing a constructive procedure for computing any n-variable symmetric function by a depth-3 threshold circuit of size $O(\sqrt{n})$. However, the lower bound result was derived by a counting argument, and it is not known which specific symmetric function requires $\Omega(\sqrt{n/\log n})$ threshold gates to be computed.

Using techniques of rational approximation, we show in Section 7.2 that any depth-3 threshold circuit with polynomially bounded weights computing the n-variable Parity function must have size $\Omega(\sqrt{n}/\log^2 n)$. We also extend this result to threshold circuits of higher depth and show that any depth-

$(d + 1)$ threshold circuit with polynomially bounded weights computing the Parity function must have size $\Omega(n^{1/d}/\log^2 n)$. A central idea of our proof is the use of a result from the theory of rational approximation, which states the following [152, 45]: The function $\text{sgn}(x)$ can be approximated with an error of $O(e^{-ck/\log(1/\epsilon)})$ by a rational function of degree k, for $0 < \epsilon < |x| < 1$. This result allows us to approximate several layers of threshold gates by a rational function of low degree when the size of the circuit is small. Then, by deriving a lower bound on the degree of the rational function that approximates the Parity function, we derive a lower bound on the size of the circuit.

We extend the above results in Section 7.3 to circuits in which additional gates other than threshold gates are allowed. In particular, we show that any constant-depth circuit of subexponentially many AND/OR gates and $n^{o(1)}$ threshold gates cannot compute the Parity function.

In Section 7.4, the lower bound results presented in Section 7.2 are generalized and shown to be valid even when the elements of the networks can assume continuous values. The class of networks for which the techniques can be applied includes networks of sigmoidal elements and radial basis elements. In particular, we generalize our results on threshold circuits to a general class of neural networks whose elements can be piecewise approximated by low-degree rational functions. We show that for symmetric Boolean functions of large strong degree (e.g., the Parity function), any depth-d network whose elements can be piecewise approximated by low-degree rational functions requires almost the same size as a depth-d threshold circuit computing the same function.

7.2   Lower Bounds on Threshold Circuits

Definition 7.1:   Let the *degree* of a rational function $R(x) = P(x)/Q(x)$ be the maximum of the degrees of its numerator polynomial $P(x)$ and its denominator polynomial $Q(x)$. Let ϵ be any constant such that $0 < \epsilon < 1$. Define

$$\| R(x) - \text{sgn}(x) \| = \sup_{|x| \in [\epsilon, 1]} |R(x) - \text{sgn}(x)|$$

and

$$E_k(\text{sgn}(x)) = \inf \{\| R(x) - \text{sgn}(x) \| : R(x) \text{ rational and } \deg R(x) \le k\}. \quad \square$$

First let us state a classical result from the theory of rational approximation which is due to Gončar [45].

Lemma 7.1:   There exists a constant $c > 0$ such that

$$E_k(\text{sgn}(x)) = O(e^{-ck/\log(1/\epsilon)}) \ . \qquad \Box$$

In other words, one can approximate exponentially close (in k) the function $\text{sgn}(x)$ by a rational function of degree k. It is worth mentioning that the rational function that achieves the approximation to $\text{sgn}(x)$ stated in Gončar's Lemma can be constructed explicitly, though the explicit form of the rational function is not needed for deriving our lower bound results. Exercises 7.4 and 7.5 outline an alternative technique due to Newman [88] for an explicit construction of a rational function for approximating $\text{sgn}(x)$.

One complexity measure of a Boolean function that we shall use later is the strong degree.

Definition 7.2:   Given any Boolean function $f(X) : \{1, -1\}^n \to \{1, -1\}$, its *strong degree*, $SD(f)$, is the smallest degree of any multilinear polynomial $F(X)$ such that $f(X) = \text{sgn}(F(X))$ and $F(X) \neq 0$ for all $X \in \{1, -1\}^n$. $\Box$

Remark 7.1:   The spectral representation (see Sections 1.5 and 8.5) of a Boolean function $f(X)$ shows that there always exists a multilinear polynomial $F(X)$ such that $f(X) = \text{sgn}(F(X))$. Hence, the strong degree is always well-defined for any Boolean function f. Moreover, if $P(X) = f(X)$ is the spectral representation, then $SD(f) \leq deg(P(X)) \leq n$, where n is the number of variables. It is easy to see that if $f(X) \in LT_1$, then $SD(f) = 1$. For example, $SD(AND) = 1$, even though the degree of the spectral representation of AND is n. On the other hand, the Parity function $PAR_n(X)$ has the largest strong degree among all n-variable Boolean functions, as stated next. $\Box$

Lemma 7.2: $SD(PAR_n(X)) = n$.

Proof: See Lemma 8.3 in Chapter 8. $\Box$

In the following discussions, we assume that each threshold gate has polynomially bounded (in n) integer weights. In other words, each threshold gate computes $\text{sgn}(\sum_i w_i f_i)$, where $\sum_i |w_i| < n^\alpha$, and α is some constant > 0. Without loss of generality, we further assume that $\sum_i w_i x_i \neq 0$ for every input $X \in \{1, -1\}^n$.

Lemma 7.3:   Let $X = (x_1, \ldots, x_n) \in \{1, -1\}^n$ and let

$$g(X) = \text{sgn}(\sum_{i=1}^{t} w_i f_i(X)),$$

where each $f_i(X)$ is a ± 1 valued function. Assume that $\sum_i |w_i| < n^\alpha$, w_i's are integers, and $\sum_{i=1}^{t} w_i f_i(X) \neq 0$. We have the following claims:

1. If $|R_{f_i}(X) - f_i(X)| < n^{-(\alpha+1)}$ for all $X \in \{1, -1\}^n$, then $g(X) = \text{sgn}(\sum_{i=1}^{t} w_i R_{f_i}(X))$ for all $X \in \{1, -1\}^n$.

2. If $R_{f_i}(X)$ satisfies the assumption in Claim 1, then

$$1 - 1/n \leq |\sum_{i=1}^{t} w_i R_{f_i}(X)| \leq n^\alpha$$

   for all $X \in \{1, -1\}^n$.

3. If $R_{f_i}(X)$ is a rational function of degree $\leq d$, for $i = 1, \ldots, t$, and satisfies the assumption in Claim 1, then there exists a rational function $R_g(X)$ of degree $O(td \log^2 n)$ such that $|R_g(X) - g(X)| < n^{-(\alpha+1)}$ for all $X \in \{1, -1\}^n$.

Proof:

1. Write $R_{f_i}(X) = f_i(X) + \epsilon_i$ with $|\epsilon_i| < n^{-(\alpha+1)}$, then $\sum_{i=1}^{t} w_i R_{f_i}(X) = \sum_{i=1}^{t} w_i(f_i(X) + \epsilon_i) = \sum_{i=1}^{t} w_i f_i(X) + \epsilon$, where $\epsilon = \sum_{i=1}^{t} w_i \epsilon_i$. Since $\sum_i |w_i| < n^\alpha$ by assumption, we have $|\epsilon| \leq \sum_{i=1}^{t} |w_i| \cdot n^{-(\alpha+1)} < n^{-1}$. Since $\sum_{i=1}^{t} w_i f_i(X)$ is a non-zero integer from our assumption, it follows that $\text{sgn}(\sum_{i=1}^{t} w_i R_{f_i}(X)) = \text{sgn}(\sum_{i=1}^{t} w_i f_i(X) + \epsilon) = \text{sgn}(\sum_{i=1}^{t} w_i f_i(X))$ for all X.

2. As above, $|\sum_{i=1}^{t} w_i R_{f_i}(X)| \leq |\sum_{i=1}^{t} w_i f_i(X)| + |\epsilon| \leq n^\alpha$. The left-hand inequality follows from $|\sum_{i=1}^{t} w_i R_{f_i}(X)| \geq |\sum_{i=1}^{t} w_i f_i(X)| - |\epsilon| \geq 1 - 1/n$.

3. Clearly, we can write the sum $\sum_{i=1}^{t} w_i R_{f_i}(X)$ as a rational function $\tilde{R}(X)$ of degree at most td. By the claim 2 above, $n^{-\alpha}/2 \leq |\tilde{R}(X)|/n^\alpha \leq 1$. If we let $y(X) = \tilde{R}(X)/n^\alpha$, then $g(X) = \text{sgn}(y(X))$.

Choose $\epsilon = n^{-\alpha}/2$ and $k = ((\alpha+1)\log n)^2/c$ for the approximating rational function to $\mathrm{sgn}(y(X))$ as in Lemma 7.1; then, $g(X) = \mathrm{sgn}(y(X))$ can be approximated by a rational function in the variable $y(X)$ of degree $O(\log^2 n)$ with an error of at most $n^{-(\alpha+1)}$ for all inputs. Substituting $y(X) = \tilde{R}(X)/n^\alpha$ into this approximating rational function, we obtain a rational function $R_g(X)$ of degree $O(td\log^2 n)$ such that $|R_g(X) - g(X)| < n^{-(\alpha+1)}$. □

Lemma 7.4: Let $F(X)$ be a function computed by a depth-3 threshold circuit such that the first layer and the second layer consist of t_1 and t_2 gates, respectively. Then, $F(X)$ can be written as a sign of a rational function of degree $O(t_1 t_2 \log^4 n)$.

Proof: The idea of the proof is to approximate the outputs of the threshold gates in the first two layers by rational functions of low degrees, such that the output of the third layer can be written as the sign of a rational function of degree $O(t_1 t_2 \log^4 n)$.

First, let us derive rational approximations for the gates in the first layer. The output f_j of a threshold gate in the first layer can be expressed as $f_j = \mathrm{sgn}(\sum_i w_i \cdot x_i + w_0)$, where, by assumption, $\sum_i |w_i| < n^\alpha$ for some constant $\alpha > 0$. Lemma 7.1 can be directly applied to obtain a rational function $R^1_{f_j}(X)$ (the superscript 1, denotes that the rational function is approximating the output of a threshold gate in the first layer) of degree $c_j \log^2 n$ (where $c_j > 0$ is some constant), such that $|R^1_{f_j} - f_j| < n^{-(\alpha+1)}$ for all X. Since the approximation error is $< n^{-(\alpha+1)}$, it follows from Lemma 7.3 that for the threshold gates in succeeding layers, one can always replace the output f_j by its rational approximation and incur no errors.

Next we show how to approximate the outputs of the threshold gates in the second layer by appropriate rational functions. In general, an output g_i in the second layer can be written as

$$g_i(X) = \mathrm{sgn}(\sum_{j=1}^{t_1} w_j f_j(X) + \sum_{j=1}^{n} \tilde{w}_j x_j + \tilde{w}_o),$$

where f_j's are the outputs of the gates in the first layer, and t_1 is the number of gates in the first layer. As observed before, one can replace each $f_j(X)$ by the corresponding rational function $R^1_{f_j}(X)$ (of degree $c_j \log^2 n$) without changing the output g_i; hence,

$$g_i(X) = \text{sgn}(\sum_{j=1}^{t_1} w_j R_{f_j}^1(X) + \sum_{j=1}^{n} \tilde{w}_j x_j + \tilde{w}_o).$$

Following the same arguments as in part 2 of Lemma 7.3, one can show that $1 - O(1/n) \le |(\sum_{j=1}^{t_1} w_j R_{f_j}^1 + \sum_{j=1}^{n} \tilde{w}_j x_j)| \le n^\alpha$. Also since $R_{f_j}^1(X)$'s are rational functions of degree $c_j \log^2 n$, we can write $g_i(X) = \text{sgn}(y(X))$, where $y(X) = (\sum_{j=1}^{t_1} w_j R_{f_j}^1(X) + \sum_{j=1}^{n} \tilde{w}_j x_j)/n^\alpha$ is a rational function of degree $c_i t_1 \log^2 n$, and $|y(X)| \in [1/n^\alpha, 1]$. Next, using the same arguments as in part 3 of Lemma 7.3, one can determine a rational function $R_{g_i}^2(y)$ of degree $c_y \log^2 n$ such that $|R_{g_i}^2 - g_i| \le n^{-(\alpha+1)}$. Since $y(X)$ is a rational function in X of degree $c t_1 \log^2 n$, it easily follows that the rational function $R_{g_i}^2(X)$ that approximates g_i has degree $c_i t_1 \log^4 n$.

Finally, the output $h(X)$ of a depth-3 threshold circuit can be written as

$$h(X) = \text{sgn}(\sum_{j=1}^{t_2} w_{1j} g_j(X) + \sum_{j=1}^{t_1} w_{2j} f_j(X) + \sum_{j=1}^{n} w_{3j} x_j + w_{3o}),$$

where g_j's and f_j's are outputs of the threshold gates in the second and first layer, respectively. Again, it follows from Lemma 7.3 that one can replace every g_j (or f_j) by the corresponding rational function without changing the value of h; hence,

$$h(X) = \text{sgn}(\sum_{j=1}^{t_2} w_{1j} R_{g_j}^2(X) + \sum_{j=1}^{t_1} w_{2j} R_{f_j}^1(X) + \sum_{j=1}^{n} w_{3j} x_j + w_{3o}),$$

where $R_{g_j}^2$ is of degree $c_{g_j} t_1 \log^4 n$ and $R_{f_j}^1$ is of degree $c_{f_j} \log^2 n$. One can now combine these rational functions and conclude

$$h(X) = \text{sgn}(R_h(X)),$$

where R_h is a rational function of degree $c_1 t_2 t_1 \log^4 n + c_2 t_1 \log^2 n + c_3$. In other words, the output of a depth-3 threshold circuit with t_1 gates in the first layer and t_2 gates in the second layer can be written as the sign of a rational function of degree $O(t_2 t_1 \log^4 n)$.   □

Theorem 7.1:   Any depth-3 threshold circuit computing $PAR_n(X)$ must have size $\Omega(\sqrt{n}/\log^2 n)$.

Proof: By Lemma 7.4, any function $F(X)$ computed by a depth-3 threshold circuit can be written as $\text{sgn}(P(X)/Q(X))$, where $P(X)$ and $Q(X)$ are polynomials of degree at most $O(t_1 t_2 \log^4 n)$. We can also write $F(X) = \text{sgn}(P(X) \cdot Q(X))$. Hence, if $F(X) = PAR_n(X)$, then by previous remarks, degree $(P(X) \cdot Q(X)) = n$. It implies that $t_1 t_2 = \Omega(n/\log^4 n)$. The size $(t_1 + t_2 + 1)$ of the circuit is minimized when $t_1 = t_2 = \Omega(\sqrt{n}/\log^2 n)$ and the result follows.   □

In fact, by applying the rational approximation repeatedly layer by layer, we can generalize the lower bound result to higher depth circuits.

Theorem 7.2: Any depth-$(d+1)$ threshold circuit computing $PAR_n(X)$ must have size $\Omega(dn^{1/d}/\log^2 n)$.

Proof: Let t_i be the number of threshold gates in the ith layer, for $i = 1, \ldots, d$. One can follow the same arguments as in the proof Lemma 7.3, and repeat the rational approximation technique from the first layer to the dth layer, and show that the output of the $(d+1)^{th}$ layer can be written as the sign of a polynomial whose degree is dominated by $O(t_1 \ldots t_d (c \log n)^{2d})$. Since the degree has to be n, it follows that the size $(1 + \sum_{i=1}^{d} t_i)$ is at least $\Omega(dn^{1/d}/\log^2 n)$.   □

We can prove a result similar to Theorem 7.2 for the Complete Quadratic (CQ) function (see Definition 5.1).

Theorem 7.3: Any depth-$(d+1)$ threshold circuit computing the CQ function must have size $\Omega(dn^{1/d}/\log^2 n)$.

Proof: By the same degree argument as in the proof of Theorem 7.2, it suffices to show that the CQ function (of n variables) has strong degree $\Omega(n)$. This fact is stated as Lemma 7.5.   □

Lemma 7.5: $SD(CQ) \geq \lfloor \frac{n}{2} \rfloor$, where n is the number of variables, i.e., if $P(X)$ is a multilinear polynomial such that $CQ(X) = \text{sgn}(P(X))$, then $\deg(P(X)) \geq \lfloor \frac{n}{2} \rfloor$.

Proof: Since $CQ(X) = 1$ if number of -1's in X is 0 or 1 mod 4 (see Exercise 7.1), it is easy to see that $PAR_n(X)$ is a simple projection of the CQ function of $2n$ variables. If $CQ(\tilde{X}) = \text{sgn}(P(\tilde{x}_1, \tilde{x}_2, \ldots, \tilde{x}_{2n}))$, then $PAR_n(x_1, x_2, \ldots, x_n) = \text{sgn}(P(x_1, x_1, \ldots, x_n, x_n))$. Since $SD(PAR_n) = n$, it implies that $SD(CQ) \geq \lfloor \frac{n}{2} \rfloor$.   □

Remark 7.2:   The only property of the functions Parity and CQ we use in proving the lower bounds in Theorem 7.2 and 7.3 is that these functions have strong degree $\Omega(n)$. Thus, these results also hold for any function with strong degree $\Omega(n)$. Note that the preceding lower bounds are nontrivial for $d = o(\log n/\log\log n)$. The upper bound results presented in Chapter 5 give a construction of depth-$(\log n)$ threshold circuits with size $\log n$ computing these functions. Below, we show that such a threshold circuit for Parity also has optimal size.   □

Theorem 7.4:   Any threshold circuit computing $PAR_n(X)$ must have $\log n$ gates.

Proof: Given any linear threshold function $f(X) = \text{sgn}(w_0 + \sum_{i=1}^{n} w_i x_i)$, there exists a subset S of the variables $\{x_1, \ldots, x_n\}$ with $|S| \leq \lceil n/2 \rceil$, such that the value of $f(X)$ can be determined once the variables in S are fixed to some values (Exercise 7.6). However, the value of a Parity function cannot be determined unless all values of its variables are known.

Consider any threshold circuit of k gates computing $PAR_n(X)$. Label the gates so that gate g_j only receives inputs from the variables or from the output of gate g_i, where $i < j$. First we restrict the values to at most half of the input variables such that the output of gate g_1 is fixed. Next restrict at most half of the remaining variables such that the output of gate g_2 is also fixed. Repeat the restriction procedure to gates g_3, g_4, ... and so on, each time restricting the values of at most half of the remaining variables. After the kth step, the circuit output will be fixed, but at least $n/2^k$ variables are still free. This is impossible unless $k \geq \log n$, thus proving Theorem 7.4.   □

7.2.1   Approximating Parity with Threshold Circuits

Suppose that the size of a threshold circuit is smaller than the lower bound in Theorems 7.2 and 7.3. How well can we approximate $PARITY$ or CQ when the inputs are random and uniformly distributed? In this section, we derive almost tight lower bounds on the size of threshold circuits that approximate these functions. We make use of the following result.

Theorem 7.5:   Let $p(X) = \text{sgn}(P_k(X))$, where $P_k(X) \neq 0$ for all $X \in \{1, -1\}^n$ is a multilinear polynomial of degree $k < n$. Suppose $\frac{n-k-1}{2}$ is an integer. Then, $p(X) \neq PAR_n(X)$ on at least

$$N_k = \sum_{i=0}^{(n-k-1)/2} \binom{n}{i}$$

inputs.   Moreover, there is a degree k polynomial $\hat{P}_k(X)$ such that $\text{sgn}(\hat{P}_k(X)) \neq PAR_n(X)$ on exactly N_k inputs.   □

For the proof of the above theorem, we need the following two lemmas.

Lemma 7.6:   Let $S \subset \{1, -1\}^n$ with $|S| < \sum_{i=0}^{k} \binom{n}{i}$. Then, there exists a polynomial $q(X)$ of degree $2k$ such that $q(X) \not\equiv 0$, $q(X) \geq 0$ for all $X \in \{1, -1\}^n$, and $q(X) = 0$ for all $X \in S$.

Proof: We first find a multilinear polynomial $r(X)$ of degree k so that $r(X) = 0$ for all $X \in S$. In general, $r(X)$ has $\sum_{i=0}^{k} \binom{n}{i}$ coefficients, say α_i's. For each $\hat{X} \in S$, $r(\hat{X}) = 0$ represents a linear homogeneous equation with $\sum_{i=0}^{k} \binom{n}{i}$ unknowns (i.e., the coefficients) α_i's. Thus, the constraints $r(X) = 0$ for all $X \in S$ represents a homogeneous system of $|S|$ linear equations in the unknowns α_i's, and has a nontrivial solution since $|S| < \sum_{i=0}^{k} \binom{n}{i}$.

Now let $q(X) = r^2(X)$. Then, $q(X)$ is a polynomial of degree $2k$ (not identically zero) that is 0 on S and non-negative elsewhere.   □

Lemma 7.7:   Let $S \subsetneq \{1, -1\}^n$. Suppose $\tilde{q}(X)$ is a multilinear polynomial with the following properties:

1. $\tilde{q}(X) = 0$ for all $X \in S$.

2. For each $X \notin S$, $\tilde{q}(X) \neq 0$ and $\text{sgn}(\tilde{q}(X)) = PAR_n(X)$.

Then, the degree of $\tilde{q}(X) = n$.

Proof: Suppose $\deg \tilde{q}(X) < n$. Observe that for every monomial term X^α of degree less than n, we have $\sum_{X \in \{1, -1\}^n} X^\alpha \cdot PAR_n(X) = 0$. Thus,

$\sum_{X\in\{1,-1\}^n} \tilde{q}(X)\cdot PAR_n(X) = 0$. On the other hand, because of the two prop-
erties of $\tilde{q}(X)$, we have $\sum_{X\in\{1,-1\}^n} \tilde{q}(X)\cdot PAR_n(X) = \sum_{X\notin S} \tilde{q}(X)\cdot PAR_n(X) >$
0, a contradiction.   □

Proof of Theorem 7.5:
We prove the first part of the theorem by contradiction. Let $\hat{S} \subset \{1,-1\}^n$
such that $|\hat{S}| < \sum_{i=0}^{(n-k-1)/2} \binom{n}{i}$ and $\text{sgn}(P_k(X)) = PAR_n(X)$ for all $X \notin \hat{S}$.

By Lemma 7.6, there exists a polynomial $\hat{q}(X)$ of degree $(n-k-1)$ such
that $\hat{q}(X) = 0$ on $\hat{S}$ and non-negative elsewhere. Let $\tilde{q}(X) = P_k(X)\hat{q}(X)$.
Then, there exists a subset S with $\hat{S} \subseteq S$ such that $\tilde{q}(X) = 0$ on S and for
each $X \notin S$, $\tilde{q}(X) \neq 0$ and $\text{sgn}(\tilde{q}(X)) = PAR_n(X)$. On the other hand,
$\deg \tilde{q}(X) < k + (n-k-1) = n$. This is impossible by Lemma 7.7.

For the second part of the proof, one can construct the desired polyno-
mial $\hat{P}_k(X)$ of the form $\tilde{P}(\sum_{i=1}^n x_i)$ whose sign is the same as $PAR_n(X)$
on a connected, center range of inputs $\sum_{i=1}^n x_i$. We leave the details of the
construction as an exercise.   □

The construction of the polynomial $\hat{P}_k(X)$ (defined in the preceding theorem) leads to a size-efficient construction of a threshold circuit that approximates the Parity function $PAR_n(X)$ with error only on N_k inputs.

Theorem 7.6:   Let $N_k = \sum_{i=0}^{(n-k-1)/2} \binom{n}{i}$ (as in Theorem 7.5). Then, for
every depth $(d+1)$, there is a threshold circuit of size $O(dk^{1/d})$ that gives
the same output as $PAR_n(X)$ except on N_k inputs.   □

The above construction of the threshold circuit that approximates
$PAR_n(X)$ is almost optimal for $k = \Omega(n^\epsilon)$ as stated in the following theorem.

Theorem 7.7:   Let $C(X)$ be computed by a depth-$(d+1)$ threshold circuit
(of n input variables) such that $C(X) \neq PAR_n(X)$ on at most N_k inputs.
Then, the circuit of $C(X)$ must have size $\Omega(dk^{1/d}/\log^2 n)$.

Proof: Let t_i be the number of threshold gates in the i-th layer of $C(X)$,
for $i = 1,\ldots,d$. Then, from the proof of Theorem 7.2, it is easy to see that
the output of the $(d+1)^{th}$ layer can be written as the sign of a polynomial
whose degree is dominated by $O(t_1\ldots t_d(c\log n)^{2d})$. But this degree has to

be at least k (by the previous theorem), and hence, the size of the circuit $(1 + \sum_{i=1}^{d} t_i)$ is at least $\Omega(dk^{1/d}/\log^2 n)$.   □

7.3 Extensions to Threshold Circuits with Various Gates

In this section, we consider circuits when additional gates other than the threshold gates are allowed. We first consider depth-2 threshold circuits with polynomially many additional gates in $\widetilde{SP}$ (to be defined). We show that for depth-2 circuits, $\Omega(n/\log^2 n)$ threshold gates are still needed to compute the CQ function even if polynomially many gates in $\widetilde{SP}$ are allowed. Next we show that any constant-depth circuit of subexponentially many AND/OR gates and $n^{o(1)}$ threshold gates (with subexponentially large weights) cannot compute the Parity function.

Definition 7.3:   The class $\widetilde{SP}$ consists of all Boolean functions that can be closely approximated by sparse polynomials, more precisely, given any $f(X) \in \widetilde{SP}$, $k > 0$, there are only polynomially many (in n) α's such that $c_\alpha \neq 0$ and

$$|f(X) - \sum c_\alpha X^\alpha| < n^{-k} \text{ for all } X \in \{1, -1\}^n .$$ □

For example, functions such as AND, OR, Comparison, Addition, and more generally, functions that have polynomially bounded spectral norms are contained in $\widetilde{SP}$ (see Section 4.2). With these additional gates, the degree argument does not apply directly. Instead, we use the following result from harmonic analysis of Boolean functions:

Lemma 7.8:   If $CQ(X) = \text{sgn}(P(X))$, where $P(X)$ is a multilinear polynomial, then the number of monomial terms in $P(X)$ must be at least $2^{n/2}$.   □

We shall postpone the proof of Lemma 7.8 until Chapter 8, where a more general result will be proved using a simple geometric argument.

Lemma 7.9:   For any $\lambda > 1$, there exists a $\delta > 0$ such that for δn an integer, we have $\sum_{i=0}^{\delta n} \binom{n}{i} < \lambda^n$.

Proof: Observe that $\sum_{i=0}^{\delta n} \binom{n}{i}/2^n$ is the probability of having δn number of heads in n independent tosses of a fair coin. Define independent Bernoulli random variables X_i's such that each $X_i = 1$ or 0 with probability $1/2$. Then, by the symmetry of the distribution,

$$Pr\{\sum_{i=0}^{n} X_i \geq n - \delta n\} = Pr\{\sum_{i=0}^{n} X_i \leq \delta n\} = \sum_{i=0}^{\delta n} \binom{n}{i}/2^n.$$

By Chernoff inequality, for $\theta > 0$,

$$Pr\{\sum_{i=0}^{n} X_i \geq n - \delta n\} \leq E(e^{\theta X_i})^n e^{-\theta(1-\delta)n} = (e^{\theta\delta} + e^{-\theta(1-\delta)})^n/2^n.$$

For any $\lambda > 1$, take $\theta = 1/\sqrt{\delta}$. It is clear that there exists a $\delta > 0$ small enough such that $e^{\theta\delta} + e^{-\theta(1-\delta)} = e^{\sqrt{\delta}}(1 + e^{-1/\sqrt{\delta}}) < \lambda$. Hence, for such choices of δ and θ, we have

$$\sum_{i=0}^{\delta n} \binom{n}{i} \leq (e^{\theta\delta} + e^{-\theta(1-\delta)})^n < \lambda^n .   \square$$

The next theorem shows that including polynomially many gates from $\widetilde{SP}$ in a depth-2 threshold circuit computing CQ will not significantly reduce the number of threshold gates.

Theorem 7.8:   Suppose in addition to threshold gates, we have polynomially many gates $\in \widetilde{SP}$ in the first layer of a depth-2 threshold circuit that computes the CQ function. Then, the number of threshold gates required in the circuit is $\Omega(n/\log^2 n)$.

Proof: Let $f_1(X), \ldots, f_s(X)$ be the $\widetilde{SP}$ gates and $g_1(X), \ldots, g_t(X)$ be the threshold gates in the first layer, where s and t are polynomially bounded. Let us denote the output function of the depth-2 circuit by

$$h(X) = h(f_1(X), \ldots, f_s(X), g_1(X), \ldots, g_t(X)) = \text{sgn}(H(X)),$$

where

$$H(X) = \sum_{i=1}^{s} w_i f_i(X) + \sum_{j=1}^{t} \tilde{w}_j g_j(X) + w_0 \neq 0 \quad \forall X \in \{1, -1\}^n,$$

$w_i, \tilde{w}_i$ are integers, and $\sum_{i=1}^{s} |w_i| + \sum_{j=1}^{t} |\tilde{w}_j| + |w_0| < n^k$ for some constant $k > 0$.

We can approximate each $f_i(X)$ by a sparse polynomial $P_{f_i}(X)$ and each $g_j(X)$ by a rational function $R_{g_i}(X)$ of degree $O(\log^2 n)$ with error $\leq n^{-k}$ (see Lemma 7.3), so that

$$
\begin{aligned}
h(X) &= h(P_{f_1}(X), \dots, P_{f_s}(X), R_{g_1}(X), \dots, R_{g_t}(X)) \\
&= \mathrm{sgn}\left(\sum_{i=1}^{s} w_i P_{f_i}(X) + \sum_{j=1}^{t} \tilde{w}_j R_{g_j}(X) + w_0\right) \\
&= \mathrm{sgn}\left(\sum_{i=1}^{s} w_i P_{f_i}(X) + \frac{P(X)}{Q(X)}\right),
\end{aligned}
$$

where $\frac{P(X)}{Q(X)}$ is a rational function of degree $d \leq ct \log^2 n$ for some constant $c > 0$ (t is the number of threshold gates in the first layer). The last expression in the preceding equations can be written as $\mathrm{sgn}(\tilde{H}(X))$, where $\tilde{H}(X) = (\sum_{i=1}^{s} w_i P_{f_i}(X)) Q^2(X) + Q(X)P(X)$. Since both $P(X)$ and $Q(X)$ are multilinear polynomials of degree-d (in n variables), each can have at most $\sum_{i=0}^{d} \binom{n}{i}$ number of monomial terms. Also note that $\sum_{i=1}^{s} w_i P_{f_i}(X)$ is a sparse polynomial, so that the number of monomial terms is $< n^\beta$ for some constant $\beta > 0$. Thus, the total number of monomial terms in $\tilde{H}(X)$ is $\leq n^\beta \sum_{i=0}^{2d} \binom{n}{i}$. By Lemma 7.9 with $\lambda < \sqrt{2}$, this number will be smaller than $2^{n/2}$ (for sufficiently large n) if $2d < \delta n$, for some $\delta > 0$. It then follows from Lemma 7.8 that $2d = 2ct \log^2 n \geq \delta n$. Therefore, the number of threshold gates, $(t+1)$, is at least $\Omega(n/\log^2 n)$. □

Remark 7.3: Using the same arguments as above, it is possible to show that the above lower bound still holds even if we allow subexponentially many $(2^{n^{o(1)}})$ AND, OR gates in the first layer of the depth-2 circuit. □

Remark 7.4: One can easily generalize the lower bound results in Theorem 7.2 and Theorem 7.3 to constant-depth threshold circuits in which additional gates with low-degree polynomial approximations are allowed. In other words, it can be shown that any depth-$(d + 1)$ polynomial-size threshold circuit that computes $PARITY$ or CQ with additional gates that can be approximated by low-degree $(\log^{O(1)} n)$ polynomials must require $\Omega(dn^{1/d}/\log^{O(1)} n)$ threshold gates. For example, the class of gates that has low-degree polynomial approximations contains as a subclass functions that can be computed by a *decision tree* of *polylogarithmic depth*. In particular, if the additional gates of the threshold circuits have $O(\log^c n)$ fan-in, then the lower bound in Theorem 7.2 applies.   □

Our difficulty in generalizing the result in Theorem 7.2 to threshold circuits with additional AND, OR gates lies in the fact that AND, OR gates do not have low-degree polynomial approximations. We prove this fact next.

Theorem 7.9:   Let $X \in \{1, -1\}^n$. Suppose $P_{AND}(X)$ and $P_{OR}(X)$ are polynomials such that $|P_{AND}(X) - AND(X)| = o(n^{-1})$ and $|P_{OR}(X) - OR(X)| = o(n^{-1})$ for all X. Then, $P_{AND}(X)$ and $P_{OR}(X)$ must have degrees $\Omega(\sqrt{n})$.

Proof: It is enough to give a proof for the OR($\bigvee$) function, since

$$AND(x_1, \ldots, x_n) = -OR(-x_1, \ldots, -x_n).$$

Recall from our convention that $\{0, 1\}$ is encoded as $\{1, -1\}$, and hence, $OR(x_1, \ldots, x_n) = 1$ if all $x_i = 1$.

Note that any symmetric function $f(X)$ can be expressed as

$$f(X) = \bigvee_{k \in S} EX_k^n(X),$$

for some $S \subset \{-n, \ldots, n\}$, where $EX_k^n(X)$, the Exact function, is 1 if $\sum_{i=1}^n x_i = k$. In particular, $PAR_n(X)$ can be written as a linear threshold function with $EX_k^n(X)$'s as inputs, with total weights bounded in magnitude by $O(n)$. If EX_k^n's all have low-degree polynomial approximations (with error $o(n^{-1})$), then the $PAR_n(X)$ function can be written as a sign of a low-degree polynomial, which is impossible. Thus, $EX_k^n(X)$ cannot have low-degree polynomial approximations for all $k \in \{-n, \ldots, n\}$. (In fact, the degree must be n for some EX_k^n.)

Suppose $f(X)$ is a symmetric function and $P(X)$ is a polynomial such that $|P(X) - f(X)| \le n^{-c}$. Then, $|\frac{1}{n!}(\sum_\sigma P(X_\sigma)) - f(X)| \le n^{-c}$, where σ denotes a permutation of the variables and ranges over all $n!$ permutations, $X_\sigma = (x_{\sigma(1)}, \ldots, x_{\sigma(n)})$. It is easy to see that $\frac{1}{n!} \sum_\sigma P(X_\sigma) = \tilde{P}(\sum_{i=1}^n x_i)$ for some polynomial $\tilde{P}$, since $\frac{1}{n!} \sum_\sigma P(X_\sigma)$ is a symmetric polynomial. Observe that each term $P(X_\sigma)$ of the sum has the same degree as $P(X)$, thus $\tilde{P}(\sum_{i=1}^n x_i)$ also has the same degree as $P(X)$. So for symmetric functions, it suffices to consider all approximating polynomials of the form $\tilde{P}(\sum_{i=1}^n x_i)$.

Let $X_n = \sum_{i=1}^{n} x_i$. Consider a polynomial approximation to the OR function with error $o(n^{-1})$, i.e.,

$$\tilde{P}(X_n) = \begin{cases} 1 + \epsilon(X_n) & \text{if } X_n = n, \\ -1 + \epsilon(X_n) & \text{otherwise,} \end{cases}$$

where $|\epsilon(X_n)| = o(n^{-1})$ for all possible values of X_n. Now consider the functions EX_k^n of $\sqrt{n}/2$ variables and observe that $\tilde{P}(n - (k - \tilde{X})^2)$ is a polynomial approximation to the function $EX_k^{\sqrt{n}/2}$, where $\tilde{X} = \sum_{i=1}^{\sqrt{n}/2} x_i$. Thus, by previous argument, for some k, $\tilde{P}(n - (k - \tilde{X})^2)$ must have degree $\Omega(\sqrt{n})$. Hence, $\tilde{P}(X)$ has degree $\Omega(n^{1/2})$. The result follows. □

The result of Theorem 7.2 states that any constant-depth threshold circuit with only $n^{o(1)}$ gates cannot compute $PAR_n(X)$. We also showed that any depth-2 threshold circuit computing the CQ function must have size $\Omega(\sqrt{n}/\log^2 n)$ even if we allow subexponentially many $(2^{n^{o(1)}})$ extra AND/OR gates in the circuit. This result was obtained using rational approximation and harmonic analysis. In the following, we employ some new techniques and previous ideas of approximation to generalize the results to multilayer threshold circuits with subexponentially many extra AND/OR gates. In particular, the result states that any constant-depth circuit of subexponentially many AND/OR gates and $n^{o(1)}$ threshold gates cannot compute the Parity function.

For the rest of the discussions in this section, we find it more convenient to switch back to the $\{0, 1\}$ notations. We make use of two key lemmas in proving our results. The first lemma states that if the Parity function of $X = (x_1, \ldots, x_n) \in \{0, 1\}^n$ can be expressed as the sign of an (integral) sum of AC^0 functions in X, then the sum of the magnitudes of the integer coefficients must be exponentially large. The second lemma is basically a restatement of Lemma 7.1 that gives the error of rational approximation to a linear threshold function in terms of the magnitudes of the coefficients in the rational function.

In order to facilitate our discussions, we introduce some more terminologies.

Definition 7.4:

1. The *norm* of a polynomial is the sum of the absolute values of its coefficients.

2. The *norm* of a rational function is the norm of its numerator polynomial plus the norm of its denominator polynomial.   □

Unless otherwise specified, we assume that all coefficients in the polynomials and the rational functions are integers. Also observe that a polynomial with variables in $\{0,1\}^n$ and norm N is equal to the sum of N conjunctions (ANDs), and the value of the polynomial is also bounded by N.

The following lemma is useful in proving our result and whose proof will be presented in Chapter 8.

Lemma 7.10: If $PAR_n(x_1, \ldots, x_n) = \text{sgn}(p(f_1, \ldots, f_\ell))$, where p is a polynomial and $f_1, \ldots, f_\ell$ are AC^0 functions (in x_i's), then the norm of $p(\cdot)$ must be exponentially large, i.e., $2^{n^{\Omega(1)}}$.   □

The next lemma follows immediately from Lemma 7.1.

Lemma 7.11: Let $g(x_1, \ldots, x_n) = \text{sgn}(\sum_{i=1}^n w_i x_i) : \{0,1\}^n \to \{0,1\}$ be a linear threshold function with integer weights w_i. Then, $g(x_1, \ldots, x_n)$ can be approximated by a rational function $R_g(x_1, \ldots, x_n)$ with integer coefficients having norm $2^{O(\log(1/\epsilon)\log^2(\sum_{i=1}^n |w_i|))}$, with an error

$$|g(x_1, \ldots, x_n) - R_g(x_1, \ldots, x_n)| \leq \epsilon$$

for every input $(x_1, \ldots, x_n) \in \{0,1\}^n$.   □

We need the following inequality in proving our theorem.

Lemma 7.12: Let m and N be positive integers, and let $\epsilon = \frac{1}{m(1+2^m N)}$. Then, $2^m N((1+\epsilon)^m - 1) < 1$.

Proof: For all real y, $1 + y \leq \exp(y)$, with equality holding if $y = 0$. Therefore, for all x,

$$\frac{1}{1 + 1/x} = \frac{x}{1+x} = 1 - \frac{1}{1+x} \leq \exp(-\frac{1}{1+x}),$$

so $1 + 1/x \geq \exp(\frac{1}{1+x})$. Letting $x = 2^m N$, we have

$$1 + 2^{-m}N^{-1} \geq \exp(\frac{1}{1+2^m N}),$$

so

$$(1 + 2^{-m}N^{-1})^{\frac{1}{m}} \geq \exp(\frac{1}{m(1 + 2^m N)}) > 1 + \frac{1}{m(1 + 2^m N)} = 1 + \epsilon.$$

Therefore,

$$2^m N((1 + \epsilon)^m - 1) \leq 1,$$

as claimed. □

Theorem 7.10: Let C be a constant-depth subexponential-size $(2^{n^{o(1)}})$ circuit of AND/OR gates and $n^{o(1)}$ threshold gates with subexponential integer weights. Then, C cannot compute PARITY. □

Before we present a formal proof of the Theorem 7.10, let us first sketch the essential ideas. Given a constant-depth circuit with $m = n^{o(1)}$ threshold gates and subexponentially many AND/OR gates, we will transform it into a constant-depth circuit (which computes the same function) such that the new circuit also has subexponentially many AND/OR gates but has a threshold gate only at the output. It then follows from Lemma 7.10 that this circuit cannot compute the Parity function. A major part of the proof is devoted to showing how the transformation of the circuit can be carried out using rational approximations.

Proof: Let $f : \{0, 1\}^n \to \{0, 1\}$ be computable by a depth-d circuit of $2^{n^{o(1)}}$ AND/OR gates and $m = n^{o(1)}$ threshold gates. Let the total size of the circuit be $S = 2^{n^{o(1)}}$. Also assume that the threshold gates have subexponentially large $2^{n^{o(1)}}$ integer weights. For convenience, let us define $\tilde{AC}$ to be the class of functions computable by constant-depth circuits of $2^{n^{o(1)}}$ AND/OR gates. We will show that $f = \text{sgn}(p(f_1, \ldots, f_\ell))$, where p is a polynomial with integer coefficients and norm $2^{m^d \log^{O(d)} S} = 2^{n^{o(1)}}$ and $f_1, \ldots, f_\ell$ are $\tilde{AC}$ functions.

We first assume that in the circuit for f, the threshold gates are only on the last k layers. We construct a polynomial p and $f_1, \ldots, f_\ell$ such that $f = \text{sgn}(p(f_1, \ldots, f_\ell))$, p is a polynomial (with integer coefficients) whose norm is bounded by some function $N_p(k)$, and $f_1, \ldots, f_\ell$ are $\tilde{AC}$ functions. For convenience, we also assume without loss of generality that p is never zero. We find a recurrence for $N_p(k)$, and show by induction that $N_p(k) = 2^{m^k \log^{O(k)} S}$. We then let $k = d$ to establish our claim.

So consider the function f and the corresponding circuit C_f which satisfy the preceding assumptions. Then, all the gates from the first layer to the $(d-k)^{th}$ layer in C_f are AND/OR gates. Let $(t_1, t_2, .., t_\alpha) \in \{0,1\}^\alpha$ be the outputs of the α threshold gates on the $(d-k+1)^{th}$ layer in C_f, where $\alpha \leq m$. Clearly, there are at most $2^\alpha \leq 2^m$ possible values of $(t_1, t_2, .., t_\alpha)$. For each possible value of $(t_1, t_2, .., t_\alpha)$, say $\mathbf{T}^i \in \{0,1\}^\alpha$, consider the circuit $C_f^{(i)}$ derived from C_f with the outputs of the threshold gates $(t_1, t_2, .., t_\alpha)$ fixed to $\mathbf{T}^i$. Then, each circuit $C_f^{(i)}$ has threshold gates only on the last $(k-1)$ levels. By the inductive hypothesis, the output of each of these circuits can be expressed as $\mathrm{sgn}(p^{(i)}(f_{i1}, \ldots, f_{i\ell}))$, where $p^{(i)}$ is a polynomial with integer coefficients whose norm is bounded by some function $N_p(k-1)$, and $f_{i1}, \ldots, f_{i\ell}$ are $\tilde{A}C$ functions.

For each $\mathbf{T}^i \in \{0,1\}^\alpha$, let $\bigwedge_{\mathbf{T}^i}(t_1, t_2, \ldots, t_\alpha)$ be the AND function such that $\bigwedge_{\mathbf{T}^i} = 1$ if and only if $(t_1, t_2, \ldots, t_\alpha) = \mathbf{T}^i$. Now the function f is the sum, over all possible values $\mathbf{T}^i \in \{0,1\}^\alpha$, of the output of $C_f^{(i)}$ (i.e., $\mathrm{sgn}(p^{(i)}(f_{i1}, \ldots, f_{i\ell}))$) multiplied by $\bigwedge_{\mathbf{T}^i}$. In other words,

$$f = \sum_{\mathbf{T}^i \in \{0,1\}^\alpha} \mathrm{sgn}(p^{(i)}(f_{i1}, \ldots, f_{i\ell})) \cdot \bigwedge_{\mathbf{T}^i}. \qquad (7.1)$$

Recall that each of the AND functions $\bigwedge_{\mathbf{T}^i}$ is the product of exactly α factors, each of which is the output of a threshold gate (or its complement) on the $(d-k+1)^{th}$ layer in C_f. Moreover, each of these threshold gates has at most $S = 2^{n^{o(1)}}$ inputs which are either the inputs to the circuit C_f, or the outputs from the AND/OR gates on the first $(d-k)^{th}$ layers. By Lemma 7.11, each of these threshold gates can be approximated with error at most $\epsilon = 1/(m(2^m N_p(k-1)+1))$ by a rational function whose norm is $2^{O(\log(1/\epsilon)\log^2(2^{n^{o(1)}}S))} = 2^{O((\log^2 S)(m+\log N_p(k-1)))}$. If a threshold gate is approximated within ϵ by a rational function r, then its complement is approximated within ϵ by the rational function $1-r$, which has the same denominator as r and norm $2^{O((\log^2 S)(m+\log N_p(k-1)))}$. Now approximate each AND term $\bigwedge_{\mathbf{T}^i}$ in Eqn. (7.1) by the product of the rational functions that approximate the corresponding threshold gates (or their complements) on the $(d-k+1)^{th}$ layer. The product of the rational functions (which itself is a rational function), denoted by $R_{\mathbf{T}^i}$, corresponding to each AND term $\bigwedge_{\mathbf{T}^i}$ has norm $2^{O((\log^2 S)(m^2+m\log N_p(k-1)))}$, and approximates each AND term $\bigwedge_{\mathbf{T}^i}$

with error

$$|R_{\mathbf{T}^i} - \bigwedge_{\mathbf{T}^i}| \leq (1+\epsilon)^\alpha - 1 \leq (1+\epsilon)^m - 1.$$

Also, observe that the denominators of the rational functions $R_{\mathbf{T}^i}$ are the same for all AND terms $\bigwedge_{\mathbf{T}^i}$.

Now consider the following expression $\tilde{f}$,

$$\tilde{f} = \sum_{\mathbf{T}^i \in \{0,1\}^\alpha} p^{(i)}(f_{i1}, \ldots, f_{i\ell}) \cdot R_{\mathbf{T}^i}. \tag{7.2}$$

In the following, we show that $f = \operatorname{sgn}(\tilde{f})$.

Let $\hat{X}$ be an input to the circuit. Denote the outputs of the threshold gates on the $(d - k + 1)^{th}$ layer corresponding to $\hat{X}$ by $\hat{\mathbf{T}} \in \{0,1\}^\alpha$. Then, in the summation of Eqn. (7.1), all the AND terms except $\bigwedge_{\hat{\mathbf{T}}}$ are zero. So

$$f = \operatorname{sgn}(\hat{p}(\hat{f}_1, \ldots, \hat{f}_\ell)) \cdot \bigwedge_{\hat{\mathbf{T}}} = \operatorname{sgn}(\hat{p}(\hat{f}_1, \ldots, \hat{f}_\ell)),$$

where $\hat{p}$ has norm bounded by $N_p(k - 1)$. Let the rational function that approximates $\bigwedge_{\hat{\mathbf{T}}}$ be denoted by $R_{\hat{\mathbf{T}}}$. Then, $|R_{\hat{\mathbf{T}}} - 1| \leq (1+\epsilon)^m - 1$, and for $\mathbf{T}^i \neq \hat{\mathbf{T}}$, we have $|R_{\mathbf{T}^i}| \leq (1+\epsilon)^m - 1$.

From Eqn. (7.2),

$$
\begin{aligned}
|\tilde{f} - \hat{p}(\hat{f}_1, \ldots, \hat{f}_\ell)| \quad &\leq \quad |\hat{p}(\hat{f}_1, \ldots, \hat{f}_\ell)| \cdot |R_{\hat{\mathbf{T}}} - 1| + \\
&\qquad \sum_{\mathbf{T}^i \neq \hat{\mathbf{T}}} |p^{(i)}(f_{i1}, \ldots, f_{i\ell})| \cdot |R_{\mathbf{T}^i}| \\
&\leq \quad 2^m N_p(k-1)((1+\epsilon)^m - 1) \\
&< \quad 1.
\end{aligned}
$$

The last inequality follows from Lemma 7.12 (where $\epsilon = 1/(m(2^m N_p(k-1) + 1)))$. Recall our assumption that $\hat{p}$ is never zero and has integer coefficients, and thus $|\hat{p}(\hat{f}_1, \ldots, \hat{f}_\ell)| \geq 1$. Hence,

$$\operatorname{sgn}(\tilde{f}) = \operatorname{sgn}(\hat{p}(\hat{f}_1, \ldots, \hat{f}_\ell))$$

and therefore,

$$\operatorname{sgn}\left(\sum_{\mathbf{T}^i \in \{0,1\}^\alpha} p^{(i)}(f_{i1}, \ldots, f_{i\ell}) \cdot R_{\mathbf{T}^i} \right) = f.$$

Recall that all the rational functions $R_{\mathbf{T}^i}$ used in the approximation have the same denominator and has norm bounded by $2^{O((\log^2 S)(m^2 + m \log N_p(k-1)))}$.

We obtain a polynomial that has the same sign as f by multiplying $\tilde{f}$ by the square of that common denominator. Moreover, this polynomial has integer coefficients and has norm $N_p(k)$ bounded by

$$N_p(k) \leq 2^m N_p(k-1) 2^{O((\log^2 S)(m^2 + m \log N_p(k-1)))}.$$

This recurrence implies that $N_p(k)$ can be chosen to be less than $2^{m^k} \log^{O(k)} S$.

Now substitute $k = d$, $S = 2^{n^{o(1)}}$, and $m = n^{o(1)}$. Hence, for every f computable by a depth-d circuit of $2^{n^{o(1)}}$ AND/OR gates and $m = n^{o(1)}$ threshold gates, we can obtain a polynomial $p(f_1, \ldots, f_\ell)$ with norm $2^{m^d} \log^{O(d)} S = 2^{n^{o(1)}}$, where $f_1, \ldots, f_\ell$ are $\tilde{AC}$ functions, such that $f = \mathrm{sgn}(p(f_1, \ldots, f_\ell))$. Therefore, by Lemma 7.10, f cannot be the Parity function.   □

7.4   Lower Bounds on Networks with Continuous-Valued Elements

In this section, we derive lower bounds on networks with continuous-valued elements. Recall from Section 1.2 the definition of computation of Boolean functions by γ networks (see Definitions 1.8 and 1.9):

Definition 7.5:   A γ network C is said to compute a Boolean function $f : \{1, -1\}^n \to \{1, -1\}$ with separation $\epsilon > 0$ if there is some $t_C \in \mathbf{R}$ such that for any input $X = (x_1, \ldots, x_n) \in \{1, -1\}^n$ to the network C, the output element of C computes a value $C(X)$ with the following property: If $f(X) = 1$, then $C(X) \geq t_C + \epsilon$. If $f(X) = -1$, then $C(X) \leq t_C - \epsilon$.   □

Remark 7.5:   Computing with γ networks without separation at the output element is less interesting because an infinitesimal change in the output of any γ element may change the output bit. We are mainly interested in computations on γ networks C_n with separation at least $\Omega(n^{-k})$ for some fixed $k > 0$. Together with the assumption of polynomially bounded weights, this notion of computing with separation makes the complexity class of constant-depth polynomial-size γ networks quite robust and more interesting to study from a theoretical point of view.   □

We now extend the lower bound results in previous sections to general γ networks. Recall that the main idea in deriving the lower bound results on

threshold circuits is that one can easily obtain a low-degree rational approximation of a linear threshold function by using Gončar's Lemma. On the
other hand, when the element in a γ network is, say, a sigmoidal function, we
do not know a direct method of obtaining a low-degree rational approximation to the element. Instead, we first obtain a piecewise low-degree rational
approximation to the sigmoidal function. Then, with the following result, we
can obtain a single rational function by joining the pieces together, without
increasing the degree significantly. The proof can be found in [102].

Lemma 7.13:   Let f be a continuous function over $\Delta = [a, b]$. Let $\Delta_1 = [a, c]$ and $\Delta_2 = [c, b]$, $a < c < b$. Denote $\parallel g \parallel_{\Delta_i} = \sup_{x \in \Delta_i} |g(x)|$. Suppose
there are rational functions r_1 and r_2 such that for some $\epsilon > 0$,

$$\parallel f - r_i \parallel_{\Delta_i} \leq \epsilon.$$

Then, for each $\tilde{\epsilon} > 0$ and $\delta > 0$, there is a rational function r such that

$$\parallel f - r \parallel_\Delta \ \leq \epsilon + \tilde{\epsilon} + \omega(f; \delta)_\Delta,$$

and

$$\deg r \leq 2 \deg r_1 + 2 \deg r_2 + C_1 \log(e + \frac{b-a}{\delta}) \log(e + \frac{\parallel f \parallel_\Delta}{\tilde{\epsilon}}), \quad (7.3)$$

where $\omega(f; \delta)_\Delta$ is the modulus of continuity of f over Δ, C_1 is a constant.

$\square$

As an application of the preceding lemma, one can easily obtain a low-
degree rational approximation to the sigmoidal function, as stated in the
following theorem. (One can also obtain similar results on networks of radial
basis elements).

Theorem 7.11:   Let $X = (x_1, \ldots, x_n) \in \{1, -1\}^n$ and

$$f(X) = \sigma(F(X)) = \frac{2}{1 + e^{-F(X)}} - 1,$$

where

$$F(X) = \sum_{i=1}^{S} w_i \cdot x_i + w_0,$$

and $\sum_{i=0}^{S} |w_i| \leq n^c$. Then, there exists a rational function $R_f(X)$ of degree $O(\log^4 n)$ such that

$$|R_f(X) - f(X)| = O(n^{-\lambda \log n})$$

for some constant $\lambda > 0$ and for all $X \in \{1, -1\}^n$.

Proof: For simplicity, we derive an approximation to $\sigma(x)$ and substitute $x = F(X)$ in the final result. First, observe that the derivative (with respect to y) of the function $1/(1 + y)$ is bounded in magnitude by 1 for all $y > 0$. Therefore, by Mean Value Theorem,

$$\left| \frac{1}{1+y} - \frac{1}{1+\tilde{y}} \right| \leq |y - \tilde{y}| \quad \text{for all } y > 0.$$

Thus, to closely approximate $\sigma(x)$, it suffices to find a good approximation to e^x.

Let $\Delta_1 = [-n^c, -\log^2 n]$, $\Delta_2 = [-\log^2 n, \log^2 n]$, $\Delta_3 = [\log^2 n, n^c]$. By Taylor series expansion, for $x \in \Delta_2$, we have $e^x = p(x) + \xi^k/k!$, where $p(x)$ is a polynomial of degree $k-1$ and the error term $|\xi^k/k!| \leq (\log^2 n)^k/k!$. We can choose $k = O(\log^3 n)$ such that

$$\| e^x - p(x) \|_{\Delta_2} = O(n^{-\lambda_2 \log n})$$

for some constant $\lambda_2 > 0$. If we let

$$r_2(x) = \frac{1}{1 + p(-x)},$$

then

$$\| r_2(x) - \sigma(x) \|_{\Delta_2} = O(n^{-\lambda_2 \log n}).$$

Observe that when $x \geq \log^2 n$, $|\sigma(x) - 1| \leq 2(1 - \frac{1}{1+e^{\log^2 n}}) \leq 2e^{-\log^2 n}$. Similarly, when $x \leq -\log^2 n$, $|\sigma(x) - (-1)| \leq 2e^{-\log^2 n}$. Thus, the functions $r_1(x) = -1$ and $r_3(x) = 1$ are good (rational) approximations to $\sigma(x)$ on Δ_1 and Δ_3, respectively. In other words, for some constants $\lambda_1 > 0$ and $\lambda_3 > 0$,

$$\| r_1(x) - \sigma(x) \|_{\Delta_1} = O(n^{-\lambda_1 \log n}),$$

and

$$\| r_3(x) - \sigma(x) \|_{\Delta_3} = O(n^{-\lambda_3 \log n}).$$

It is easy to check that the derivative $|\sigma'(x)| \leq 2$ for all x. Hence, the modulus of continuity of the sigmoidal function $\omega(\sigma; \delta)_\Delta \leq 2\delta$. Now apply Lemma 7.13 twice, first obtaining a rational approximation to $\sigma(x)$ on $\tilde{\Delta} = \Delta_1 \cup \Delta_2$ and then on $\Delta = \tilde{\Delta} \cup \Delta_3$, and choose δ and $\tilde{\epsilon}$ in Lemma 7.13 to be both bounded by $O(n^{-\lambda \log n})$ for some constant $\lambda > 0$. We obtain a rational function $r(x)$ such that

$$\| r(x) - \sigma(x) \|_\Delta = O(n^{-\lambda \log n})$$

and $\deg r(x) = O(\log^4 n)$.   □

Suppose instead of the sigmoidal function, we consider a continuous function $f(x)$ on $[a, b]$ which can be piecewise approximated by low-degree (i.e., $\log^{O(1)} n$) rational functions. Using the same technique as in the proof of Theorem 7.11, it is not hard to show that a similar result holds for the function f, provided that the number of consecutive intervals for which the approximation holds is bounded by a constant. We would like to generalize the result to the case when the number of intervals over which the function is approximated can increase with the size of the inputs. In fact, we can extend the result of Lemma 7.13 and use this extended result to prove the following theorem.

Theorem 7.12:   Let f be a continuous function with $\| f \|_\Delta \leq n^{O(1)}$ and modulus of continuity $\omega(f; \delta) = O(n^\alpha \delta)$ on $\Delta = [-n^c, n^c]$ for some fixed constants $c > 0$ and $\alpha > 0$. Suppose there exist $m \leq \log^{O(1)} n$ subintervals $\Delta_{i+1} = [c_i, c_{i+1}]$, where $-n^c = c_0 < c_1 < \ldots < c_m = n^c$, $i = 0, \ldots, m - 1$, and rational functions r_i of degree $\leq \log^{O(1)} n$ such that

$$\| f - r_i \|_{\Delta_i} = O(n^{-\lambda \log n})$$

for some constant $\lambda > 0$. Then, there is a rational function $r(x)$ such that

$$\| f - r \|_\Delta = O(n^{-\tilde{\lambda} \log n}) \, ,$$

and

$$\deg r \leq \log^{O(1)} n$$

for some constant $\tilde{\lambda} > 0$.   □

We shall prove Theorem 7.12 using a key lemma. Before presenting a formal proof of the general result, it is helpful to give an informal discussion of the difficulty involved in applying the previous techniques. In order to apply Lemma 7.13, first we require that the modulus of continuity $\omega(f;\delta)$ to be small. Second, the upper bound on the degree of the resulting rational function is doubled (see Eqn. (7.3)) every time we apply the lemma to 'join together' two rational functions over adjacent subintervals. Thus, for the resulting upper bound on the degree to be small ($\leq \log^{O(1)} n$), the number of subintervals m must be bounded by $O(\log \log n)$. To relax the second requirement, we need to use a more subtle argument that would lead to a generalization of Lemma 7.13. In fact, the proof of Lemma 7.13 is based on the following result (whose proof can be found in [102]):

Lemma 7.14:   Let f be a continuous function over $[a, b]$. Let there exist compact subintervals Δ_1 and Δ_2 such that $\Delta = \Delta_1 \cup \Delta_2$ and $|\Delta_1 \cap \Delta_2| > 0$. Suppose there are rational functions r_1 and r_2 such that

$$\| f - r_i \|_{\Delta_i} \leq \epsilon,$$

and

$$\| r_i \|_{(-\infty,\infty)} \leq A$$

for $i = 1, 2$, where $\epsilon > 0$, $A > 0$. Then, for each $\tilde{\epsilon} > 0$ there is a rational function r such that

$$\| f - r \|_{\Delta} \leq \epsilon + \tilde{\epsilon},$$

$$\| r \|_{(-\infty,\infty)} \leq A,$$

and

$$\deg r \leq \ \deg r_1 + \ \deg r_2 + C \log(e + \frac{|\Delta|}{|\Delta_1 \cap \Delta_2|}) \log(e + \frac{A}{\tilde{\epsilon}}) \ ,$$

where $C > 1$ is a constant, $|\Delta|$ is the length of Δ. $\square$

The following result gives a generalization of Lemma 7.13 and is the key lemma for the proof of Theorem 7.12.

Lemma 7.15:   Let f be a continuous function over $\Delta = [a, b]$. Let there exist m subintervals $\Delta_{i+1} = [c_i, c_{i+1}]$, where $a = c_0 < c_1 < \ldots < c_m = b$, $i = 0, \ldots, m - 1$, and rational functions r_i such that

$$\| f - r_i \|_{\Delta_i} \leq \epsilon, \tag{7.4}$$

where $\epsilon > 0$. Then, for each $\tilde{\epsilon} > 0$, there is a rational function r such that

$$\| f - r \|_\Delta \le (m-1)(\epsilon + \tilde{\epsilon} + \omega(f; \delta)_\Delta),$$

where $0 < \delta < \min_i(c_{i+1} - c_i)$,

$$\deg r \le 2 \sum_{i=1}^{m} \deg r_i + C_1(m-1) \log(e + \frac{b-a}{\delta}) \log(e + \frac{\| f \|_\Delta}{\tilde{\epsilon}}), \quad (7.5)$$

and $C_1 > 0$ is a constant.

Proof: Our main tool is the result stated in Lemma 7.14; however, there are two difficulties in applying the result for our proof of Eqn. (7.5). First, Lemma 7.14 assumes that the subintervals Δ_i be overlapped by a small subinterval of length δ. Second, the lemma also assumes that the rational functions r_i all be bounded by a constant over the entire real line.

To overcome the first difficulty, define subintervals $\tilde{\Delta}_1 = [a, c_1 + \delta/2]$ and for $i = 1, \ldots, m-2$, $\tilde{\Delta}_{i+1} = [c_i - \delta/2, c_{i+1} + \delta/2]$, and $\tilde{\Delta}_m = [c_{m-1} - \delta/2, b]$. Let $l_1(x) = (c_1 - a)(x - a)/(c_1 + \delta/2 - a) + a$, $l_{m-1}(x) = (b - c_{m-1})(x - c_{m-1} + \delta/2)/(b - c_{m-1} + \delta/2) + c_{m-1}$, and for $i = 2, \ldots, m-2$, $l_i(x) = (c_{i+1} - c_i)(x - c_i + \delta/2)/(c_{i+1} - c_i + \delta) + c_i$. In other words, each l_i is a linear function that maps $\tilde{\Delta}_i$ onto Δ_i. Note that $\| x - l_i(x) \|_{\tilde{\Delta}_i} = \delta$ for every i.

Define $\tilde{r}_i(x) = r_i(l_i(x))$ for $i = 1, \ldots, m$. Then, from Eqn. (7.4) we have

$$\begin{aligned} \| f(x) - \tilde{r}_i(x) \|_{\tilde{\Delta}_i} &\le \| f(x) - f(l_i(x)) \|_{\tilde{\Delta}_i} \\ &\quad + \| f(l_i(x)) - \tilde{r}_i(x) \|_{\Delta_i} \\ &\le \omega(f; \delta) + \epsilon. \end{aligned} \quad (7.6)$$

Next we show how to overcome the second difficulty. Given $\tilde{\epsilon} > 0$, define for $i = 1, \ldots, m$,

$$\hat{r}_i = \frac{\tilde{r}_i}{1 + \zeta \tilde{r}_i^2},$$

where $\zeta = \frac{\tilde{\epsilon}}{\| f \|_\Delta^3} > 0$, $\| f \|_\Delta > 0$. (The case where $\| f \|_\Delta = 0$ is trivial.)

Clearly, $\deg \hat{r}_i \le 2 \deg \tilde{r}_i = 2 \deg r_i$. Moreover, it is easy to show from elementary calculus that

$$\| \hat{r}_i \|_{(-\infty, \infty)} \le \frac{1}{2\sqrt{\zeta}} = \frac{\| f \|_\Delta^{3/2}}{2\tilde{\epsilon}^{1/2}}. \quad (7.7)$$

Let us consider the approximation error of f by $\hat{r}_i$. There are two cases to be considered. If $\| \tilde{r}_i \|_{\tilde{\Delta}_i} \leq 2 \| f \|_{\tilde{\Delta}_i}$, then from Eqn. (7.6) we have

$$
\begin{aligned}
\| f - \hat{r}_i \|_{\tilde{\Delta}_i} &= \| \frac{f - \tilde{r}_i + f\zeta\tilde{r}_i^2}{1 + \zeta\tilde{r}_i^2} \|_{\tilde{\Delta}_i} \\
&\leq \| f(x) - \tilde{r}_i(x) \|_{\tilde{\Delta}_i} + \zeta \| f \|_\Delta \| \tilde{r}_i(x) \|_{\tilde{\Delta}_i}^2 \\
&\leq \omega(f;\delta) + \epsilon + 4\zeta \| f \|_\Delta^3 \\
&= \omega(f;\delta) + \epsilon + 4\tilde{\epsilon}.
\end{aligned}
$$

If $\| \tilde{r}_i \|_{\tilde{\Delta}_i} > 2 \| f \|_{\tilde{\Delta}_i}$, then from Eqn. (7.6) we have

$$
\| f \|_{\tilde{\Delta}_i} \leq \| \tilde{r}_i \|_{\tilde{\Delta}_i} - \| f \|_{\tilde{\Delta}_i} \leq \| f - \tilde{r}_i \|_{\tilde{\Delta}_i} \leq \epsilon + \omega(f;\delta).
$$

In other words, the rational function $\hat{r}_i = 0$ approximates f on $\tilde{\Delta}_i$ with error $\epsilon + \omega(f;\delta)$. Hence, for each $i = 1,\ldots,m$, we can find a rational function $\hat{r}_i$ such that $\hat{r}_i$ satisfies Eqn. (7.7) and

$$
\| f - \hat{r}_i \|_{\tilde{\Delta}_i} \leq \omega(f;\delta) + \epsilon + 4\tilde{\epsilon}. \tag{7.8}
$$

We can now apply Lemma 7.14 $(m-1)$ times and recall that the degree of each $\hat{r}_i$ is bounded by $2 \deg r_i$. Thus, we conclude that there is a rational function r such that

$$
\| f - r \|_\Delta \leq (m-1)(\omega(f;\delta) + \epsilon + 4\tilde{\epsilon})
$$

and

$$
\begin{aligned}
\deg r &\leq 2\sum_{i=1}^{m} \deg r_i + C\sum_{i=1}^{m-1} \log(e + \frac{|\Delta|}{|\Delta_i \cap \Delta_{i+1}|}) \log(e + \frac{\| f \|_\Delta^{3/2}}{2\tilde{\epsilon}^{3/2}}) \\
&\leq 2\sum_{i=1}^{m} \deg r_i + C_1(m-1) \log(e + \frac{b-a}{\delta}) \log(e + \frac{\| f \|_\Delta}{\tilde{\epsilon}}),
\end{aligned}
$$

where $C_1 > 1$ is a constant. $\square$

Theorem 7.12 follows as a consequence of Lemma 7.15.

Proof of Theorem 7.12:   In Lemma 7.15, choose $[a,b] = [-n^c, n^c]$ and let ϵ, $\tilde{\epsilon}$, and δ all be bounded by $O(n^{-\lambda\log n})$. Then, from the assumption of Theorem 7.12, $\omega(f;\delta) = O(n^{-\tilde{\lambda}\log n})$. Now, the conclusion of Theorem 7.12 follows from the fact that $\deg r_i = \log^{O(1)} n$ for each $i = 1,\ldots,m$, where $m = \log^{O(1)} n$. $\square$

Our objective is to determine whether the size of the network can be reduced by using γ elements that assume continuous values. As we shall see, the answer is negative when the γ elements can be piecewise approximated by low-degree rational functions and the Boolean function to be computed is highly oscillating, such as Parity.

One of the arguments in the proof of the lower bound result on threshold circuits is that if the Parity function of n variables $PAR_n(X)$ can be represented as a sign of a polynomial, then the polynomial must have degree n. We have a similar statement when the sign function is replaced by a sigmoidal function.

Lemma 7.16: Let $F(X)$ be a multilinear polynomial in X $=$ $(x_1, \ldots, x_n) \in \{1, -1\}^n$ such that for some fixed θ, the sigmoidal function $\sigma(F(X)) > \theta$ only if the Parity of X is 1, then the degree of $F(X)$ is n.

Proof: Observe that since $\sigma(x)$ is a strictly increasing function, its inverse σ^{-1} exists and is also strictly increasing. Thus, $\sigma((F(X)) > \theta$ if $F(X) > \sigma^{-1}(\theta)$. Then, by assumption, the multilinear polynomial $P(X) = F(X) - \sigma^{-1}(\theta) > 0$ whenever the Parity of X is 1. The conclusion follows from the fact that any such polynomial $P(X)$ must have degree n. $\square$

The result of Theorem 7.12 motivates the following definition:

Definition 7.6: The class $\tilde{R}$ consists of all functions $f(x)$ such that for every constant $c > 0$, $\| f \|_\Delta = n^{O(1)}$ on $\Delta = [-n^c, n^c]$. Moreover, there exists a partition of Δ into $m = \log^{O(1)} n$ subintervals $\Delta_{i+1} = [c_i, c_{i+1}]$, where $-n^c = c_0 < c_1 < \ldots < c_m = n^c$, $i = 0, \ldots, m - 1$, such that the modulus of continuity $\omega(f; \delta) = O(n^\alpha \delta)$ on each Δ_i for some constant $\alpha > 0$, and there exist rational functions r_i of degree $= \log^{O(1)} n$ such that

$$\| f - r_i \|_{\Delta_i} = O(n^{-\lambda \log n})$$

for some constant $\lambda > 0$. $\square$

Roughly speaking, $\tilde{R}$ is the class of polynomially bounded functions that are piecewise continuous and can be piecewise approximated by low-degree rational functions. Most functions that are commonly used to implement the elements in neural networks belong to this class of functions. Theorem 7.12 implies that every function in $\tilde{R}$ can be closely approximated by a single

low-degree rational function over a large domain. In particular, it includes
the common sigmoidal functions and radial basis functions. We shall generalize the result of Lemma 7.16 to every function in $\tilde{R}$. Note that the proof
of Lemma 7.16 only requires that the function be strictly monotone and not
necessarily be closely approximated by low-degree rational functions. On
the other hand, a function in $\tilde{R}$ is no longer guaranteed to be strictly monotone. A similar result still holds when the output value is computed with
(polynomially small) separation.

Lemma 7.17:   Let $g \in \tilde{R}$ and $R(X)$ be a rational function in $X = (x_1, \ldots, x_n) \in \{1, -1\}^n$ such that $|R(X)| = n^{O(1)}$ for all X. If for some
fixed θ, $g(R(X)) \geq \theta + \epsilon$ when the parity of X is 1, and $g(R(X)) \leq \theta - \epsilon$
otherwise, where $0 < \epsilon = \Omega(n^{-k})$ for some constant $k > 0$. Then, the degree
of $R(X)$ must be $\Omega(n/\log^{O(1)} n)$.

Proof: Let $c > 0$ be a constant such that $|R(X)| \leq n^c$. Since $g \in \tilde{R}$, by
Theorem 7.12 there is a rational function r of degree $\log^{O(1)} n$ such that
$|g(x) - r(x)| = O(n^{-\lambda \log n})$ for all $x \in [-n^c, n^c]$ and for some $\lambda > 0$.
Therefore, $|g(R(X)) - r(R(X))| = O(n^{-\lambda \log n})$ for all $X \in \{1, -1\}^n$.
Let $\delta = \sup_{X \in \{1,-1\}^n} |g(R(X)) - r(R(X))|$. For sufficiently large n, we
have $\delta < 2\epsilon$. Then, $r(R(X)) > \theta - \delta$ when the parity of X is 1, and
$r(R(X)) < \theta - \epsilon + \delta$ otherwise. If we let $\frac{p(X)}{q(X)} = r(R(X)) - \theta + \delta$, then the
Parity of X is 1 if and only if $\frac{p(X)}{q(X)} > 0$ or equivalently, $p(X)q(X) > 0$. Since
the degree of $p(X)q(X)$ is bounded by $O(\log^{O(1)} n \cdot \deg R(X))$, we must have
$\deg R(X) = \Omega(n/\log^{O(1)} n)$.   □

The preceding lemmas provide us with the tools for proving an almost
tight lower bound on the size of neural networks computing Parity. The class
of neural networks for which the lower bound results apply is quite large;
each element in the network can compute any function in the class $\tilde{R}$.

Theorem 7.13:   Let W be any depth-$(d+1)$ neural network in which each
element v_j computes a function $f^j(\sum_i w_i x_i)$, where $f^j \in \tilde{R}$ and $\sum_i |w_i| = n^{O(1)}$ for each element. If the network W computes the Parity function of n
variables with separation δ, where $0 < \delta = \Omega(n^{-k})$ for some $k > 0$, then for
any fixed $\epsilon > 0$, W must have size $\Omega(dn^{1/d-\epsilon})$.   □

We shall first prove two related lemmas (see Lemmas 7.18 and 7.19) and
then provide a proof for Theorem 7.13.

In particular, since the sigmoidal function and the radial basis function belong to the class $\tilde{R}$, the lower bound results of Theorem 7.13 apply to sigmoidal networks and radial basis networks.

Remark 7.6:   We proved in Section 5.2.1 that the n-variable parity function can be computed by a depth-$(d+1)$ threshold circuit of size $O(dn^{1/d})$. One can easily show that such threshold circuits can be simulated by sigmoidal networks of the same size. Thus, the lower bound in the preceding theorem is almost optimal. $\square$

We shall divide the proof of Theorem 7.13 into two lemmas. The arguments are similar to the proofs of the lower bounds on threshold circuits.

Lemma 7.18:   Let $g(X) = h(\sum_{i=1}^{t} w_i f_i(X))$, where f, g, and h are functions in $\tilde{R}$. Assume that $\sum_i |w_i| = n^{O(1)}$. We have the following claims:

1. If $|R_{f_i}(X) - f_i(X)| = O(n^{-\lambda_1 \log n})$ for all $X \in \{1, -1\}^n$, then $|g(X) - h(\sum_{i=1}^{t} w_i R_{f_i}(X))| = O(n^{-\lambda_2 \log n})$.

2. If $R_{f_i}(X)$ is a rational function of degree $\leq d$, for $i = 1, \ldots, t$, and satisfies the assumption in Claim 1, then there exists a rational function $R_g(X)$ of degree $O(td \log^{O(1)} n)$ such that $|R_g(X) - g(X)| = O(n^{-\lambda_2 \log n})$.

Proof:

1. One can rewrite $R_{f_i}(X) = f_i(X) + \epsilon_i$, where $|\epsilon_i| = O(n^{-\lambda_1 \log n})$, then

$$\sum_{i=1}^{t} w_i R_{f_i}(X) = \sum_{i=1}^{t} w_i(f_i(X) + \epsilon_i) = \sum_{i=1}^{t} w_i f_i(X) + \epsilon,$$

where $\epsilon = \sum_{i=1}^{t} w_i \epsilon_i$. Since $\sum_i |w_i| = n^{O(1)}$ by assumption, we have $|\epsilon| = O(\sum_{i=1}^{t} |w_i| \cdot n^{-\lambda_1 \log n}) = O(n^{-\hat{\lambda}_1 \log n})$. In other words,

$$|\sum_{i=1}^{t} w_i R_{f_i}(X) - \sum_{i=1}^{t} w_i f_i(X)| = O(n^{-\hat{\lambda}_1 \log n}).$$

Since $h \in \tilde{R}$, the modulus of continuity $\omega(h; \delta) = n^{O(1)}\delta = O(n^{-\lambda_2 \log n})$ when $\delta = O(n^{-\hat{\lambda}_1 \log n})$. It follows that

$$|h(\sum_{i=1}^{t} w_i R_{f_i}(X)) - h(\sum_{i=1}^{t} w_i f_i(X))| \leq \omega(h; \delta) = O(n^{-\lambda_2 \log n})$$

for all $X \in \{1, -1\}^n$.

2. Clearly, we can write the sum $\sum_{i=1}^{t} w_i R_{f_i}(X)$ as a rational function $\tilde{R}(X)$ of degree at most td. Since

$$|\sum_{i=1}^{t} w_i R_{f_i}(X)| \le |\sum_{i=1}^{t} w_i f_i(X)| + |\epsilon| = n^{O(1)},$$

we can apply Theorem 7.12 to $h(\tilde{R}(X))$ and obtain a rational function $R_g(X)$ of degree $O(td \log^{O(1)} n)$ such that $|R_g(X) - h(\tilde{R}(X))| = O(n^{-\hat{\lambda}_2 \log n})$. Then,

$$
\begin{aligned}
|R_g(X) - g(X)| \;\le\; & |h(\sum_{i=1}^{t} w_i R_{f_i}(X)) - g(X)| \\
& + |h(\sum_{i=1}^{t} w_i R_{f_i}(X)) - R_g(X)| \\
=\; & O(n^{-\lambda_3 \log n}).
\end{aligned}
$$
$\qquad\square$

Lemma 7.19:   Let $X \in \{1, -1\}^n$ and $F(X)$ be the output of a depth-$(d + 1)$ neural network in which each element computes a function in $\tilde{R}$ and the weights of each element are polynomially bounded (in n). Let t_i denote the number of elements in the ith layer and $h \in \tilde{R}$ be the function corresponding to the output element. Then, there is a rational function $R(X)$ of degree $O(t_1...t_d(\log^{O(1)} n)^d)$ such that $|R(X)| = n^{O(1)}$ for all X and $|h(R(X)) - F(X)| = O(n^{-\lambda \log n})|$ for some constant $\lambda > 0$.

Proof: We prove the lemma by induction on d. The case $d = 1$ has been proved in claim 1 of Lemma 7.18 by noting that each function in the first layer can be approximated by a rational function of degree $= \log^{O(1)} n$. For a general $d > 1$, let the output functions in the dth layer be $f_1(X)$, $f_2(X)$, ..., $f_{t_d}(X)$. Then, $F(X) = h(\sum_{i=1}^{t_d} w_i f_i(X))$. By the inductive hypothesis, for each $f_i(X)$, there is a rational function $R_{f_i}(X)$ of degree $O(t_1...t_{d-1}(\log^{O(1)} n)^{d-1})$ such that $|R_{f_i}(X) - f_i(X)| = O(n^{-\lambda_1 \log n})$ for all X. The conclusion follows from claim 2 of Lemma 7.18. $\qquad\square$

Proof of Theorem 7.13:   Using the same notation as in Lemma 7.19, the assumption of Theorem 7.13 implies that for some fixed θ, $F(X) \ge \theta + \delta$ when the Parity of X is 1, and $F(X) < \theta - \delta$ otherwise, where $0 < \delta = \Omega(n^{-k})$ for some $k > 0$. We can replace $F(X)$ by $h(R(X))$ as in Lemma 7.19 without changing the assertion in the previous statement, where the degree of $R(X)$ is bounded by $O(t_1...t_d(\log^{O(1)} n)^d)$. But the degree of $R(X)$ is $\Omega(n/\log^{O(1)} n)$ by Lemma 7.17. It follows that the network size $(1 + \sum_{i=1}^{d} t_i)$ is at least $\Omega(dn^{1/d}/\log^{O(1)} n) = \Omega(dn^{1/d-\epsilon})$ for any fixed $\epsilon > 0$. $\square$

Exercises

7.1: [*An alternative definition of $CQ(X)$*]
Using the $\{1, -1\}$ notation, show that

$$CQ(X) = \begin{cases} 1 & \text{if number of } -1\text{'s in } X \bmod 4 = 0 \text{ or } 1, \\ -1 & \text{otherwise.} \end{cases}$$

7.2: [*Stratified circuit*]
A circuit is said to be *stratified* if the gates at the kth layer receive inputs only from the outputs at the $(k-1)^{th}$ layer. Show that given any depth-d threshold circuit with node complexity $\Omega(n)$, one can always find an equivalent stratified circuit with an increase in node complexity and edge complexity by a factor of at most d.

7.3: [*Relating node complexity to edge complexity for depth-2 threshold circuits computing Parity*]
Let C_n be a depth-2 threshold circuit of minimal node complexity $N(n)$ computing the Parity of n inputs PAR_n. Let $E(n)$ denote the minimal edge complexity of any depth-2 threshold circuit computing PAR_n.

(a) Let x_i be any input variable. Let C' be a circuit derived from C_n by assigning the value 1 to x_i. Let C'' be a circuit derived from C_n by assigning the value 0 to x_i and negating the output threshold and the weights of all incoming edges to the output node. Show that C' and C'' compute the same function, which is the negation of the Parity of the rest $(n-1)$ inputs.

(b) Let $\text{sgn}(\sum w_i f_i' - t)$ and $\text{sgn}(\sum -w_i f_i'' + t)$ be the output nodes of C' and C'', respectively. Obtain a new circuit $\hat{C}$ by merging the two output nodes of C' and C'', i.e., the output node of $\hat{C}$ computes $\text{sgn}(\sum w_i f_i - w_i f_i'')$. Show that $\hat{C}$ computes the same function as C' and C''. Now argue that all the nodes in the first layer of $\hat{C}$ that are not connected to x_i can be deleted without changing the function computed by $\hat{C}$.

(c) Show that the node complexity of $\hat{C}$ is bounded above from $(2\deg(x_i)+1)$, where $\deg(x_i)$ is the number of edges incident on x_i.

(d) Show that $E(n) \geq [2N(n) + nN(n-1) - n - 2]/2$.

(e) Using the lower bound on the node complexity (Theorem 7.2), conclude that any depth-2 threshold circuit (with polynomially bounded integer weights) computing PAR_n must have $\Omega(n^2/\log^2 n)$ edges.

7.4:   [*A good rational approximation of $|x|$, Newman's Theorem*]

(a) Given $0 < \epsilon < 1$ and $n \geq 2$, define $p(x) = \prod_{k=1}^{n-1}(x + \xi^k)$, where $\xi = \epsilon^{1/n}$.

Let $x \in [\epsilon, 1]$ and let $x \in [\xi^{i+1}, \xi^i]$, $0 \leq i \leq (n-1)$. Then, using the fact that $(a-x)/(a+x)$ is monotone decreasing for $a > 0$, show that

$$\left|\frac{p(-x)}{p(x)}\right| = \prod_{k=1}^{i}\frac{\xi^k - x}{\xi^k + x} \prod_{k=i+1}^{n-1}\frac{x - \xi^k}{x + \xi^k} \leq \prod_{k=1}^{n-1}\frac{1 - \xi^k}{1 + \xi^k}.$$

Using the inequality $(1-t)/(1+t) \leq e^{-2t}$, $\forall t \geq 0$, show that for $x \in [\epsilon, 1]$

$$\left|\frac{p(-x)}{p(x)}\right| \leq \exp\{-\frac{2(\epsilon^{1/n} - \epsilon)}{1 - \epsilon^{1/n}}\}.$$

(b) If $\epsilon = e^{-\sqrt{n}}$, and $n \geq 5$, then show that for all $x \in [e^{-\sqrt{n}}, 1]$

$$\left|\frac{p(-x)}{p(x)}\right| \leq e^{-\sqrt{n}}. \tag{7.9}$$

(c) Let $R(x) = x\dfrac{p(x) - p(-x)}{p(x) + p(-x)}$, where $n \geq 5$, $\epsilon = e^{-\sqrt{n}}$, and $\xi = \epsilon^{1/n} = e^{-1/\sqrt{n}}$.

Show that if $x \in [0, e^{\sqrt{n}}]$, then

$$|x - R(x)| = x - R(x) \leq e^{-\sqrt{n}}. \tag{7.10}$$

Using Eqn. (7.9) show that for all $x \in [e^{-\sqrt{n}}, 1]$

$$|x - R(x)| \leq \frac{2|p(-x)/p(x)|}{1 - |p(-x)/p(x)|} < 3e^{-\sqrt{n}}. \tag{7.11}$$

(d) Since $|x|$ is an *even* function, conclude from Eqns. (7.10) and (7.11) that for all $x \in [-1, 1]$, and $n \geq 5$ $||x| - R(x)| < 3e^{-\sqrt{n}}$.

7.5: [*A good rational approximation of* $\text{sgn}(x)$]

(a) Given a rational function $R(x)$ of degree-d approximating $|x|$ with an error ϵ, i.e., $||x| - R(x)| \leq \epsilon$ for $x \in [-1, 1]$, determine a rational function $R'(x)$ of degree at most $d+1$ such that $|\text{sgn}(x) - R'(x)| \leq \epsilon/\delta$ for $|x| \in [\delta, 1]$.

(b) Using the preceding result and the results stated in Exercise 7.4, construct a rational function $R'(x)$, such that for any constant $c > 0$ $\deg(R'(x)) = 2c \ln^2 n + 1 = O(\log^2 n)$, and $|\text{sgn}(x) - R'(x)| \leq n^{-c}$, for all $|x| \in [n^{-c}, 1]$, and $n \geq 5$.

7.6: [*Critical assignment of a linear threshold function*]

Let $f(X) = \text{sgn}(w_0 + \sum_{i=1}^{n} w_i x_i)$ be any linear threshold function, where $x_i \in \{1, -1\}$. Show that there exists a subset S of the variables $\{x_1, \ldots, x_n\}$ with $|S| \leq \lceil n/2 \rceil$ such that there is a value assignment to the variables in S that fixes the value of f.

7.7: [*Polynomials approximating Parity*]

Complete the proof of Theorem 7.5 by giving explicit constructions of polynomials of degree k, which approximate Parity with errors on at most

$$N_k = \sum_{i=0}^{(n-k-1)/2} \binom{n}{i} \text{ inputs.}$$

7.8: [*A generalization of Theorem 7.8*]

Show that the statement of Theorem 7.8 remains true even if polynomially many additional gates computing functions equivalent to decision trees of polylogarithmic depths are included in the first layer of the depth-2 circuit.

7.9: [*Depth-2 sigmoidal networks computing XOR-of-Majority*]

Let $F(X, Y) = XOR(MAJ(x_1, \ldots, x_n), MAJ(y_1, \ldots, y_n))$, where $X = (x_1, \ldots, x_n)$, $Y = (y_1, \ldots, y_n) \in \{0, 1\}^n$. Let $\gamma : \mathbf{R} \to \mathbf{R}$ such that $\gamma(x)$ is differentiable on some open interval containing $x = 0$, and $\gamma''(0)$ exists and is greater than zero.

(a) Consider $\theta(x) = \gamma(x) + \gamma(-x)$. Show that $\theta(x)$ is an even function, and there exist some $\epsilon > 0$ and $c > 0$ such that $\theta(a+h) - \theta(a) \geq ch^2$ for all $a, h \in [0, \epsilon]$.

(b) Show that for any $a, b \in [-\epsilon, \epsilon]$, $\theta(a) > \theta(b)$ if and only if $|a| > |b|$. Conclude that any two non-zero real numbers $u, v \in [-\epsilon/2, \epsilon/2]$ have opposite sign if and only if $\theta(u-v) - \theta(u+v) > 0$.

(c) Using the above results, construct a depth-2 γ network of 5 elements that computes $F(X,Y)$ with separation $\Omega(1/n^2)$.

(d) Derive an upper bound on the size of a depth-2 threshold circuit that computes $F(X,Y)$.

Bibliographic Notes

The techniques of rational approximation are used in Paturi and Saks [100] to derive an $\Omega(n/\log^2 n)$ almost tight lower bound on the size of depth-2 threshold circuit computing Parity. Siu, Roychowdhury, and Kailath [126] generalize the techniques to higher depth and provide both $\Omega(dn^{1/d}/\log^2 n)$ almost-tight lower bound on the size of depth-$(d+1)$ threshold circuits and a $O(dn^{1/d})$-size construction for Parity. Prior to the results in [126], all known lower bound results on the size only applied to depth-2 threshold circuits. The rational approximation is the first-known technique that can be applied to threshold circuits with depth more than 2. While the techniques employed in [100, 126] only seem to apply to threshold circuits with polynomially bounded integer weights, probabilistic methods are used in Impagliazzo, Paturi, and Saks [62] to show that any constant-depth threshold circuit with arbitrary weights computing the n-variable Parity function must have $n^{\Omega(1)}$ gates.

It is shown in [126] that a depth-2 polynomial-size threshold circuit with additional gates in $\widetilde{SP}$ in the first layer still requires $\Omega(n/\log^2 n)$ threshold gates to compute the CQ function. Beigel, Reingold, and Speilman [14], and Beigel [13] refine the techniques of rational approximation, and it is shown in [13] that any constant-depth circuit with subexponential AND/OR gates and $n^{o(1)}$ threshold gates cannot compute Parity (Theorem 7.10). Our proof follows closely the presentation in [13].

The techniques in proving Lemma 7.6 and Theorem 7.5 are due to Aspnes, Beigel, Furst, and Rudich [8]. In fact, Theorem 7.5 is a special case of a more general result in [8].

A stronger version of Theorem 7.9 can be derived using a deeper algebraic technique developed by Szegedy [135]. It is shown in [135] that if the error of approximation is $\leq 1/2 - \epsilon$ for some fixed $\epsilon > 0$, then the approximating polynomial to the AND function or the OR function must have degree $\Omega(\sqrt{n})$. Paturi [99] extends this result to more general symmetric functions.

For lower bound results on networks with continuous-valued elements, it is shown in Maass, Schnitger, and Sontag [75] that when the depth of the network is restricted to be 2, then there is a Boolean function of n variables that can be computed by a depth-2 sigmoidal network with a fixed number of elements, but requires a depth-2 threshold circuit with size that increases at least logarithmic in n. The lower bound techniques in [75] are refinements of the correlation techniques first developed in [48]. These results are improved in DasGupta and Schnitger [34], which shows that there exists an n-variable Boolean function computable by a fixed number of sigmoidal elements, but requires $\Omega(\log n)$ linear threshold elements, even if unrestricted depth circuits are allowed. The results in [34] imply that for certain functions, one can reduce the size of the network by at least a logarithmic factor by using continuous elements such as the sigmoidal elements instead of threshold elements with binary output values.

The first-known almost-optimal lower bound results for networks of continuous-valued elements with depth more than 2 can be found in Siu, Roychowdhury, Kailath [128, 129]. The proofs of Lemmas 7.13 and 7.14 upon which the key lemma (Lemma 7.15) is based, are from Petrushev and Popov [102].

Chapter 8

Geometric Framework and Spectral Analysis

8.1   Introduction

We introduce a number of concepts and techniques for relating the output function of a threshold gate with its set of input functions. A diverse set of tools from disciplines such as linear programming, linear algebra, and harmonic analysis are introduced and their applications to threshold circuits are demonstrated.

In Sections 8.2–8.4, we introduce a geometric approach for investigating the power of threshold circuits. Viewing n-variable Boolean functions as vectors in $\mathcal{R}^{2^n}$, we invoke tools from linear algebra to derive results on the realizability of Boolean functions using threshold gates. Geometric concepts such as correlation and generalized spectrum of Boolean functions are introduced. The geometric framework and the related results provide the set of concepts and tools necessary for the rest of this chapter.

Section 8.5 introduces the spectral/polynomial representations of Boolean functions. In Chapters 4–7, we have already made key use of the polynomial representation of Boolean functions. For example, in Chapter 4, the L_1 spectral norms of Comparison and Addition are used to show that these functions can be computed by depth-2 polynomial-size threshold circuits with polynomially bounded integer weights. Similarly, the concept of strong degree (SD(f)) associated with a Boolean function f is used in

Chapter 7 to derive almost tight lower bounds on the number of threshold elements required for computing Parity. In Section 8.5, we formally study the spectral representation of Boolean functions using the set of basic results derived in Sections 8.2–8.4.

In Section 8.7, we further investigate methods for deriving lower bounds on the number of input functions to a threshold gate using the concept of correlation of Boolean functions. We refer to such analytical techniques as the method of correlations. One of the applications of the method of correlations shows that the Inner Product Mod 2 function is not in $\widehat{LT_2}$. This leads to an important separation result in threshold circuit complexity: $\widehat{LT_2} \subsetneq \widehat{LT_3}$. We also investigate the limitations of the method of correlations and in the process develop constructive procedures showing that every n-variable Boolean function f is a threshold function of polynomially many input functions, none of which is significantly correlated with f. Such procedures for decomposing a given function f are referred to as threshold-decomposition procedures. The approach used in this section integrates the geometric framework of Section 8.2 with some basic tools used in linear programming.

8.2   Geometric Concepts and Definitions

In this chapter, we use the $\{1, -1\}$ notation for Boolean variables and consider an n-variable Boolean function as a mapping $f : \{1, -1\}^n \to \{1, -1\}$. Unless otherwise mentioned, all Boolean functions considered in this chapter are functions of n input variables. We view f as a (column) vector in $\mathcal{R}^{2^n}$. Each of f's 2^n components is either -1 or $+1$ and represents $f(x)$ for a distinct value assignment $x \in \{1, -1\}^n$ of the n Boolean variables. As introduced in Chapter 2, we view the S weights of an S input threshold gate as a weight vector $\mathbf{w} = [w_1, \ldots, w_S]^T$ in $\mathcal{R}^S$.

Let the functions $f_1, \ldots, f_S$ be the inputs of a threshold gate with weight vector $\mathbf{w}$. The gate computes a function f (or f is the output of the gate) if the following vector equation holds:

$$f = \text{sgn}\left(\sum_{i=1}^{S} f_i w_i\right). \tag{8.1}$$

We assume without loss of generality that all coordinates of $\sum_{i=1}^{S} f_i w_i$ be non-zero; see Chapter 2 for justification. Recall that in the context of $\{1, -1\}$ notation, $\text{sgn}(x) = 1$ if $x > 0$, $\text{sgn}(x) = -1$ if $x < 0$. It is convenient to write Eqn. (8.1) in a matrix form:

$$f = \text{sgn}(Y\mathbf{w}),$$

where the input matrix

$$Y = [f_1 \cdots f_S]$$

is a $2^n \times S$ matrix whose columns are the input functions. We assume that Y has full column rank (see Exercise 2.3). The function f is a *threshold function* of $f_1, \ldots, f_S$ if there exists a threshold gate with weights $\mathbf{w} = [w_1, \ldots, w_s]^T$ and with inputs $f_1, \ldots, f_S$ that computes f.

Geometrically, each function f, being a ± 1 vector in $\mathcal{R}^{2^n}$, determines an *orthant* in $\mathcal{R}^{2^n}$ — i.e., the set of vectors whose non-zero coordinates agree in sign with the corresponding coordinates of f. The orthant's *interior* consists of all vectors in the orthant with non-zero coordinates.

In this interpretation, f is the output of a threshold gate whose input functions are $f_1, \ldots, f_S$ if and only if the linear combination $Y\mathbf{w} = \sum_{i=1}^{S} f_i w_i$ defined by the gate lies in the interior of f's orthant. This view forms the basis of many of the results in this chapter. Combined with some basic observations, it enables us to prove various results on threshold circuits. This is demonstrated by Lemma 8.1.

If two vectors lie in the same orthant, their inner product is non-negative. If one of the vectors is in the orthant's interior, and the other is non-zero, then the inner product is positive. We therefore have the following result:

Lemma 8.1: If f is orthogonal to a set of functions $f_1, \ldots, f_S$ (i.e., $C_{fY} = 0$), then f is not a threshold function of $f_1, \ldots, f_S$.

Proof: If f is orthogonal to all of $f_1, \ldots, f_S$, then f is orthogonal to any linear combination $Y\mathbf{w}$ of these functions. By the observations preceding the lemma, $Y\mathbf{w}$ cannot be in f's orthant, and hence, f cannot be a threshold function of $f_1, \ldots, f_S$. $\square$

We next introduce the key notion of correlation.

Definition 8.1: The *correlation* of two n-variable Boolean functions f_1 and f_2 is

$$C_{f_1 f_2} = (f_1^T f_2)/2^n.$$

The two functions are *uncorrelated or orthogonal* if $C_{f_1 f_2} = 0$. $\qquad\square$

Note that $C_{f_1 f_2} = 1 - 2d_H(f_1, f_2)/2^n$, where $d_H(f_1, f_2)$ is the Hamming distance between f_1 and f_2; thus, the correlation can be interpreted as a measure of how 'close' the two functions are.

Fix the input functions $f_1, \ldots, f_S$ to a threshold gate. The *correlation vector* of a function f with the input functions is defined as

$$C_{fY} = (Y^T f)/2^n = [C_{ff_1}\ C_{ff_2}\ \cdots\ C_{ff_S}]^T.$$

We define $\hat{C}$ as the maximum in magnitude among the correlation coefficients, i.e.,

$$\hat{C} = \max\left\{|C_{ff_i}| \ : \ 1 \leq i \leq S\right\}.$$

8.3 Uniqueness

The correlation between two n-variable Boolean functions is a multiple of $2^{-(n-1)}$, bounded between -1 and 1; hence, it can assume $2^n + 1$ values. Given $Y = [f_1, \ldots, f_S]$, the correlation vector $C_{fY} = [C_{ff_1}, \ldots, C_{ff_S}]^T$ can therefore assume at most $(2^n + 1)^S$ different values over all Boolean functions. There are 2^{2^n} n-variable Boolean functions; hence, many share the same correlation vector. However, the next theorem shows that a threshold function of $f_1, \ldots, f_S$ does not share its correlation vector with any other function.

Theorem 8.1: **[Uniqueness Theorem]** Let f be a threshold function of $f_1, \ldots, f_S$. Then, for all Boolean functions $g \neq f$,

$$C_{gY} \neq C_{fY},$$

where $Y = [f_1 \cdots f_S]$.

Proof: Let $(v)_i$ denote the ith entry of a vector v. For any Boolean function g and for all $i \in \{1, \ldots, 2^n\}$,

$$(f - g)_i = \begin{cases} 0 & \text{if } (f)_i = (g)_i, \\ 2(f)_i & \text{if } (f)_i \neq (g)_i. \end{cases}$$

By assumption, there is a weight vector $\mathbf{w}$ such that $f = \mathrm{sgn}(Y\mathbf{w})$. Hence, whenever $(f)_i \neq (g)_i$,

$$\mathrm{sgn}((f-g)_i) = \mathrm{sgn}((f)_i) = \mathrm{sgn}((Y\mathbf{w})_i).$$

Moreover, if $f \neq g$, then there exists an index i such that $(f)_i \neq (g)_i$, and by definition $(Y\mathbf{w})_i \neq 0$ for all i. Hence, $(f-g)^T Y\mathbf{w} > 0$, which implies $f^T Y \neq g^T Y$.   □

The proof has a simple geometric interpretation. If f is a threshold function of $f_1, \ldots, f_S$, then some linear combination $Y\mathbf{w}$ of these functions lies in the interior of the orthant in $\mathcal{R}^{2^n}$ determined by f. But for any $g \neq f$, the non-zero vector $(f-g)$ lies in the same orthant. Hence, $(f-g)^T Y\mathbf{w} > 0$, which implies that $f^T Y \neq g^T Y$.

In Section 8.5, we apply Theorem 8.1 to obtain results on the uniqueness of spectral coefficients. As an immediate consequence of Theorem 8.1, we derive an upper bound on the number of threshold functions of any set of input functions. A similar bound can be derived using the upper bound results presented in Chapter 2 (see Exercise 2.9); however, Theorem 8.1 provides a much simpler derivation.

Corollary 8.1:   There are at most $(2^n + 1)^S$ threshold functions of any set of S input functions.

Proof: Given any set of S input functions, there are at most $(2^n + 1)^S$ different correlation vectors.   □

Remark 8.1:   Theorem 8.1 holds even if the entries of Y (and hence, of f_i) are real valued and not restricted to ± 1. It also holds if the domains of the output and the input functions are restricted, i.e., f and f_i can be m-dimensional vectors, where m is not necessarily a power of 2. Of course, the correlations are then defined over the restricted domains.   □

The following theorem shows that the converse of the Uniqueness Theorem (Theorem 8.1) is not true in general.

Theorem 8.2:   Fix a set $f_1, \ldots, f_S$ of input functions, and let f be a Boolean function satisfying $C_{fY} \neq C_{gY}$ for all Boolean functions $g \neq f$. Then, f is not necessarily a threshold function of $f_1, \ldots, f_S$.

Proof: A proof can be given by constructing an example. Let f be the Parity function $PAR_n(x_1, \ldots, x_n)$ of all the input variables, and let the input functions be the Parity functions of all subsets of the variables x_i's except f. Thus, $S = 2^n - 1$. We shall show in Section 8.5 that f is orthogonal to all the input functions, i.e., $Y^T f = 0$. Hence, it follows from Lemma 8.1 that f cannot be a threshold function of the input functions. However, it is easy to verify that $g^T Y \neq 0$ (see Exercise 8.9) for every Boolean function $g \neq f$. Thus, even though C_{fY} is unique, f cannot be computed as a threshold function of the functions in Y. $\square$

8.4   Generalized Spectrum

As in Section 8.2, we let $Y = [f_1 \; f_2 \; \cdots \; f_S]$ be the $2^n \times S$ input matrix whose columns are the input functions to a threshold gate and has full column rank. A basic linear algebraic result guarantees that any function f can be expressed as

$$f = Y\beta + Z, \tag{8.2}$$

where $Z^T Y = 0$ and $\beta = [\beta_1 \; \beta_2 \; \cdots \; \beta_S]^T$ is an S-dimensional column vector. Then, β can be computed as follows:

$$\beta = (Y^T Y)^{-1} Y^T f = 2^n (Y^T Y)^{-1} C_{fY}, \tag{8.3}$$

where $C_{fY} = [C_{ff_1} \; C_{ff_2} \; \cdots \; C_{ff_S}]^T$ is the correlation vector defined in Section 8.2.

Geometrically, $Y\beta$ represents the orthogonal projection of the vector f onto the subspace spanned by the input vectors f_i, and Z is the error term that is orthogonal to that subspace. For an interpretation of the β vector, consider the case where the input functions form an orthogonal basis for $\mathcal{R}^{2^n}$. Such a basis is formed, for example, by the 2^n Parity functions of subsets of the n input variables (this special case is discussed in Section 8.5). Then, $f_i^T f_j = 0$ for $i \neq j$ and $f_i^T f_i = 2^n$. Hence, $2^n (Y^T Y)^{-1} = I_{S \times S}$, and from Eqn. (8.3) it follows that $\beta = C_{fY}$, or β_i's are the corresponding spectral coefficients. In general, if the input functions f_i are mutually orthogonal, then there exists an orthogonal basis $\mathcal{B} = \{f_1, f_2, \ldots, f_S, \tilde{f}_{S+1}, \ldots, \tilde{f}_{2^n}\}$, where $\tilde{f}_k \in \mathcal{R}^{2^n}$ (not necessarily restricted to have ± 1 entries). Since f_i's are mutually orthogonal, we again get $\beta = C_{fY}$, or β_i's are some of the spectral coefficients corresponding to the orthogonal basis $\mathcal{B}$. With this motivation, we define β as the *generalized spectrum* of an output function f with respect to $f_1, \ldots, f_S$, where f_i's are not required to be mutually orthogonal.

The next theorem uses the generalized spectral coefficients to derive a lower bound on the number of input functions required by a threshold gate computing a function.

Theorem 8.3: **[Spectral Bound Theorem]** If f is a linear threshold function of $f_1, \ldots, f_S$, then

$$\sum_{i=1}^{S} |\beta_i| \geq 1.$$

Hence,

$$S \geq 1/\hat{\beta},$$

where $\hat{\beta} = \max\{|\beta_i| : 1 \leq i \leq S\}$.

Proof: Consider the decomposition $f = Y\beta + Z$ as described in Eqn. (8.2). Suppose $\sum_{i=1}^{S} |\beta_i| < 1$. Then, $(Y\beta)_i < 1$ for all i. Thus, $\text{sgn}(Z) = \text{sgn}(f - Y\beta) = f$. However, by assumption, $\text{sgn}(Y\mathbf{w}) = f$ for some $\mathbf{w}$, which implies that $Z^T Y \mathbf{w} > 0$. This is a contradiction since $Z^T Y = 0$. $\square$

Geometrically, if f is a threshold function of $f_1, \ldots, f_S$, then some linear combination $Y\mathbf{w}$ of these functions lies in the interior of the orthant in $\mathcal{R}^{2^n}$ determined by f. If $(Y\beta)_i < 1$ for all i, then the vector Z lies in the interior of f's orthant as well. However, this leads to a contradiction as, on the one hand, Z is orthogonal to $f_1, \ldots, f_S$, while, on the other hand, it lies in the same orthant as the linear combination $Y\mathbf{w}$ of $f_1, \ldots, f_S$.

If the input functions are mutually orthogonal, i.e., $C_{f_i f_j} = 0$, $i \neq j$, then $\beta_i = C_{f f_i}$. Hence, $\hat{\beta} = \max\{|C_{f f_i}| : 1 \leq i \leq S\} = \hat{C}$, and we can state the next corollary.

Corollary 8.2: If f is a threshold function of $f_1, \ldots, f_S$, and if the f_i's are mutually orthogonal, then

$$S \geq \frac{1}{\max\{|C_{f f_i}| : 1 \leq i \leq S\}}. \qquad \square$$

Thus, if the correlation of f with every input function is exponentially small and the input functions are mutually orthogonal, then exponentially many input functions are required for computing f. In Section 8.5, we shall present applications of Theorem 8.3 for the special case where the input functions are restricted to be Parity functions of subsets of the input variables.

Next we prove a variation of Corollary 8.2 for the case where the input functions are asymptotically orthogonal. For every integer n, let f be a desired output function, let $\{f_1, \ldots, f_{S_n}\}$ (all Boolean functions of n Boolean variables — the index n is omitted from the function notation) be a set of input functions, and let

$$\hat{v}_n = \max\{|C_{f_i f_j}| : 1 \le i < j \le S_n\}.$$

We shall define the set of input functions $\{f_1, \ldots, f_{S_n}\}$ to be *strongly* asymptotically orthogonal if there exists an ϵ, $0 \le \epsilon < 1$, such that for sufficiently large n,

$$\hat{v}_n < \epsilon/S_n.$$

That is, the input functions are asymptotically mutually orthogonal, and their mutual correlation is bounded above by ϵ/S_n.

Define

$$\hat{C}_n \stackrel{\text{def}}{=} \max\{|C_{ff_i}| : 1 \le i \le S_n\}.$$

Theorem 8.4:   If for every n, f is a linear threshold function of the strongly asymptotically orthogonal functions $f_1, \ldots, f_{S_n}$, then

$$S_n = \Omega(\frac{1}{\hat{C}_n}).$$

Proof: One can represent $(Y^T Y)/2^n$ as follows:

$$\frac{1}{2^n} Y^T Y = I_{S_n} - E_{S_n},$$

where I_{S_n} is an $S_n \times S_n$ identity matrix and E_{S_n} is an $S_n \times S_n$ matrix, such that $(E_{S_n})_{ii} = 0$ and $\max |(E_{S_n})_{ij}| \le \hat{v}_n$. Since $\{f_1, \ldots, f_{S_n}\}$ is strongly asymptotically orthogonal, we have for sufficiently large n, $\max |(E_{S_n})_{ij}| \le \hat{v}_n \le \epsilon/S_n$. Since $0 \le \epsilon < 1$, one can now show that for sufficiently large n:

1. $\max |(E_{S_n}^k)_{ij}| \le \epsilon^k/S_n$, and

2. $(I_{S_n} - E_{S_n})^{-1} = \sum_{k=0}^{\infty} E_{S_n}^k = (I_{S_n} - F_{S_n})$, where

$$\max |(F_{S_n})_{ij}| \leq \frac{1}{S_n} \times \frac{\epsilon}{1 - \epsilon}. \tag{8.4}$$

Now from Eqn. (8.3) we get

$$\beta = (\frac{1}{2^n} Y^T Y)^{-1} C_{fY} = (I_{S_n} - F_{S_n}) C_{fY}.$$

If $\hat{\beta}_n = \max \{|\beta_i| : 1 \leq i \leq S_n\}$, then using the upper bound on $\max |(F_{S_n})_{ij}|$ in Eqn. (8.4), it follows that

$$\hat{\beta}_n \leq \hat{C}_n (1 + \frac{\epsilon}{S_n (1 - \epsilon)}).$$

The result of the theorem follows from the fact that $\hat{\beta}_n \geq 1/S_n$ (Theorem 8.3).   □

8.4.1 Characterizations with Generalized L_1 Spectral Norms

In Chapter 4, we show that if the spectral norm of a function is polynomially bounded, then it can be expressed as a threshold function of polynomially many Parity functions (or equivalently, monomials). We present here a generalization of these results in terms of the generalized spectral coefficients.

Let $g_i(X) : \{1, -1\}^n \rightarrow \{1, -1\}$ for $i = 1, \ldots, 2^n$ be a basis (not necessarily orthogonal) in R^{2^n}. Then, every n-variable Boolean function $f(X)$ can be expressed uniquely as

$$f(X) = \sum_{i=1}^{2^n} \beta_i g_i(X),$$

where β_i are the generalized spectral coefficients. Exercise 8.15 shows that a nonorthogonal basis always exists.

Definition 8.2: The generalized L_1 spectral norm of f with respect to the basis $\{g_i : i = 1, \ldots, 2^n\}$ is defined as $\|f\|_{\mathcal{F}}^g = \sum_{i=1}^{2^n} |\beta_i|$.   □

The following theorem is an immediate generalization of the result in Chapter 4. Since the proof is very similar to that presented for Lemma 4.1, it is omitted here.

Theorem 8.5:   If the generalized L_1 spectral norm of $f(X)$ is bounded by a polynomial, i.e., $\|f\|_{\mathcal{F}}^g \leq n^c$ for some $c > 0$, then for any $k > 0$, there exists an $S \subset \{1, \ldots, 2^n\}$ such that $|S| \leq n^p$ (i.e., S has only polynomially many elements in it) and

$$|f(X) - \frac{1}{N} \sum_{i \in S} w_{g_i} g_i(X)| \leq n^{-k},$$

where N and w_{g_i} are polynomially bounded integers. As a consequence, $f(X) = \text{sgn}(\sum_{i \in S} w_{g_i} g_i(X))$. $\square$

8.5 Harmonic Analysis and Spectral/Polynomial Representation of Boolean Functions

The spectral or polynomial representation of an n-variable Boolean function is obtained by expressing it in terms of an orthogonal basis formed by the 2^n Parity functions of subsets of the n input variables.

Let $x_i \in \{1, -1\}$ represent the input variables. Recall that the Parity function of n variables, $x_1, x_2, \ldots, x_n$, equals -1 only if an odd number of input variables equals -1. Hence, the Parity function of all the input variables equals the product $x_1 x_2 \cdots x_n$. In general, any Parity function of a subset of the n input variables is given by a product of the form

$$x_1^{\alpha_1} x_2^{\alpha_2} \cdots x_n^{\alpha_n},$$

where $\alpha_i \in \{0, 1\}$. For example, if $n = 5$, then the Parity function of the inputs x_1, x_3 and x_5 is given by the product $x_1 x_3 x_5$; hence, $\alpha_1 = \alpha_3 = \alpha_5 = 1$ and $\alpha_2 = \alpha_4 = 0$.

For notational convenience, the product $\prod_{i=1}^{n} x_i^{\alpha_i}$ will be represented as X^α, where $\alpha = (\alpha_1, \alpha_2, \ldots, \alpha_n) \in \{0, 1\}^n$. Each such X^α is referred to as a *monomial*. Thus, the monomials $X^{(0,0,\ldots,0)}$, $X^{(1,0,\ldots,0)}$ and $X^{(1,1,0,\ldots,0)}$ equal the constant function 1, the input variable x_1, and the Parity function of x_1 and x_2, respectively.

The results in this section will be derived from the results in Sections 8.2-8.4, and it is convenient to have a vector representation for each monomial. For every $\alpha \in \{0,1\}^n$, let $\phi_\alpha \in \mathcal{R}^{2^n}$ be the vector representation of the Parity function X^α.

The following two properties establish that Parity functions are mutually orthogonal:

1. $\displaystyle\sum_{X \in \{1,-1\}^n} X^\alpha = \phi_{(0,\dots,0)}^T \phi_\alpha = 0$ for all $\alpha \neq (0,\dots,0)$. This follows from the observation that every Parity function equals -1 and 1 for an equal number of inputs.

2. $\displaystyle\sum_{X \in \{1,-1\}^n} X^{\alpha_1} X^{\alpha_2} = \phi_{\alpha_1}^T \phi_{\alpha_2} = 0$ for all $\alpha_1 \neq \alpha_2$. This follows easily from the property 1 by observing that $x_i^2 = 1$. In particular, $X^{\alpha_1} X^{\alpha_2} = X^{\alpha'}$, where $\alpha' \in \{0,1\}^n$ and has a 1 entry only if exactly one of the corresponding entries of α_1 and α_2 is 1.

Thus, the set of vectors $\{\phi_\alpha : \alpha \in \{0,1\}^n\}$ forms an orthogonal basis in $\mathcal{R}^{2^n}$. If $f \in \mathcal{R}^{2^n}$ represents the vector form of a given n-variable Boolean function $f(X)$, then f can be uniquely represented as

$$f = \sum_{\alpha \in \{0,1\}^n} a_\alpha \phi_\alpha. \tag{8.5}$$

Equivalently, any Boolean function $f(X)$ admits a unique representation in terms of the monomials:

$$f(X) = \sum_{\alpha \in \{0,1\}^n} a_\alpha X^\alpha. \tag{8.6}$$

Eqn. (8.6) defines the *polynomial* representation of any Boolean function $f(X)$, and a_α's are referred to as *spectral* coefficients.

The following discussion explains why a_α's are referred to as Fourier or spectral coefficients. If $f(k_1,\dots,k_n)$ is an n-variable function from $GF(p)^n \to GF(p)$ (i.e., $k_i \in \{0,1,\dots,p-1\}$ and $f(k_1,\dots,k_n) \in \{0,1,\dots,p-1\}$), then the n-dimensional Fourier transform of f is given as:

$$\mathcal{F}(\omega_1,\dots,\omega_n) = \frac{1}{p^n} \sum_{K \in P^n} \exp^{-\frac{j2\pi}{p} \sum_{i=1}^n \omega_i k_i} f(K),$$

where $\omega_i \in \{0, 1, \ldots, p - 1\}$, $K = \{k_1, \ldots, k_n\}$, and $P = \{0, 1, \ldots, p - 1\}$. Equivalently, the inverse Fourier transform is given as

$$f(K) = \sum_{(\omega_1, \ldots, \omega_n) \in P^n} \exp^{\frac{j 2\pi}{p} \sum_{i=1}^{n} \omega_i k_i} \mathcal{F}(\omega_1, \ldots, \omega_n) \, .$$

For our case let $p = 2$. Then, $k_i, \omega_i \in \{0, 1\}$ and

$$
\begin{aligned}
f(K) &= \sum_{(\omega_1, \ldots, \omega_n) \in \{0,1\}^n} \exp^{j\pi \sum_{i=1}^{n} \omega_i k_i} \mathcal{F}(\omega_1, \ldots, \omega_n) \\
&= \sum_{(\omega_1, \ldots, \omega_n) \in \{0,1\}^n} (-1)^{\sum_{i=1}^{n} \omega_i k_i} \mathcal{F}(\omega_1, \ldots, \omega_n) \\
&= \sum_{(\omega_1, \ldots, \omega_n) \in \{0,1\}^n} \prod_{i=1}^{n} (-1)^{\omega_i k_i} \mathcal{F}(\omega_1, \ldots, \omega_n) \, . \quad (8.7)
\end{aligned}
$$

Since $\omega_i \in \{0, 1\}$, it can be denoted by α_i in our notation. Moreover,

$$
(-1)^{k_i} = \begin{cases} 1 & \text{if } k_i = 0, \\ -1 & \text{if } k_i = 1, \end{cases}
$$

and thus $(-1)^{k_i} = x_i$ in our notation. One can then write Eqn. (8.7) as

$$f(X) = \sum_{\alpha \in \{0,1\}^n} X^\alpha \mathcal{F}(\alpha_1, \ldots, \alpha_n).$$

By the uniqueness property of the representation in Eqn. (8.6), we get that a_α equals the Fourier/spectral coefficients $\mathcal{F}(\alpha_1, \ldots, \alpha_n)$.

Note that the spectral and the polynomial representation of f are equivalent. When the spectral/harmonic terminology is used, the parity functions are referred to as the basis functions, and the a_α's are called the spectral coefficients. When the polynomial terminology is used, the Parity functions (X^α) are referred to as monomials and the coefficients are again referred to as spectral coefficients.

We next relate the polynomial/spectral representation to the geometric framework. We can rewrite Eqn. (8.5) in matrix notation as

$$f = YA, \quad (8.8)$$

where

$$Y = [\phi_{(0, \ldots, 0)}, \cdots, \phi_{(1, \ldots, 1)}]$$

is an $2^n \times 2^n$ matrix whose columns are the Parity functions ϕ_α, and

$$A = [a_{(0,\ldots,0)}, \ldots, a_{(1,\ldots,1)}]^T$$

is the transform vector comprising all the spectral coefficients.

Remark 8.2:   Given the orthogonality of the Parity functions, we have

$$a_\alpha = \frac{1}{2^n} \sum_{X \in \{-1,1\}^n} X^\alpha f(X) = C_{f\phi_\alpha}.$$

Thus, each a_α equals the correlation of f with the Parity function ϕ_α, or equivalently with the monomial X^α. Moreover, the orthogonality property of Parity functions implies that the generalized spectral coefficients defined in Eqn. (8.3) are the same as the spectral coefficients a_α's, i.e.,

$$\beta^T = [a_{(0,\ldots,0)}, \ldots, a_{(1,\ldots,1)}].$$

The preceding two relationships make the results in Section 8.2 applicable to the particular case where the input functions are Parity functions, and such applications are discussed in Sections 8.5.2 and 8.6.   □

Since $Y^T Y = 2^n I_{2^n \times 2^n}$, Y is a Hadamard matrix. Moreover, if the rows and columns of Y (see Eqn. (8.8)) are appropriately permuted, then the matrix equals the so-called *Sylvester-type Hadamard matrix*, H_{2^n}. The required ordering of the rows and columns is as follows:

1. Each column of Y is a Parity function ϕ_α. In the desired permutation, the columns are ordered from left to right lexicographically based on the values of α. Thus, $\phi_{(0,\ldots,0)}$ is the first column, $\phi_{(1,0,0,\ldots,0)}$ is the second column, and so on.

2. Similarly each row of Y can be considered as a Parity function ϕ_α, and in the desired permutation, the rows are ordered lexicographically from top to bottom based on the values of α.

One advantage of these permutations is that the resultant Sylvester-type Hadamard matrix can be recursively defined as follows:

$$
\begin{aligned}
H_1 &= [1], \\
H_2 &= \begin{bmatrix} 1 & 1 \\ 1 & -1 \end{bmatrix}, \\
H_{2^{n+1}} &= \begin{bmatrix} H_{2^n} & H_{2^n} \\ H_{2^n} & -H_{2^n} \end{bmatrix}.
\end{aligned}
\tag{8.9}
$$

The following equation shows an 8×8 Hadamard matrix and the function to which each row corresponds. Since H_{2^n} is a symmetric matrix, the ith row is the transpose of the ith column, and for convenience we have identified only the rows.

$$H_8 = H_8^T = \begin{bmatrix} 1 & 1 & 1 & 1 & 1 & 1 & 1 & 1 \\ 1 & -1 & 1 & -1 & 1 & -1 & 1 & -1 \\ 1 & 1 & -1 & -1 & 1 & 1 & -1 & -1 \\ 1 & -1 & -1 & 1 & 1 & -1 & -1 & 1 \\ 1 & 1 & 1 & 1 & -1 & -1 & -1 & -1 \\ 1 & -1 & 1 & -1 & -1 & 1 & -1 & 1 \\ 1 & 1 & -1 & -1 & -1 & -1 & 1 & 1 \\ 1 & -1 & -1 & 1 & -1 & 1 & 1 & -1 \end{bmatrix} \begin{matrix} 1 \\ x_1 \\ x_2 \\ x_2 \oplus x_1 \\ x_3 \\ x_3 \oplus x_1 \\ x_3 \oplus x_2 \\ x_3 \oplus x_2 \oplus x_1. \end{matrix}$$

8.5.1 Computing Spectral Coefficients

The degree of a monomial X^α is the number of 1's in $\alpha = (\alpha_1, \ldots, \alpha_n) \in \{0,1\}^n$; hence, $deg(X^\alpha) = |\alpha| = \sum \alpha_i$. The degree of a multilinear polynomial is the largest degree of any monomial included in it.

Linear spectral coefficients of a Boolean function are the spectral coefficients of the linear terms in the polynomial representation, i.e., the spectral coefficients corresponding to the monomials $x_1, x_2, \ldots, x_n$. The linear spectral coefficients are also referred to as Chow parameters, and some of the related issues are discussed in Section 8.5.2.

We shall make repeated use of the polynomial representation of the 2-variable AND function in computing the spectral coefficients of different functions.

$$\begin{aligned} AND(x_1, x_2) &= \tfrac{1}{2}(1 + x_1 + x_2 - x_1 x_2) \\ &= \tfrac{1}{2}(1 + x_1 + x_2(1 - x_1)) . \end{aligned}$$

Theorem 8.6:   For the n-variable Inner Product Mod 2 function,

$$|a_\alpha| = 2^{-n/2}$$

for all $\alpha \in \{0,1\}^n$.

Proof: In fact we can explicitly determine the polynomial representation. Recall that the Inner Product Mod 2 function of n-variables is defined as

$$IP(X, Y) = \bigoplus_{i=1}^{n/2} (x_i \wedge y_i).$$

By expressing the 2-variable AND functions by their respective polynomial representations, and representing the Parity as a product, we get

$$IP(X,Y) = \frac{1}{2^{n/2}} \prod_{i=1}^{n/2} (1 + x_i + y_i - x_i y_i) .$$

One can easily show that when $\prod_{i=1}^{n/2} (1 + x_i + y_i - x_i y_i)$ is expanded, every possible monomial term appears and has a coefficient of magnitude equal to 1. $\qquad\square$

We next compute the spectral coefficients of the Complete Quadratic function, $CQ(X)$, which was introduced in Chapter 5. Recall that $CQ(x_1, x_2, \ldots, x_n)$ equals the Parity function of all $\binom{n}{2}$ possible AND's between pairs of variables, and can be written as

$$\begin{aligned}
CQ(x_1, \ldots, x_n) = \; & x_n \wedge (x_1 \oplus x_2 \oplus \cdots \oplus x_{n-1}) \oplus \\
& x_{n-1} \wedge (x_2 \oplus \cdots \oplus x_{n-2}) \oplus \\
& \cdots\cdots\cdots\cdots \oplus \\
& \cdots\cdots\cdots \oplus \\
& x_2 \wedge x_1 .
\end{aligned}$$

Theorem 8.7:   For the n-variable Complete Quadratic function:

1. $|a_\alpha| = 2^{-n/2}$ $\forall \alpha \in \{0,1\}^n$ if n is even, and

2. $|a_\alpha| = 0$ or $2^{-(n-1)/2}$ $\forall \alpha \in \{0,1\}^n$ if n is odd.

Proof: We provide an explicit construction for the polynomial representation and use induction to prove the theorem for even values of n. The proof for odd values of n is left as an exercise.

Theorem 8.7 is valid when $n = 2$:

$$CQ(x_1, x_2) = AND(x_1, x_2) = \frac{1}{2}(1 + x_1 + x_2 - x_1 x_2).$$

Let us assume that it is true for n variables, i.e.,

$$CQ(x_1, \ldots, x_n) = \frac{1}{2^{n/2}} \sum_{\alpha \in \{0,1\}^n} s_\alpha X^\alpha,$$

where $s_\alpha \in \{1, -1\}$. We next show that the theorem is also true for $(n+2)$ variables. If we add two more variables, x_{n+1} and x_{n+2}, then the augmented function can be written as

$$
\begin{aligned}
CQ(x_1, \ldots, x_{n+2}) \;=\; & CQ(x_1, \ldots, x_n) \oplus \\
& x_{n+2} \wedge (x_1 \oplus \cdots \oplus x_n \oplus x_{n+1}) \oplus \\
& x_{n+1} \wedge (x_1 \oplus \cdots \oplus x_n).
\end{aligned} \tag{8.10}
$$

Substituting the polynomial representation of the 2-variable AND function and recognizing that Parity represents multiplication, one can show the following:

$$
\begin{aligned}
& x_{n+2} \wedge (x_1 \oplus \cdots \oplus x_n \oplus x_{n+1}) \\
&= \frac{1}{2}(1 + x_{n+2} + (\prod_{i=1}^{n} x_i)(x_{n+1} - x_{n+1}x_{n+2})) \,,
\end{aligned}
$$

and

$$
\begin{aligned}
& x_{n+1} \wedge (x_1 \oplus \cdots \oplus x_n) \\
&= \frac{1}{2}(1 + x_{n+1} + (\prod_{i=1}^{n} x_i)(1 - x_{n+1})) \,.
\end{aligned}
$$

Substituting these expressions in Eqn. (8.10) and simplifying, one gets

$$
\begin{aligned}
CQ(x_1, \ldots, x_{n+2}) \;=\; & [\frac{1}{2^{n/2}} \sum_{\alpha \in \{0,1\}^n} s_\alpha X^\alpha] \cdot \\
& \frac{1}{2}[x_{n+1} + x_{n+2} + (\prod_{i=1}^{n} x_i)(1 - x_{n+1}x_{n+2})] \,. \tag{8.11}
\end{aligned}
$$

For any $\alpha \in \{0,1\}^n$, $(\prod_{i=1}^{n} x_i)X^\alpha = X^{\overline{\alpha}}$, where $\overline{\alpha}$ is the complement of α. Hence, multiplying out the expressions in Eqn. (8.11) one gets

$$
\begin{aligned}
CQ(x_1, \ldots, x_{n+2}) \;=\; & \frac{1}{2^{(n+2)/2}}[(x_{n+1} + x_{n+2})(\sum_{\alpha \in \{0,1\}^n} s_\alpha X^\alpha) \\
& + (1 - x_{n+1}x_{n+2})(\sum_{\alpha \in \{0,1\}^n} s_{\overline{\alpha}} X^\alpha)] \,.
\end{aligned}
$$

One can verify that every monomial X^α, $\alpha \in \{0,1\}^{n+2}$, appears in the preceding expression.   □

In the $\{1, -1\}$ notation, the n-variable Majority function is given as $MAJ_n(X) = \text{sgn}(\sum_{i=1}^{n} x_i)$. The following theorem derives the spectral coefficients of Majority.

Theorem 8.8: For $MAJ_n(X)$, spectral coefficient $a_\alpha = 0$ if $|\alpha| = \sum_{i=1}^{n} \alpha_i$ is even and

$$a_\alpha = \frac{2(-1)^{(|\alpha|-1)/2}}{2^n} \frac{(|\alpha|-1)!}{\left(\frac{|\alpha|-1}{2}\right)!} \frac{(n-|\alpha|)!}{\left(\frac{n-|\alpha|}{2}\right)! \left(\frac{n-1}{2}\right)!}$$

if $|\alpha|$ is odd.

Proof: We derive the spectral coefficients for the case where $|\alpha| \leq (n-1)/2$. The proof for the case where $|\alpha| \geq (n+1)/2$ is similar and is left as an exercise (see Exercise 8.3). By definition of the spectral coefficients,

$$2^n a_\alpha = \sum_{X \in \{1,-1\}^n} MAJ_n(X)X^\alpha . \tag{8.12}$$

Let $m = |\alpha|$ in the rest of this proof. Since α has m 1's, without loss of generality one can consider α as

$$\alpha = (\overbrace{1, \ldots, 1}^{m}, \overbrace{0, \ldots, 0}^{n-m}) .$$

For any $X \in \{1, -1\}^n$, let k be the number of -1's in X for which the corresponding entries in α are 1's, and let l be the number of -1's in X for which the corresponding entries in α are 0's. Thus, X has $(k+l)$ number of -1's and can be visualized as

$$X = (\underbrace{\overbrace{-1, \ldots, -1}_{k}, 1, \ldots, 1}^{m}, \underbrace{\overbrace{-1, \ldots, -1}_{l}, 1, \ldots, 1}^{n-m}).$$

Clearly $X^\alpha = (-1)^k$. Since the summation in Eqn. (8.12) is over all $X \in \{1, -1\}^n$, one can write it as

$$2^n a_\alpha = \sum_{k=0}^{m} \sum_{l=0}^{n-m} \binom{m}{k} \binom{n-m}{l} (-1)^k g(k+l),$$

where

$$g(k+l) = \begin{cases} 1 & \text{if } k+l \le (n-1)/2, \\ -1 & \text{if } k+l \ge (n+1)/2 . \end{cases}$$

Thus,

$$
\begin{aligned}
2^n a_\alpha &= \sum_{k=0}^{m} \binom{m}{k} (-1)^k \left[\sum_{l=0}^{\frac{n-1}{2}-k} \binom{n-m}{l} - \sum_{l=\frac{n+1}{2}-k}^{n-m} \binom{n-m}{l} \right] \\
&= \sum_{k=0}^{\frac{m-1}{2}} \binom{m}{k} (-1)^k \left[\sum_{l=0}^{\frac{n-1}{2}-k} \binom{n-m}{l} - \sum_{l=\frac{n+1}{2}-k}^{n-m} \binom{n-m}{l} \right] \\
&\quad + \sum_{k=\frac{m+1}{2}}^{m} \binom{m}{k} (-1)^k \left[\sum_{l=0}^{\frac{n-1}{2}-k} \binom{n-m}{l} - \sum_{l=\frac{n+1}{2}-k}^{n-m} \binom{n-m}{l} \right] . \quad (8.13)
\end{aligned}
$$

These expressions can be simplified by observing that $k \le \dfrac{m-1}{2}$ if and only if $\dfrac{n-m}{2} \le \dfrac{n-1}{2} - k$, and $k \ge \dfrac{m+1}{2}$ if and only if $\dfrac{n-m}{2} \ge \dfrac{n+1}{2} - k$. Thus, we get

$$
\begin{aligned}
& \sum_{k=0}^{\frac{m-1}{2}} \binom{m}{k} (-1)^k \left[\sum_{l=0}^{\frac{n-1}{2}-k} \binom{n-m}{l} - \sum_{l=\frac{n+1}{2}-k}^{n-m} \binom{n-m}{l} \right] \\
&= \sum_{k=0}^{\frac{m-1}{2}} \binom{m}{k} (-1)^k \sum_{l=\frac{n+1}{2}-m+k}^{\frac{n-1}{2}-k} \binom{n-m}{l} , \quad (8.14)
\end{aligned}
$$

and

$$
\begin{aligned}
& \sum_{k=\frac{m+1}{2}}^{m} \binom{m}{k} (-1)^k \left[\sum_{l=0}^{\frac{n-1}{2}-k} \binom{n-m}{l} - \sum_{l=\frac{n+1}{2}-k}^{n-m} \binom{n-m}{l} \right] \\
&= - \sum_{k=\frac{m+1}{2}}^{m} \binom{m}{k} (-1)^k \sum_{l=\frac{n+1}{2}-k}^{\frac{n-1}{2}-m+k} \binom{n-m}{l} . \quad (8.15)
\end{aligned}
$$

Substituting the preceding identities in Eqn. (8.13), one gets

$$
2^n a_\alpha = \sum_{k=0}^{\frac{m-1}{2}} \binom{m}{k} (-1)^k \sum_{l=\frac{n+1}{2}-m+k}^{\frac{n-1}{2}-k} \binom{n-m}{l}
$$

$$- \sum_{k=\frac{m+1}{2}}^{m} \binom{m}{k} (-1)^k \sum_{l=\frac{n+1}{2}-k}^{\frac{n-1}{2}-m+k} \binom{n-m}{l}$$

$$= \sum_{k=0}^{\frac{m-1}{2}} \binom{m}{k} (-1)^k \sum_{j=-(\frac{m-1}{2}-k)}^{\frac{m-1}{2}-k} \binom{n-m}{\frac{n-m}{2}+j}$$

$$- \sum_{k=\frac{m+1}{2}}^{m} \binom{m}{k} (-1)^k \sum_{j=-(k-\frac{m+1}{2})}^{k-\frac{m+1}{2}} \binom{n-m}{\frac{n-m}{2}+j}$$

$$= 2 \sum_{k=0}^{\frac{m-1}{2}} \binom{m}{k} (-1)^k \sum_{j=-(\frac{m-1}{2}-k)}^{\frac{m-1}{2}-k} \binom{n-m}{\frac{n-m}{2}+j} \quad \text{(since } m \text{ is odd)}$$

$$= 2 \sum_{j=-\frac{m-1}{2}}^{\frac{m-1}{2}} \binom{n-m}{\frac{n-m}{2}+j} \sum_{k=0}^{\frac{m-1}{2}-|j|} (-1)^k \binom{m}{k}.$$

Using the identity $\sum_{k=0}^{a} (-1)^k \binom{b}{k} = (-1)^a \binom{b-1}{a}$, one gets

$$2^n a_\alpha = 2 \sum_{j=-\frac{m-1}{2}}^{\frac{m-1}{2}} \binom{n-m}{\frac{n-m}{2}+j} \binom{m-1}{\frac{m-1}{2}-|j|} (-1)^{\frac{m-1}{2}-|j|}$$

$$= 2(-1)^{\frac{m-1}{2}} \sum_{j=-\frac{m-1}{2}}^{\frac{m-1}{2}} \binom{n-m}{\frac{n-m}{2}+j} \binom{m-1}{\frac{m-1}{2}-j} (-1)^j.$$

Using another identity (see Exercise 8.4)

$$\sum_{j=-a}^{a} \binom{2a}{a+j} \binom{2b}{b-j} (-1)^j = \frac{(2a)!(2b)!}{a!\,b!\,(a+b)!}$$

one gets

$$2^n a_\alpha = 2(-1)^{\frac{m-1}{2}} \frac{(n-m)!\,(m-1)!}{(\frac{n-m}{2})!\,(\frac{m-1}{2})!\,(\frac{n-1}{2})!}.$$

This derivation holds for $m \leq (n-1)/2$ and odd. Next, we show that if $m = \sum_{i=1}^{n} \alpha_i$ is even, then $a_\alpha = 0$. The initial discussions are the same when

m is even and Eqn. (8.13) implies that

$$
\begin{aligned}
2^n a_\alpha &= \sum_{k=0}^{m} \binom{m}{k}(-1)^k \left[\sum_{l=0}^{\frac{n-1}{2}-k} \binom{n-m}{l} - \sum_{l=\frac{n+1}{2}-k}^{n-m} \binom{n-m}{l} \right] \\
&= \sum_{k=0}^{\frac{m}{2}-1} \binom{m}{k}(-1)^k \left[\sum_{l=0}^{\frac{n-1}{2}-k} \binom{n-m}{l} - \sum_{l=\frac{n+1}{2}-k}^{n-m} \binom{n-m}{l} \right] \\
&\quad + \sum_{k=\frac{m}{2}+1}^{m} \binom{m}{k}(-1)^k \left[\sum_{l=0}^{\frac{n-1}{2}-k} \binom{n-m}{l} - \sum_{l=\frac{n+1}{2}-k}^{n-m} \binom{n-m}{l} \right] \\
&\quad + \binom{m}{\frac{m}{2}}(-1)^{\frac{m}{2}} \left[\sum_{l=0}^{\frac{n-m-1}{2}} \binom{n-m}{l} - \sum_{l=\frac{n+m+1}{2}}^{n-m} \binom{n-m}{l} \right].
\end{aligned}
$$

Now,

$$
\sum_{l=0}^{\frac{n-m-1}{2}} \binom{n-m}{l} - \sum_{l=\frac{n+m+1}{2}}^{n-m} \binom{n-m}{l} = 0
$$

and using the identities in Eqns. (8.14) and (8.15), one gets

$$
\begin{aligned}
2^n a_\alpha &= \sum_{k=0}^{\frac{m}{2}-1} \binom{m}{k}(-1)^k \sum_{l=\frac{n+1}{2}-m+k}^{\frac{n-1}{2}-k} \binom{n-m}{l} \\
&\quad - \sum_{k=\frac{m}{2}+1}^{m} \binom{m}{k}(-1)^k \sum_{l=\frac{n+1}{2}-k}^{\frac{n-1}{2}-m+k} \binom{n-m}{l} \\
&= \sum_{k=0}^{\frac{m}{2}-1} \binom{m}{k}(-1)^k \sum_{j=-(\frac{m-1}{2}-k)}^{\frac{m-1}{2}-k} \binom{n-m}{\frac{n-m}{2}+j} \\
&\quad - \sum_{k=\frac{m}{2}+1}^{m} \binom{m}{k}(-1)^k \sum_{j=-(k-\frac{m+1}{2})}^{k-\frac{m+1}{2}} \binom{n-m}{\frac{n-m}{2}+j}. \quad (8.16)
\end{aligned}
$$

Now by a shift in the summation index, one gets

$$
\sum_{k=\frac{m}{2}+1}^{m} \binom{m}{k}(-1)^k \sum_{j=-(k-\frac{m+1}{2})}^{k-\frac{m+1}{2}} \binom{n-m}{\frac{n-m}{2}+j}
$$

$$= (-1)^m \sum_{k=0}^{\frac{m}{2}-1} \binom{m}{k} (-1)^k \sum_{j=-(\frac{m-1}{2}-k)}^{\frac{m-1}{2}-k} \binom{n-m}{\frac{n-m}{2}+j}. \qquad (8.17)$$

Since m is even, $(-1)^m = 1$, and substituting Eqn. (8.17) in Eqn. (8.16), one gets that $a_\alpha = 0$. $\qquad\qquad\qquad\qquad\qquad\qquad\qquad\qquad\qquad\qquad\qquad\square$

8.5.2 Polynomial Threshold Functions and Related Results

So far we have discussed an exact representation of a Boolean function by a polynomial or equivalently by a linear combination of the Parity functions. In the context of threshold gates, it is natural to consider a representation in which the sign of the polynomial corresponds to that of the function for every input $X \in \{1, -1\}^n$, i.e.,

$$f(X) = \text{sgn}(\sum_{\alpha \in Q} w_\alpha X^\alpha), \qquad (8.18)$$

where $Q \subset \{0, 1\}^n$.

Definition 8.3: The set of functions that can be expressed as in Eqn. (8.18) such that $|Q| \le n^c$ for some constant c (i.e., the number of monomials are polynomially bounded in n) is defined as PT_1. If the weights are also restricted to be integers that are polynomially bounded in n, then the class of functions is referred to as $\widehat{PT_1}$. $\qquad\qquad\qquad\qquad\square$

Note that there is no restriction on the degree of the polynomials used in representing functions in PT_1 or $\widehat{PT_1}$.

 The following lemma underscores the motivation for studying functions in PT_1. In particular, if a function is in PT_1, then it can be computed by a depth-2 polynomial-size threshold circuit. In Chapter 4, for example, depth-2 circuits for Comparison and Addition are constructed by showing that these functions are in $\widehat{PT_1}$.

Lemma 8.2:
$$PT_1 \subseteq LT_2 \text{ and } \widehat{PT_1} \subseteq \widehat{LT_2}.$$

Proof: If a function $f(X)$ is in PT_1, then $f(X) = \text{sgn}(\sum_{\alpha \in Q} w_\alpha X^\alpha)$, where $|Q| \le n^c$ for some integer c. Now each monomial X^α is a Parity function of an appropriate number of variables. In Chapter 3, we show that an n-variable Parity function can be written as a sum of at most n functions in $\widehat{LT_1}$. Substituting X^α by the corresponding sum, we get $f(X) = \text{sgn}(\sum_{i=1}^{N} w_i f_i(X))$, where $f_i(X) \in \widehat{LT_1}$ and $N \leq n^{c+1}$. Hence, $f(X) \in LT_2$. If the weights, w_α, are restricted to be polynomially bounded integers, then the above arguments also show that if $f(X) \in \widehat{PT_1}$, then it is in $\widehat{LT_2}$.　　　$\square$

If $f(X) = \text{sgn}(\sum_{\alpha \in Q} w_\alpha X^\alpha)$, then in our terminology, $f(X)$ is the output of a threshold gate in which the input functions are restricted to be the parity functions or monomials. Thus, the results of Section 8.2 can be applied with the substitution (see Remark 8.2): $a_\alpha = C_{f\phi_\alpha} = \beta_\alpha$ for all $\alpha \in \{0,1\}^n$.

A direct application of the Uniqueness Theorem yields the next corollary.

Corollary 8.3:　　Let $f(X) = \text{sgn}(\sum_{\alpha \in Q \subseteq \{0,1\}^n} w_\alpha X^\alpha)$, then the set of spectral coefficients of $f(X)$ over Q, represented as $\{a_\alpha^f : \alpha \in Q\}$, is unique. That is, for any other Boolean function $g(X) \neq f(X)$, there exists at least one $\alpha \in Q$ such that $a_\alpha^f \neq a_\alpha^g$.　　　$\square$

If f is a linear threshold function, i.e., $f(X) = \text{sgn}(\sum_{i=1}^{n} w_i x_i + w_0)$, then Corollary 8.3 implies that the set of linear spectral coefficients and the constant spectral coefficient (i.e., $a_{(0,\ldots,0)}$), of the function f is unique. Hence, linear threshold functions can be uniquely characterized by the linear terms and the constant term in the spectrum. Since the linear spectral coefficients are also referred to as Chow parameters, Corollary 8.3 shows that Chow parameters are unique for functions in LT_1.

As one application of the Spectral Bound Theorem, we state the following lower bound on the number of monomials required in a polynomial so that the sign of the polynomial equals the given function.

Corollary 8.4:　　Let $f(X) = \text{sgn}(\sum_{\alpha \in Q \subseteq \{0,1\}^n} w_\alpha X^\alpha)$, then

$$\sum_{\alpha \in Q} |a_\alpha| \geq 1 .$$

Hence,

$$|Q| \geq \frac{1}{\max\{|a_\alpha| : \alpha \in Q\}} .$$ □

This corollary can be used to show the following separation results.

Theorem 8.9:

$$PT_1 \subsetneq LT_2 \text{ and } \widehat{PT_1} \subsetneq \widehat{LT_2}.$$

Proof: Lemma 8.2 establishes that PT_1 and $\widehat{PT_1}$ are contained in LT_2 and $\widehat{LT_2}$, respectively. Next we use the $CQ(X)$ function to show the separation between the classes. Since the spectral coefficients of $CQ(X)$ (as computed in Theorem 8.7) satisfy $|a_\alpha| \leq 2^{-n/2}$, it follows from Corollary 8.4 that $CQ(X) \notin PT_1$ or $\widehat{PT_1}$. However, $CQ(X)$ is a symmetric function, and hence, it is in $\widehat{LT_2}$.   □

Remark 8.3:   Theorem 8.9 does not address the relationship between the classes PT_1 and $\widehat{LT_2}$. A result by Goldmann, Håstad, and Razborov [40], however, shows that there is a function $f(X)$ that is in PT_1 but not in $\widehat{LT_2}$. Thus, $PT_1 \not\subseteq \widehat{LT_2}$. Moreover, since $CQ(X)$ is in $\widehat{LT_2}$ but not in PT_1, we have $\widehat{LT_2} \not\subseteq PT_1$. Hence, neither class is a subset of the other.   □

Next we present some basic results on the degree of polynomials used in representing functions in PT_1. Recall that the strong degree of a Boolean function $f(X) : \{1, -1\}^n \to \{1, -1\}$, $SD(f)$, is the smallest degree of any multilinear polynomial $F(X)$ such that $f(X) = \text{sgn}(F(X))$ and $F(X) \neq 0$ for all $X \in \{1, -1\}^n$.

In the literature, the concept of a weak degree of a Boolean function has also been introduced.

Definition 8.4:   The *weak degree* of $f(X) : \{1, -1\}^n \to \{1, -1\}$, $WD(f)$, is defined as the smallest degree of any multilinear polynomial $F(X)$ (not identically zero) such that for some nonempty $S \subset \{1, -1\}^n$, $f(X) = \text{sgn}(F(X))$ for all $X \in S$, whereas $F(X) = 0$ for $X \notin S$.   □

It is obvious that for any n-variable Boolean function f, we have $WD(f) \leq SD(f) \leq n$. Recall that PAR_n denotes the n-variable Parity function. The following lemma states that the Parity function has the largest strong degree among all Boolean functions.

Lemma 8.3: $WD(PAR_n) = SD(PAR_n) = n.$

Proof: It suffices to prove that $WD(PAR_n) \geq n$. Suppose $PAR_n(X) = \text{sgn}(F(X))$ for $X \in S$, where $F(X) = \sum w_\alpha X^\alpha$ (not identically zero) has degree $< n$ and $F(X) = 0$ for $X \notin S$. Then, $PAR_n^T F = 0$, since $PAR_n^T X^\alpha = 0$ for all $\alpha \neq (1, \ldots, 1)$. But $PAR_n^T F = \sum_{X \in S} P(X)F(X) = \sum_{X \in S} |F(X)| > 0$, a contradiction.   □

Lemma 8.4:   Let $f(X) \neq PAR_n(X)$ denote any Boolean function of n variables. Then, $SD(f) < n$.

Proof: Using the polynomial representation, every Boolean function f can be written as a polynomial $f(X) = \sum_{\alpha \in \{0,1\}^n} a_\alpha X^\alpha$, where $a_\alpha = f^T \phi_\alpha / 2^n = C_{f\phi_\alpha}$ (see Remark 8.2). Since $f(X) \neq PAR_n(X)$, $|a_{1,1,\ldots,1}| < 1$. Thus, $f(X) = \text{sgn}(\sum_{\alpha \neq (1,\ldots,1)} a_\alpha X^\alpha)$, and hence, $SD(f) < n$.   □

8.6   Spectral Approximation of Boolean Functions

We present results on approximating Boolean functions using a linear combination of basis functions. A lower bound on the approximation error is derived in terms of the generalized spectral coefficients. We then apply this lower bound to the special case of approximating Boolean functions by polynomials and show that relatively simple functions, such as Majority, cannot be closely approximated by sparse polynomials.

Definition 8.5:   Given a Boolean function, f, and a set of S other Boolean functions $Y = [f_1, \ldots, f_S]$, the *approximation error*, $0 \leq \epsilon_{fY}$, is defined as

$$\epsilon_{fY} = \min\{\|f - Y\mathbf{w}\|_\infty \ : \ \mathbf{w} \in \mathcal{R}^S\} \ .$$

The set of functions $Y = [f_1, \ldots, f_S]$ is said to *approximate* f, if $\epsilon_{fY} < 1$. □

Note that a set of functions $Y = [f_1 \cdots f_S]$ approximates a function f (i.e., $0 \leq \epsilon_{fY} < 1$), if and only if f is a threshold function of $f_1, \ldots, f_S$.

Theorem 8.10:   If an n-variable Boolean function f is approximated by a set of S other n-variable Boolean functions, $f_1, f_2, \ldots, f_S$, then

$$4\epsilon_{fY} \geq 1 - \frac{\|Y\beta\|^2}{2^n},$$

where $Y = [f_1 \cdots f_S]$ and β is the generalized spectrum as defined in Eqn. (8.3) and $|| \cdot ||$ is the L_2 norm.

Proof: Let $X = Y\mathbf{w}$, such that $|(f - X)_i| \leq \epsilon_{fY} < 1$. Then, $1 - \epsilon_{fY} \leq |(X)_i| \leq 1 + \epsilon_{fY}$. Since $2^n C_{fY} = (Y^T Y)\beta$, we obtain

$$f^T X = f^T Y\mathbf{w} = 2^n C_{fY}^T \mathbf{w} = \beta^T Y^T Y\mathbf{w} = (Y\beta)^T X.$$

Using Cauchy-Schwarz inequality we get,

$$f^T X \leq ||Y\beta|| \, ||X||.$$

Hence,

$$||Y\beta|| \geq f^T X / ||X||.$$

However, we know that $f^T X = \sum_{i=1}^{2^n} |(X)_i| \geq (1 - \epsilon_{fY})2^n$, and $||X||^2 = \sum_{i=1}^{2^n} x_i^2 \leq 2^n(1 + \epsilon_{fY})^2$. Hence, we obtain

$$\frac{||Y\beta||^2}{2^n} \geq (1 - \epsilon_{fY})^2 / (1 + \epsilon_{fY})^2.$$

Now the theorem follows by observing that

$$4\epsilon_{fY} \geq 1 - \frac{(1 - \epsilon_{fY})^2}{(1 + \epsilon_{fY})^2}. \qquad \square$$

If we restrict to the case where columns of Y are orthogonal, then we have $\beta = C_{fY}$, and $||Y\beta||^2 = 2^n ||C_{fY}||^2$. Thus, if f_i's are mutually orthogonal, then

$$4\epsilon_{fY} \geq 1 - ||C_{fY}||^2. \tag{8.19}$$

We now apply Theorem 8.10 to the more specific case of polynomial approximation of Boolean functions. As discussed in Section 8.5, any Boolean function, $f(X)$, of n variables $x_1, \ldots, x_n$, can be written as $f(x_1, \ldots, x_n) = \sum_{\alpha \in \{0,1\}^n} a_\alpha X^\alpha$, where a_α are the spectral coefficients and X^α are the monomials. Each monomial X^α is a Parity function of the appropriate number of input variables (i.e., those variables for which $\alpha_i = 1$). Since the Parity functions are mutually orthogonal, we have $a_\alpha = C_{f\phi_\alpha}$, where $\phi_\alpha \in \mathcal{R}^{2^n}$ denotes the Parity function corresponding to the monomial X^α.

If $f(X)$ is approximated by a polynomial, then given the preceding equivalence, the following corollary is directly implied by Theorem 8.10.

Corollary 8.5:   If a function f is approximated by a polynomial in which the monomials are chosen from the set $\{X^\alpha \; : \; \alpha \in Q, \; Q \subseteq \{0,1\}^n\}$, then the approximation error $0 \le \epsilon < 1$ satisfies the following:

$$4\epsilon \ge 1 - \sum_{\alpha \in Q} a_\alpha^2 = \sum_{\alpha \notin Q} a_\alpha^2 . \qquad \square$$

Thus, the approximation error is bounded below by the total *spectral power* concentrated in the monomials that are not included in approximation.

A case of particular interest is when only polynomially many monomials are used to approximate a given function. Recall that a polynomial of n variables is a sparse polynomial if the number of monomials in it is polynomially bounded in n. Also, the class $\widetilde{SP}$ consists of all Boolean functions that can be closely approximated by sparse polynomials, more precisely, given any $f(X) \in \widetilde{SP}$ and $k > 0$, there exist polynomially many (in n) c_α's such that

$$\left| f(X) - \sum c_\alpha X^\alpha \right| < n^{-k}$$

for all $X \in \{1, -1\}^n$.

We apply the lower bound result in Corollary 8.5 to show that Majority does not belong to $\widetilde{SP}$. Recall that in the $\{1, -1\}$ notation, the n-variable Majority function, for odd n, can be represented as

$$MAJ_n(x_1, \dots, x_n) \;=\; \mathrm{sgn}(\sum_{i=1}^{n} x_i).$$

Theorem 8.11:   If MAJ_n is approximated by a sparse polynomial, then the approximation error ϵ_M is $\Omega(1/(\log \log n)^{3/2})$.

Proof: For any given $\alpha \in \{0,1\}^n$, the degree of the corresponding monomial, X^α, is given by $|\alpha| = \sum_{i=1}^{n} \alpha_i$. The spectral coefficients of MAJ_n, as computed in Theorem 8.8, are as follows (since $MAJ_n(X)$ is a symmetric function, the monomials with the same degree have the same spectral coefficients):

$$a_\alpha = \begin{cases} 0 & \text{if } |\alpha| \text{ is even,} \\ \frac{2(-1)^{(|\alpha|-1)/2}}{2^n} \frac{(|\alpha|-1)!}{\left(\frac{|\alpha|-1}{2}\right)!} \frac{(n-|\alpha|)!}{\left(\frac{n-|\alpha|}{2}\right)! \left(\frac{n-1}{2}\right)!} & \text{if } |\alpha| \text{ is odd.} \end{cases}$$

Consider the spectral coefficients for which $|\alpha| = \log \log n$, and assume without loss of generality that $\log \log n$ is odd. There are $\binom{n}{\log \log n}$ of such spectral coefficients and they all have the same value. Hence, the spectral power concentrated in these coefficients is

$$\sum_{\alpha \ni |\alpha| = \log \log n} a_\alpha^2 = \binom{n}{\log \log n} a_{|\alpha| = \log \log n}^2 = \Omega(1/(\log \log n)^{3/2}).$$

Suppose that a sparse polynomial $R(X)$, approximating MAJ_n, has fewer than n^c monomials for some constant c. Then, applying Corollary 8.5, we get

$$4\epsilon_M \geq \sum_{X^\alpha \notin R} a_\alpha^2 \geq \sum_{\alpha \ni |\alpha| = \log \log n; \, X^\alpha \notin R} a_\alpha^2.$$

There are $\binom{n}{\log \log n}$ monomials of degree $\log \log n$, and at least

$$(1 - \frac{n^c}{\binom{n}{\log \log n}})$$

fraction of these are not included in R; hence,

$$4\epsilon_M \geq \Omega(1/(\log \log n)^{3/2}(1 - \frac{n^c}{\binom{n}{\log \log n}})) = \Omega(1/(\log \log n)^{3/2}).$$

$\square$

Theorem 8.11 uses the spectrum of MAJORITY to show that it is not in $\widetilde{SP}$. Exercise 8.5 outlines an indirect proof of a weaker result that shows that the approximation error is $\Omega(1/n)$.

8.7 The Method of Correlations

In Section 8.4, we showed that the correlations of the desired output function with the set of input functions can be used to derive lower bounds on the number of inputs required to compute the output function using a threshold gate (see Theorem 8.3 and Corollaries 8.2 and 8.4). In particular, we showed that if the input functions are mutually orthogonal (or strongly asymptotically orthogonal) and if the correlation of the output function with each of the input functions is exponentially small, then the set of input functions must be exponentially large. A particular application of this result in Section 8.5.2 showed that if the spectral coefficients of a function are all exponentially small, then it is not in PT_1. In this section, we further investigate this method of correlations and study the case where the input functions are not always mutually orthogonal.

8.7.1   Separation Results

The following lemma shows that for any set of input functions, if the gate's weights are polynomially bounded integers (in the number of input variables), then exponentially small correlations will always imply that exponentially many input functions are required.

Lemma 8.5:   Let f and $f_1, \ldots, f_S$ be Boolean functions such that

$$f = \text{sgn}(\sum_{i=1}^{S} f_i w_i),$$

where the weights are integers. Then,

$$S \geq \frac{1}{\hat{w}\hat{C}},$$

where $\hat{w} = \max\{w_i : 1 \leq i \leq S\}$ and $\hat{C} = \max\{C_{ff_i} : 1 \leq i \leq S\}$.

Proof: Since all weights are integers, $Y\mathbf{w}$ is an integer vector with $|Y\mathbf{w}_i| \geq 1$, agreeing in sign with f. Hence,

$$
\begin{aligned}
2^n \leq f^T(Y\mathbf{w}) &= (f^T Y)\mathbf{w} = 2^n C_{fY}^T \mathbf{w} \\
&= 2^n \sum_{i=1}^{S} C_{ff_i} w_i \\
&\leq 2^n S \hat{C} \hat{w}
\end{aligned}
$$

implying the result.   □

We next apply Lemma 8.5 to show that the Inner Product Mod 2 function, $IP(X,Y) = \overset{n/2}{\underset{i=1}{\bigoplus}} x_i \wedge y_i$, is not in $\widehat{LT_2}$. In order to apply the lemma, we first show that the correlation of IP with any function in LT_1 is exponentially small. For our purposes, it is convenient to represent the Inner Product Mod 2 function as a $2^{n/2} \times 2^{n/2}$ matrix, say M_{IP}. Each row of M_{IP} is indexed by a distinct assignment to the variables $x_1, \ldots, x_{n/2} \in \{1, -1\}^{n/2}$. Similarly, each column of M_{IP} is indexed by a distinct assignment to the variables $y_1, \ldots, y_{n/2} \in \{1, -1\}^{n/2}$. Hence, the value of an entry (x, y),

$x, y \in \{1, -1\}^{n/2}$, of the matrix is given as $M_{IP}(x, y) = IP(x, y)$. Such a matrix M_{IP} is called a function matrix, and further properties of such matrices are derived in Chapter 10.

One can verify that each column (or row) of M_{IP} is a Parity function ϕ_α, $\alpha \in \{0, 1\}^{n/2}$, and that the columns (rows) comprise all the Parity functions of subsets of $n/2$ variables. This observation follows directly from the definition of $IP(X, Y)$. For example, if $y_1 = y_2 = -1$ and $y_3 = \cdots = y_{n/2} = 1$, then $IP(X, Y)$ reduces to the Parity function of the variables x_1 and x_2. Hence, the corresponding column of M_{IP} indexed by the assignment $y_1 = y_2 = -1$ and $y_3 = \cdots = y_{n/2} = 1$ is the Parity function $\phi_{(1,1,0,..,0)}$. Thus, the columns (rows) of the matrix are mutually orthogonal and M_{IP} is a Hadamard matrix, i.e., $M_{IP}^T M_{IP} = 2^{n/2} I_{2^{n/2} \times 2^{n/2}}$.

If $A \subseteq \{1, -1\}^{n/2}$ and $B \subseteq \{1, -1\}^{n/2}$, then the Cartesian product $A \times B$ is called a rectangle of M_{IP}; Chapter 10 has more detailed discussions on rectangles and their applications in communication complexity. The following lemma, which is a special case of a result on Hadamard matrices due to Lindsey (see Spencer [134]), establishes an upper bound on the sum of $IP(X, Y)$ over any such rectangle.

Lemma 8.6:   For every $A, B \subseteq \{1, -1\}^{n/2}$

$$\left| \sum_{(x,y) \in A \times B} M_{IP}(x, y) \right| \leq \sqrt{|A||B|2^{n/2}},$$

where $|A|$ and $|B|$ represent the cardinality of the respective sets.   □

Given an n-variable Boolean function $f(X, Y) \in LT_1$, let M_f be the function matrix defined by partitioning the set of input variables into sets $\{x_1, \ldots, x_{n/2}\}$ and $\{y_1, \ldots, y_{n/2}\}$. That is, each row of M_f is indexed by an assignment to $\{x_1, \ldots, x_{n/2}\}$ and each column is indexed by an assignment to $\{y_1, \ldots, y_{n/2}\}$, and $M_f(X, Y) = f(X, Y)$.

In Chapter 10, we define a matrix to be *strictly triangular* if the entries in each row and column forms a nondecreasing sequence. In a strictly triangular Boolean matrix (with ± 1 notation), the sets of 1's and -1's resemble a (possibly truncated) triangle, and hence, the name. A matrix is *triangular* if its rows and columns can be permuted so that the resulting matrix is strictly triangular. We show in Chapter 10 (see Example 10.3 and Exercise 10.3) that if $f(X, Y)$ is in LT_1, then the corresponding matrix M_f (defined by

partitioning the input variables into sets $\{x_1, \ldots, x_{n/2}\}$ and $\{y_1, \ldots, y_{n/2}\}$)
is triangular. In the following lemma we assume that the rows and columns
of M_f have been appropriately permuted so that it is strictly triangular.

Lemma 8.7:   For every $f(X, Y) \in LT_1$, the correlation of $f(X, Y)$ with
$IP(X, Y) \leq 4 \cdot 2^{-n/6}$.

Proof: Consider the strictly triangular matrix M_f and divide it into $2^{n/3}$
square submatrices of size $2^{n/3} \times 2^{n/3}$. Given the distribution of 1's and -1's
in M_f, one can show the following (Exercise 8.14):

1. There are at most $2 \times 2^{n/6}$ square submatrices that contain both 1's
   and -1's. Let R_1 be the set of such square submatrices.

2. The 1's that are not in R_1 can be covered by at most $2^{n/6}$ rectangles,
   (which are concatenation of the square submatrices) each of width $2^{n/3}$
   and height $\leq 2^{n/2}$. Let R_2 be the set of such rectangles.

3. Similarly, the -1's that are not in R_1 can be covered by at most $2^{n/6}$
   rectangles, each of width $2^{n/3}$ and height $\leq 2^{n/2}$. Let R_3 be the set of
   such rectangles.

Given this partitioning of M_f, the correlation $C_{f,IP}$ can be expressed as

$$
\begin{aligned}
|C_{f,IP}| &= \frac{1}{2^n}| \sum_{\text{all } x,y} M_{IP}(x,y) M_f(x,y)| \\
&= \frac{1}{2^n}|[\sum_{(x,y) \in R_1} M_{IP}(x,y) M_f(x,y) + \\
&\qquad \sum_{(x,y) \in R_2} M_{IP}(x,y) - \sum_{(x,y) \in R_2} M_{IP}(x,y)]| .
\end{aligned}
$$

Applying the result in Lemma 8.6, we get

$$
\begin{aligned}
|C_{f,IP}| &\leq \frac{1}{2^n}[2 \cdot 2^{n/6} \cdot 2^{2n/3} + \\
&\qquad 2^{n/6}\sqrt{2^{n/3} \cdot 2^{n/2} \cdot 2^{n/2}} + 2^{n/6}\sqrt{2^{n/3} \cdot 2^{n/2} \cdot 2^{n/2}}] \\
&= 4 \cdot 2^{-n/6} .
\end{aligned}
$$

$\square$

Theorem 8.12:

$$\widehat{LT_2} \subsetneq \widehat{LT_3}.$$

Proof: Lemma 8.7 shows that Inner Product Mod 2 has an exponentially small correlation with every output function LT_1. Lemma 8.5 then implies that the Inner Product Mod 2 function is not in $\widehat{LT_2}$. The function, however, can be computed by a depth-3 circuit with only a linear number of gates and polynomially bounded weights (see Exercise 3.12); hence, $\widehat{LT_2} \subsetneq \widehat{LT_3}$.   □

Another application of Lemma 8.5 can be found in the context of implementing the Parity function using a threshold gate and AC^0 functions. Recall that a function is defined to be in AC^0 if it can be realized by a constant-depth polynomial-size AND-OR circuit. The following result states that a threshold function with polynomially bounded integer weights and with inputs from AC^0 cannot compute Parity unless exponentially many AC^0 functions are used.

Theorem 8.13:   Let $f_i(X) \in AC^0$ for $i = 1, \ldots, n^c$ ($c > 0$) be a set of n-variable Boolean functions. Suppose $g(X)$ be another n-variable Boolean function such that

$$g(X) = \text{sgn}(\sum_{i=1}^{n^c} w_i f_i(X)),$$

where w_i's are integers and $|w_i| \leq n^k$ for some constant $k > 0$, then

$$g(X) \neq PAR_n(X).$$

Proof: It can be shown [50, 71] that any AC^0 function must have an exponentially small correlation with Parity. Hence, the theorem follows immediately from Lemma 8.5.   □

A stronger version of Theorem 8.13 is proved in Chapter 7 (see Theorem 7.10) using the preceding result and the techniques of rational approximation.

8.7.2   Limitations of the Method of Correlations

Here we present a constructive procedure (see Theorem 8.14) showing that any function f is a threshold function of $2n$ input functions such that the correlation of f with each of the input functions is exponentially small; the results in Section 8.2 imply that the input functions are not mutually orthogonal. In fact, the results of the constructive procedure are quite surprising; e.g., in Theorem 8.14, the output function f of the threshold gate is uncorrelated with all but one of the input functions, and the correlation with the last input function is the smallest non-zero value possible (i.e., $2^{-(n-1)}$). These results have the following implication: The method of correlations as discussed in Lemma 8.5 does not necessarily imply an exponential lower bound on the number of inputs if exponential weights are allowed. Note, however, that the method of correlations is always effective (Corollary 8.2) if the input functions are mutually orthogonal.

The following lemma introduces some of the linear programming tools used in this section. The analysis uses the same linear programming framework introduced in Chapter 2.

Lemma 8.8:   Given a Boolean function f and a set of S other functions, $f_1, \ldots, f_S$, let $Y = [f_1 \ f_2 \ \cdots f_S]$ and let Y^f be the matrix obtained by negating those rows of matrix Y that correspond to the -1 entries of the vector f (i.e., $y_{ij}^f = f_i y_{ij}$). Then, the following statements are true:

1. $f^T Y = \mathbf{1}^T Y^f$, where $\mathbf{1}$ is the all 1 vector.

2. f can be written as the output of a threshold gate with $f_1, f_2, \ldots, f_S$ as the input functions (i.e., $f = \text{sgn}(Y\mathbf{w})$ for some vector $\mathbf{w}$), if and only if any positive linear combination of the rows of Y^f does not equal zero, i.e., if $\mathbf{q}^T Y^f = 0$ and $\mathbf{q} \geq 0$, then $\mathbf{q} = 0$.

Proof: By definition, the $(i,j)^{th}$ entry of Y^f is given as $y_{ij}^f = f_i y_{ij}$. Hence,

$$\mathbf{1}^T Y^f \ = \ [\sum_{i=1}^{2^n} y_{i1}^f \ \sum_{i=1}^{2^n} y_{i2}^f \cdots \sum_{i=1}^{2^n} y_{iS}^f]$$

$$= \ [\sum_{i=1}^{2^n} f_i y_{i1} \ \sum_{i=1}^{2^n} f_i y_{i2} \cdots \sum_{i=1}^{2^n} f_i y_{iS}]$$

$$= \ f^T Y \ .$$

Also, note that $(Y^f)^f = Y$.

It is easy to verify that $f = \text{sgn}(Y\mathbf{w})$ for some integer weight vector $\mathbf{w}$ if and only if $\mathbf{1} = \text{sgn}(Y^f\mathbf{w})$ (or equivalently $Y^f\mathbf{w} \geq \mathbf{1}$). Thus, given a function f and a set of input functions represented by the columns of the matrix Y, f can be written as a threshold function of the columns of Y, if and only if there is a feasible solution to the following Linear Program (LP):

$$\text{Minimize } 0, \text{ such that}$$
$$Y^f\mathbf{w} \geq [1\ 1 \cdots \ 1]^T . \tag{8.20}$$

The dual of this LP is given by

$$\text{Maximize } \mathbf{1}^T\mathbf{q}, \text{ such that}$$
$$\mathbf{q}^T Y^f = 0, \quad \mathbf{q} \geq 0 . \tag{8.21}$$

A well-known result from duality theory states the following: A primal LP (e.g., Eqn. (8.20)) is feasible if and only if its dual (e.g., Eqn. (8.21)) has a bounded objective function. Now, Eqn. (8.21) has a bounded objective function ($= 0$) if and only if its only feasible solution is $\mathbf{q} = 0$. This is because if there is a feasible solution $\mathbf{q_o} \neq 0$, then $\alpha\mathbf{q_o}$ is also a feasible solution $\forall\ \alpha > 0$; hence, the objective function $\mathbf{1}^T\mathbf{q}$ becomes unbounded.

Thus, the LP in Eqn. (8.20) has a feasible solution if and only if $(\mathbf{q}^T Y^f = 0$ and $\mathbf{q} \geq 0) \Rightarrow \mathbf{q} = 0$.   □

Theorem 8.14:   Every Boolean function f of n variables (for n even) can be expressed as a threshold function of $2n$ Boolean functions $f_1, f_2, \ldots, f_{2n}$, such that (1) $C_{ff_i} = 0$, for all i, $1 \leq i \leq 2n - 1$, $i \neq 4$, and (2) $C_{ff_4} = 2^{-(n-1)}$.

Proof: The proof will be developed by constructing a $2^n \times 2n$ (n even) matrix A with ± 1 entries that satisfies the following properties:

1. Every positive linear combination (PLC) of a nonempty subset of the rows of A has at least one positive coordinate. Thus, it follows from the duality theory of Linear Programming (see the proof of Lemma 8.8) that $A\mathbf{w} \geq \mathbf{1}$, for some $\mathbf{w}$.

2. Every column, except the fourth one, sums to 0 and the fourth column sums to 2, i.e., $\mathbf{1}^T A = [0\ 0\ 0\ 4\ 0 \cdots 0]$.

For any given function f, set $Y^f = A$, and define $Y = (Y^f)^f = A^f$ (i.e., the Y matrix is derived by negating those rows of A that correspond to the -1 entries of the given function f). A proof of Theorem 8.14 then follows as a direct consequence of Lemma 8.8 and from the fact that $f = \text{sgn}(Y\mathbf{w})$ for some $\mathbf{w}$, i.e., the given function is a threshold function of the columns of Y.

The construction of the matrix A is recursive and is based on the following 6×4 array (partitioned into two sections for reasons to be clarified later); the first two rows will be referred to as r_a and r_b, respectively and the bottom four rows will be referred to as r_1, r_2, r_3, and r_4, respectively:

$$
\begin{array}{rrrr}
1 & 1 & -1 & -1 \qquad r_a \\
-1 & -1 & 1 & 1 \qquad r_b \\
\\
1 & 1 & 1 & -1 \qquad r_1 \\
1 & -1 & -1 & 1 \qquad r_2 \\
-1 & 1 & -1 & 1 \qquad r_3 \\
-1 & -1 & 1 & -1 \qquad r_4 \\
\end{array}
$$

Following are the properties of the rows of the array:

P1: Rows r_1, r_2, r_3, and r_4 sum to 0.

P2: Rows r_a and r_b sum to 0.

P3: Any positive linear combination of the rows that includes r_b but not r_a has at least one positive coordinate in the last two columns. This is because r_b possesses two 1's in the coordinates 3 and 4, while all other rows $r_1, \ldots, r_4$ contain one 1 and one -1 in those coordinates.

P4: Let S be any subset of the six rows in the array containing at least one row from among r_1, r_2, r_3, and not containing r_4. Then, every positive linear combination of S has at least one positive coordinate. This can be proved by enumeration of all possible cases: Assume that S contains the row r_2 but does not contain the row r_4. Then, one can verify that at least coordinates 1 or 4 of any PLC of the rows of S must be positive. Similarly, if S contains row r_3 but not r_4, then coordinates 2 or 4 of any PLC will be positive; finally if S contains r_1 but not r_4, then coordinates 1, 2, 3, or 4 of any PLC will be positive.

$$
\begin{array}{ll}
r_1 & r_a \\
r_1 & r_b \\
r_1 & r_a \\
r_1 & r_b \\
r_2 & r_a \\
r_2 & r_b \\
r_2 & r_a \\
r_2 & r_b \\
r_3 & r_a \\
r_3 & r_b \\
r_3 & r_a \\
r_3 & r_b \\
r_4 & r_1 \\
r_4 & r_2 \\
r_4 & r_3 \\
r_4 & r_4
\end{array}
$$

$r_4 \quad r_1 \qquad \leftarrow$ The last quarter starts here

$r_4 \quad r_4 \qquad \leftarrow$ Change the -1 entry in coordinate 4 to $+1$

Figure 8.1 Structure of the matrix A for $n = 4$ ($N = 16$).

Let $N = 2^n$ where n is even; therefore, N is a power of 4. Create an $N \times 4$ array where: Each of the first $N/4$ rows is r_1; each of the second $N/4$ rows is r_2; each of the third $N/4$ rows is r_3; and the last $N/4$ rows equal r_4.

Consider any subset S of the rows: If it does not contain any row from the bottom $N/4$ rows, then it doesn't contain r_4 and by property P4, discussed above, any PLC of S has at least one positive coordinate.

The construction of A can be continued by adding a second set of four columns to A (thus defining the first eight columns) as follows: the first $3N/4$ rows are alternately r_a and r_b; the last $N/4$ rows are divided into four groups of $N/16$ rows each, where the first $N/16$ rows are r_1, the second $N/16$ rows are r_2, the third $N/16$ rows are r_3, and the last $N/16$ rows are r_4. Fig. 8.1 shows the construction for $n = 4$ ($N = 16$).

A PLC of the rows of the array that does not contain any row from the bottom $N/16$ rows will always have at least one positive coordinate. This is because: (1) if the rows are chosen only from the top $3N/4$ rows, then it follows from the previous discussion that at least one coordinate in the first

four coordinates of the PLC will have a positive coordinate, (2) if the PLC
has at least one row from among other rows (i.e., from row number $3N/4+1$
to row number $(N - N/16) - 1$), then consider the last four coordinates: it
is a PLC of rows that contains at least one row from among r_1, r_2, and r_3
and does not contain r_4; hence, by property P4, at least one coordinate will
be positive.

Continue the procedure by adding $\log_4 N$ $(= n/2)$ layers, each of which
is four columns wide, thus constructing $2n$ columns. By then, any subset S
that does not contain the very bottom row has the following property: Any
PLC of the rows has at least one positive coordinate. Now change the -1 in
the fourth coordinate of the last row to a $+1$; this completes the construction
of A. It follows from property P3 that any PLC that contains the last row
of A must have at least one positive coordinate in the first four columns.

As for the column sums, properties $P1$ and $P2$ ensure that they are all
0, except for the fourth column that we modified in the last row, which sums
to 2. $\square$

Remark 8.4:

1. The matrix A in the preceding construction has identical rows. If one
   wants to construct an A with distinct rows, then one can add n extra
   columns to obtain an $2^n \times 3n$ matrix as follows: Add n extra columns,
   such that each row has a unique identifier in these columns. Since we
   use all 2^n identifiers, each of these n columns sums to 0.

2. Since any PLC of the rows of A has at least one positive coordinate,
   it implies that $A\mathbf{w} \geq \mathbf{1}$ for some $\mathbf{w}$. For the matrix A constructed in
   the proof of Theorem 8.14, a closed form description of such a $\mathbf{w}$ is not
   apparent.

3. Another decomposition of any given function, where a closed form
   description of the corresponding $\mathbf{w}$ can be given, is outlined in Exercise 8.16. The decomposition has the following properties: Every Boolean function f of n variables can be expressed as a threshold function of $2n$ Boolean functions: $f_1, f_2, \ldots, f_{2n}$ such that (1)
   $C_{ff_i} = 0$, $\forall\, 1 \leq i \leq 2n - 2$, and (2) $C_{ff_{2n-1}} = C_{ff_{2n}} = 2^{-(n-1)}$. $\square$

These constructions and the results in Section 8.4 thus provide a comprehensive understanding of the method of correlations. In particular: (1)

If the input functions are mutually orthogonal (or strongly asymptotically orthogonal), then the method of correlations is effective even if exponential weights are allowed, i.e., if a function is exponentially small correlated with every function from a pool of possible input functions, then one would require exponentially many input functions to implement the given function using a threshold gate; (2) If the weights are polynomially bounded integers and if the function is exponentially small correlated with every function from a set of (not necessarily orthogonal) input functions, then again exponentially many input functions are needed to implement the given function using a threshold gate; (3) If the input functions are not mutually orthogonal, then the method of correlations may not be applicable, i.e., one can construct examples in which the output function is correlated exponentially small with every input function, and yet it can be computed as a threshold function of polynomially many input functions.

Exercises

8.1: *[Spectrum of AND and OR]*
For the n-variable AND and OR functions show that the spectral coefficients satisfy the following:
$$|a_{(0,\ldots,0)}| = 1 - 2^{1-n},$$
and for all $\alpha \neq (0,\ldots,0)$
$$|a_\alpha| = 2^{1-n}.$$

8.2: *[Spectrum of $CQ(X)$]*
Give a proof of Theorem 8.7 for odd values of n.

8.3: *[Spectrum of Majority]*
Complete the proof of Theorem 8.8 by computing the spectral coefficients a_α of $MAJ_n(X) = \text{sgn}(\sum_{i=1}^{n} x_i)$, $x_i \in \{1, -1\}$, when $|\alpha| = \sum_{i=1}^{n} \alpha_i \geq \dfrac{n+1}{2}$.

8.4: *[A combinatorial identity]*

(a) Given non-negative integers a and b ($a \leq b$), let
$$S(a,b) = \sum_{j=-a}^{a} (-1)^j \binom{2a}{a+j}\binom{2b}{b+j} = \sum_{j=-a}^{a} (-1)^j \binom{2a}{a+j}\binom{2b}{b-j}.$$

Show that $(2b+1)S(a+1,b) = (2a+1)S(a,b+1)$.

(b) Using the above result and by choosing an appropriate inductive hypothesis, show that

$$S(a,b) = \frac{\binom{2a}{a}\binom{2b}{b}}{\binom{a+b}{a}} = \frac{(2a)!\,(2b)!}{a!\,b!\,(a+b)!}\, .$$

8.5:   [*On approximating the Majority function*]
Show that if the Majority function of n variables, $MAJ_n(X)$, can be approximated by a sparse polynomial with $o(n^{-1})$ error, then it will lead to the contradiction that the CQ function can be expressed as the sign of a sparse polynomial (i.e., $CQ \in PT_1$). This argument gives an indirect proof of a weaker version of Theorem 8.11. (*Hint:* First realize a depth-2 circuit for implementing $CQ(X)$, and then, use the hypothesis to approximate each threshold gate in the first layer by a sparse polynomial.)

8.6:   [*Reducing approximation error*]
Let $\{f_n : n = 1, 2, \ldots\}$ be a family of functions (in n variables) such that for any $n_2 > n_1$, f_{n_1} can be obtained from f_{n_2} via projection. Suppose that for some fixed $\epsilon > 0$, f_n can be approximated by a sparse polynomial with an error bounded by $O(n^{-\epsilon})$. Show that f_n can also be approximated by a sparse polynomial with an error bounded by $O(n^{-c})$, for any constant $c > \epsilon$.

8.7:   [*Strong degree of Parity*]
Let $P(x_1, \ldots, x_n)$ be a multilinear polynomial in $\{0,1\}^n$ such that

$$PAR_n(x_1, \ldots, x_n) = \text{sgn}(P(x_1, \ldots, x_n)).$$

Factorize P into

$$P(x_1, \ldots, x_n) = x_1 \cdot Q(x_2, \ldots, x_n) + R(x_2, \ldots, x_n).$$

First show that $\text{sgn}(R) = PAR_n(x_2, \ldots, x_n)$, and then, argue that $\text{sgn}(Q)$ is the negation of $PAR_n(x_2, \ldots, x_n)$. Now use induction to show that P must have degree n. This provides an alternative approach for showing that $SD(PAR_n\ (x_1, \ldots, x_n)) = n$.

8.8: *[An alternative proof of Corollary 8.4]*

Let $f(X) : \{1, -1\}^n \to \{1, -1\}$ and let $F(X) = \sum\limits_{\alpha \in Q} w_\alpha X^\alpha$ for some $Q \subset \{0, 1\}^n$ such that $F(X) \neq 0$. Denote the spectral representation of $f(X)$ by $\sum\limits_{\alpha \in \{0,1\}^n} a_\alpha X^\alpha$.

(a) Show that $f(X) = \text{sgn}(F(X))$ if and only if

$$\sum_{X \in \{1,-1\}^n} |F(X)| = \sum_{X \in \{1,-1\}^n} f(X)F(X).$$

(b) Show that $f(X) = \text{sgn}(F(X))$ if and only if

$$\sum_{X \in \{1,-1\}^n} |F(X)| = 2^n \sum_{\alpha \in Q} w_\alpha a_\alpha.$$

(c) Show that

$$\sum_{X \in \{1,-1\}^n} |F(X)| \geq 2^n w_{0...0}.$$

Hence, show that for all $\alpha \in Q$,

$$\sum_{X \in \{1,-1\}^n} |F(X)| \geq 2^n |w_\alpha|.$$

(d) Using the results in (b) and (c), show that if $f(X) = \text{sgn}(F(X))$, then

$$\sum_{\alpha \in Q} |w_\alpha| \leq |S| \sum_{\alpha \in Q} |w_\alpha||a_\alpha| .$$

(e) Conclude from (d) that

$$|Q| \geq \frac{1}{\max_{\alpha \in S} |a_\alpha|} .$$

This gives an alternative proof of Corollary 8.4.

8.9: *[Spectral norm and the Parity function]*

Let $\sum\limits_{\alpha \in \{0,1\}^n} a_\alpha X^\alpha$ be the polynomial representation of $f(X) : \{1, -1\}^n \to \{1, -1\}$. Suppose that the L_1 spectral norm of f is 1.

(a) Show that for each $X \in \{1, -1\}^n$, either $a_\alpha X^\alpha \geq 0$ for all $\alpha \in \{0, 1\}^n$, or $a_\alpha X^\alpha \leq 0$ for all $\alpha \in \{0, 1\}^n$.

(b) Choose any $a_{\hat{\alpha}} \neq 0$. Show that when $f(X) = 1$, $\mathrm{sgn}(a_{\hat{\alpha}}) = X^{\hat{\alpha}}$, and when $f(X) = -1$, $\mathrm{sgn}(a_{\hat{\alpha}}) = -X^{\hat{\alpha}}$. Hence, $f(X) = X^{\hat{\alpha}}$ if $a_{\hat{\alpha}} > 0$ and $f(X) = -X^{\hat{\alpha}}$ if $a_{\hat{\alpha}} < 0$.

(c) Conclude that for any Boolean function $f(X) : \{1, -1\}^n \rightarrow \{1, -1\}$, the L_1 spectral norm $\|f\|_{\mathcal{F}} = 1$ if and only if $f(X) = X^{\hat{\alpha}}$ or $f(X) = -X^{\hat{\alpha}}$ for some $\hat{\alpha} \in \{0, 1\}^n$. Hence, the monomials are the only Boolean functions with spectral norms equal to 1. Can you generalize this result to other bases?

8.10:   [*Depth-2 circuits with symmetric gates*]
Show that the Inner Product Mod 2 function cannot be computed by a polynomial-size depth-2 circuit with generalized symmetric gates in the first layer and a threshold gate with polynomially bounded integer weights in the second layer.

8.11:   [*Generalization of Uniqueness Theorem*]
Let $g(x)$ be a strictly increasing function, i.e., $g(x_1) > g(x_2)$ if $x_1 > x_2$. Let $\{f_i : i = 1, \ldots, S\}$ be a set of S (column) vectors in R^m and let $Y = [f_1 \cdots f_S]$. Given $\mathbf{w_1}$ and $\mathbf{w_2}$ in R^S, define two vectors h_1 and h_2 in R^m by $(h_1)_i = g((Y\mathbf{w_1})_i)$ and $(h_2)_i = g((Y\mathbf{w_2})_i)$. The correlations are defined as $C_{h_1 Y} = (Y^T h_1)/m$ and $C_{h_2 Y} = (Y^T h_2)/m$.

(a) Show that $[g(x_1) - g(x_2)](x_1 - x_2) > 0$ whenever $x_1 \neq x_2$.

(b) Show that $(h_1 - h_2)^T Y(\mathbf{w_1} - \mathbf{w_2}) > 0$ if $h_1 \neq h_2$. Hence, conclude that $C_{h_1 Y} \neq C_{h_2 Y}$ if $h_1 \neq h_2$.

8.12:   [$\widehat{PT_1}$ *and functions computed by depth-2 circuits*]

(a) Show that a function computed by a depth-2 circuit with polynomially many AND/OR gates in the first layer and a threshold gate with polynomially bounded weights in the second layer is in $\widehat{PT_1}$.

(b) In general, show that a function computed by a polynomial-size depth-2 circuit with gates computing functions of polynomially bounded spectral norms in the first layer and a threshold gate with polynomially bounded integer weights in the second layer is in $\widehat{PT_1}$.

8.13:   [*Lindsey's Lemma*]

Let M be an $n \times n$ matrix with entries in $\{1, -1\}$. Suppose the columns of M are mutually orthogonal. Given any two binary (column) vectors x and y in $\{0, 1\}^n$, show that

$$|x^T M y| \leq \sqrt{\|x\| \cdot \|y\| \cdot n}.$$

Hence, conclude the result of Lindsey's Lemma (Lemma 8.6).

8.14:   [*Covering a strictly triangular matrix*]

Let M be an $m \times m$ strictly triangular matrix with entries from $\{1, -1\}$. If M is divided into m^2/l^2 square submatrices (where l divides m) each of size $l \times l$, then show that:

(a) There are at most $2(m/l)$ square submatrices that contain both 1's and -1's; let R be the set of such submatrices. Give an example of a strictly triangular matrix in which there are $2(m/l)$ such square submatrices.

(b) The 1's that are not in R can be covered by at most (m/l) rectangles each of width l and height $\leq m$.

(c) The -1's that are not in R can be covered by at most (m/l) rectangles each of width l and height $\leq m$.

8.15:   [*A nonorthogonal set of basis functions*]

(a) Let $f_\wedge \in \mathcal{R}^{2^n}$ be the vector representation of the n-variable AND function $x_1 \wedge \cdots \wedge x_n$, $x_i \in \{1, -1\}$. Show that $f_\wedge$ has only one coordinate that equals -1 and rest of its coordinates equal $+1$.

(b) Let $\{f_{\wedge,i}\}$ be the set of all the 2^n n-variable AND functions obtained by complementing different sets of input variables. Show that the set of vectors $\{f_{\wedge,i}\}$ forms a basis for $\mathcal{R}^{2^n}$.

8.16:   [*Another threshold decomposition procedure*]

Consider the following 4×2 array:

$$
\begin{array}{ccc}
1 & -1 & r_a \\
-1 & 1 & r_b \\
 & & \\
1 & 1 & r_1 \\
-1 & -1 & r_2 \\
\end{array}
$$

Form a $2^n \times 2n$ matrix as follows (let $N = 2^n$):

1. For the first two columns: set the first $N/2$ rows equal to r_1, and the next $N/2$ rows equal to r_2. For the second two columns: set the first $N/2$ rows alternately to rows r_a and r_b; in the bottom $N/2$ rows, set the first $N/4$ rows to row r_1 and the second $N/4$ rows to row r_2. This pattern is continued until $2n$ columns are created, i.e., columns i and $i + 1$ (for odd i) have half of their last $2^{n-(i-1)}$ rows equal to r_1 and the other half equal to r_2, and the rest of the rows are alternately r_a and r_b.

2. After the above construction, the last row has -1 in its last two columns; change these two to $+1$.

(a) Show that the weight vector

$$\mathbf{w}^T = \frac{1}{2}[1\ 1\ 2\ 2\ 4\ 4\ \cdots\ 2^n\ 2^n]$$

satisfies the following properties: (1) $A\mathbf{w} = \mathbf{1}$, and (2) $(\mathbf{1}^T A)/2^n = [0 \cdots 0\ 2^{-(n-1)}\ 2^{-(n-1)}]$.

(b) Conclude that every Boolean function f of n variables can be expressed as a threshold function of $2n$ Boolean functions: $f_1, f_2, \ldots, f_{2n}$ such that (1) $C_{ff_i} = 0$, $\forall\ 1 \le i \le 2n - 2$, and (2) $C_{ff_{2n-1}} = C_{ff_{2n}} = 2^{-(n-1)}$. Give an explicit construction of the corresponding weight vector.

(c) In the preceding construction, there are two columns of A that sum to $+2$ (instead of just one column as in the construction for Theorem 8.14). Using the technique used in the proof of Theorem 8.14, modify the construction so that there is only one uncorrelated column.

Bibliographic Notes

The spectral/polynomial representation of Boolean functions, introduced in Section 8.5, was first introduced in Muller [82]; see also [44, 69, 89]. Such polynomial representations have extensive applications in coding theory and detailed discussions can be found in standard textbooks such as

[76]. The application of the spectral approach to linear threshold functions is shown in Chow [27], and further discussions can be found in books such as [70, 86]. Recent applications of this representation to prove lower bounds on the depth/size of circuits are described in [17, 18, 100, 126]. Chapter 7 discusses some of these applications in more detail. The spectral representation has also been used in Linial, Mansour, and Nisan [71] to characterize the complexity class AC^0.

The geometric framework and the results in Sections 8.2–8.4 (such as Theorems 8.1 and 8.3 in Section 8.2) are derived in Roychowdhury, Siu, Orlitsky, and Kailath [114, 115]. These results provide a unified and simple framework for generalizing and deriving other results based on spectral techniques. The spectral coefficients of Inner Product Mod 2 and Complete Quadratic functions (as derived in Section 8.5.1) can be found in [76] and [18], respectively. The spectrum of the Majority function is from Brandman, Orlitsky, and Hennessy [17].

The representation of a Boolean function as the sign of a polynomial of the input variables is studied in Baldi [10]. The classes of PT_1 and $\widehat{PT_1}$ functions are introduced in Bruck [18]. Corollary 8.3 is derived in Chow [27] for the case where each input function is either an input variable (x_i or $-x_i$) or a constant function (1 or −1). Bruck [18] uses the same techniques as Chow and generalizes it to the case where, in addition to the input variables, the input functions can be parities of the input variables. The results in [18, 27] are special cases of the more general results (Theorem 8.1) proved in [114]. Corollary 8.4 is shown in [18] and the technique used is outlined in Exercise 8.8. We have presented an alternative simpler proof and have shown that it follows as a special case of the more general result stated in Theorem 8.3 [114]. Lemma 8.2 is stated in [18], which is in fact a special case of an earlier result shown in Hajnal, Maass, Pudlák, Szegedy, and Turán [48]. Theorem 8.9 is derived in [18].

The lower bound results in Section 8.6 on the approximation error when a Boolean function is approximated by a linear combination of basis functions are taken from Roychowdhury, Siu, Orlitsky, and Kailath [114]. An indirect proof showing that the Majority function is not in $\widetilde{SP}$ is also presented in [123].

The method of correlations as presented in Section 8.7 is due to [48]. However, the concept of ϵ-discriminators is used in [48] without explicitly introducing the notion of correlation. Lemmas 8.5 and 8.7 and Theorem 8.12 are also derived in the pioneering work of [48]. In [114], some of the limitations of the method of correlations are established, and the results in Section 8.7.2 and Theorem 8.14 are derived therein.

Chapter 9

Limitations of AND-OR Circuits

9.1 Introduction

In this chapter, we derive lower bound results to demonstrate the limitation of constant-depth polynomial-size AND-OR circuits (AC^0 circuits). In particular, we prove that any constant-depth AND-OR circuit with unbounded fan-in that computes the n-variable Parity function must have size exponential in n. This is probably the most well-known result in circuit complexity theory and several proofs based on different techniques have been found.

Here we adopt an algebraic approach that also allows us to prove a more general result. More precisely, we show in Section 9.2 that any depth-d circuit with gates AND, OR, NOT, and MOD_p (where p is a prime) requires at least $2^{\Omega(n^{1/2d})}$ gates to compute the MOD_r function of n variables, when r is not a power of p. For simplicity, we only give a proof for the special case when $p = 3$ and $r = 2$. The proof for this special case contains the essential ideas of the underlying techniques; the exercises at the end of the chapter outline the proof for the general case.

In Section 9.3, we derive another result that also illustrates the computational limitation of AC^0 circuits. In particular, any function computable by AC^0 circuits can also be computed by depth-3 threshold circuits of superpolynomial-size ($2^{O(\log^k n)}$). The proof of this result follows essentially the same algebraic and probabilistic approach as in Section 9.2.

9.2   Lower Bounds on AC^0 Circuits

Before we present the formal proof of the lower bound results on AC^0 circuits, it is helpful to sketch the essential idea. We work with polynomials over the finite field Z_3. We consider 0, 1 as elements in Z_3, and a Boolean function as $f : \{0,1\}^n \to Z_3$, where f only takes on values 0 or 1 in Z_3. First we show that AND, OR, NOT and MOD_3 gates can be represented as low-degree polynomials over a large domain $D \subset \{0,1\}^n$. Given a small-depth circuit of these gates, we show that the function computed by the circuit can be represented by a low-degree polynomial over a large domain if the circuit size is small. On the other hand, we can show that the Parity (MOD_2) function cannot have low-degree polynomial representation over a large domain. As a result, we conclude that if the circuit computes the Parity function, the size of the circuit must be large.

As usual, we use the symbols $\bigvee$ and $\bigwedge$ to denote the OR and AND functions, respectively.

Lemma 9.1:   Let $y_i \in \{0,1\}$ for $i = 1, ..., t$. Assume that the arithmetic is defined over Z_3. For every integer $k > 0$, we have the following:

1. $\text{NOT}(y_i) = 1 - y_i$.

2. $\text{MOD}_3(y_1, y_2, ..., y_t) = 1 - (\sum_{i=1}^{t} y_i)^2$.

3. $\bigwedge(y_1, y_2, ..., y_t) = \prod_{i=1}^{t} y_i$.

4. $\bigvee(y_1, y_2, ..., y_t) = 1 - \prod_{i=1}^{t}(1 - y_i)$.

5. There is a polynomial $P_{OR}(y_1, y_2, ..., y_t)$ of degree $\leq 2k$ such that $P_{OR} = OR$ over a domain $D \subset \{0,1\}^t$ with $|D| \geq 2^t(1 - \frac{1}{3^k})$.

6. There is a polynomial $P_{AND}(y_1, y_2, ..., y_t)$ of degree $\leq 2k$ such that $P_{AND} = AND$ over a domain $\tilde{D} \subset \{0,1\}^t$ with $|\tilde{D}| \geq 2^t(1 - \frac{1}{3^k})$.

Proof:

1. Claim 1 is easily verified by direct substitution.

2. Observe that for any element x in Z_3, $x^2 = 1$ if $x \neq 0$, and $x^2 = 0$ if $x = 0$. So $1 - (\sum_{i=1}^{t} y_i)^2$ equals 0 if $(\sum_{i=1}^{t} y_i) \neq 0 \bmod 3$ and is 1 otherwise.

3. Note that $\prod_{i=1}^{t} y_i = 1$ if and only if all $y_i = 1$.

4. $\prod_{i=1}^{t}(1-y_i) = 1$ if all $y_i = 0$ and is 0 otherwise. Thus, $1-\prod_{i=1}^{t}(1-y_i) = 1$ if some $y_i = 1$.

5. Consider the polynomial $P_{OR}(y_1, y_2, ..., y_t) = \bigvee_{i=1}^{k} (\sum_{j=1}^{t} c_{ij} y_j)^2$, where each $c_{ij} \in Z_3$. It follows from Claims 2 and 4 that $P_{OR}(y_1, y_2, ..., y_t)$ can be expressed as a polynomial of degree $\leq 2k$. It remains to show that there exists a choice of the elements $c_{ij} \in Z_3$ such that $P_{OR}(y_1, y_2, ..., y_t) = \bigvee(y_1, y_2, ..., y_t)$ for all $(y_1, y_2, ..., y_t)$ over a large domain. First assume some fixed values of the inputs $(y_1, y_2, ..., y_t)$. Choose each c_{ij} independently with $Pr\{c_{ij} = 0\} = Pr\{c_{ij} = 1\} = Pr\{c_{ij} = 2\} = 1/3$. There are two cases to be considered.

   (a) If $\bigvee(y_1, y_2, ..., y_t) = 0$, then all $y_i = 0$. Thus, $\sum_{j=1}^{t} c_{ij} y_j = 0$ and $P_{OR}(y_1, y_2, ..., y_t) = 0$.

   (b) If $\bigvee(y_1, y_2, ..., y_t) = 1$, then some $y_j = 1$. For $i = 1, ..., k$, let $S_i = \sum_{j=1}^{t} c_{ij} y_j = \sum_{j: y_j=1} c_{ij}$. It can be shown that the S_i's are independent and uniformly distributed over Z_3 (see Exercise 9.1). Therefore, $Pr\{P_{OR} = \bigvee_{i=1}^{k} S_i^2 = 0\} = Pr\{S_i = 0 \text{ for } i = 1, ..., k\} = 1/3^k$. In other words, $Pr\{P_{OR}(y_1, y_2, ..., y_t) \neq \bigvee(y_1, y_2, ..., y_t)\} = 1/3^k$.

Thus, for every $(y_1, y_2, ..., y_t) \in \{0, 1\}^t$,

$$Pr\{P_{OR}(y_1, y_2, ..., y_t) \neq \bigvee(y_1, y_2, ..., y_t)\} \leq 1/3^k.$$

Hence, it follows from the pigeonhole principle that there are (fixed) c_{ij}'s such that

$$P_{OR}(y_1, y_2, ..., y_t) \neq \bigvee(y_1, y_2, ..., y_t)$$

for $\leq 1/3^k$ of all possible inputs $(y_1, y_2, ..., y_t)$. Equivalently, there is a domain $D \subset \{0, 1\}^t$ with $|D| \geq 2^t(1-\frac{1}{3^k})$ such that $P_{OR}(y_1, y_2, ..., y_t) = \bigvee(y_1, y_2, ..., y_t)$ for all $(y_1, y_2, ..., y_t) \in D$.

6. One can simply let $P_{AND}(y_1, y_2, ..., y_t) = 1 - P_{OR}(1 - y_1, 1 - y_2, ..., 1 - y_t)$. Then, the polynomial P_{AND} will have the same degree as P_{OR}, and will equal $AND(y_1, y_2, ..., y_t)$ for all $(y_1, y_2, ..., y_t)$ such that $(1 - y_1, 1 - y_2, ..., 1 - y_t)$ is in the domain D of claim 5 above.   □

Lemma 9.2:   Let C be a circuit of AND, OR, NOT, and MOD$_3$ gates and $X \in \{0,1\}^n$ be the input to the circuit C. Let $G(y_1, y_2, ..., y_t)$ be any gate in the circuit, where the y_i's are inputs to the gate G and are functions of the circuit inputs X. For every integer $k > 0$, there is a polynomial $P(y_1, y_2, ..., y_t)$ of degree $\leq 2k$ such that $P = G$ for every circuit input $X \in \hat{D}$, and $|\hat{D}| \geq 2^n(1 - \frac{1}{3^k})$.

Proof: From the results in Lemma 9.1, we only need to prove the lemma for the cases when the gate G is $\bigvee$ or $\bigwedge$. First consider the case of $G = \bigvee$ gate. From the proof of Claim 5 in Lemma 9.1, it follows that for each possible input $(y_1, y_2, ..., y_t)$ to the G gate, there exists a random polynomial P_{OR} of degree $\leq 2k$ such that $Pr\{P_{OR}(y_1, y_2, ..., y_t) \neq \bigvee(y_1, y_2, ..., y_t)\} \leq 1/3^k$. Clearly, since each of the y_i's is a function of the circuit input X, we conclude that for each possible circuit input X, $Pr\{P_{OR}(y_1, y_2, ..., y_t) \neq G(y_1, y_2, ..., y_t)\} \leq 1/3^k$. Again by the pigeonhole principle, there are (fixed) c_{ij}'s such that $P_{OR}(y_1, y_2, ..., y_t) \neq G(y_1, y_2, ..., y_t)$ for $\leq 1/3^k$ of all possible circuit inputs X. Equivalently, there is a domain $\hat{D} \subset \{0,1\}^n$ with $|\hat{D}| \geq 2^n(1 - \frac{1}{3^k})$ such that $P_{OR}(y_1, y_2, ..., y_t) = \bigvee(y_1, y_2, ..., y_t)$ for all $X \in \hat{D}$. For the case of $\bigwedge$ gate, apply the previous argument to $\bigvee(1 - y_1, 1 - y_2, ..., 1 - y_t)$ and the result follows.   □

Theorem 9.1:   Let $C(X)$ denote the output of a depth-d circuit C of AND, OR, NOT, and MOD$_3$ gates with input $X \in \{0,1\}^n$. If the size of C is S, then there is a domain $D \subset \{0,1\}^n$ and a polynomial $\hat{P}(X)$ of degree $\leq (2k)^d$ such that $\hat{P}(X) = C(X)$ for all $X \in D$, and $|D| \geq 2^n(1 - \frac{S}{3^k})$.

Proof: For each gate $G(y_1, y_2, ..., y_t)$ in the circuit C, apply Lemma 9.2 to obtain a polynomial $P_G(y_1, y_2, ..., y_t)$ of degree $\leq 2k$ such that $P_G \neq G$ for $\leq 2^n/3^k$ possible circuit inputs X. Starting from the gates in the first layer and then proceeding layer by layer, we can rewrite each P_G as a polynomial in X. Since the depth of the circuit is d, the polynomial $\hat{P}(X)$ corresponding to the output gate of the circuit has degree $\leq (2k)^d$. Moreover, since

$$\{X : \hat{P}(X) \neq C(X)\} \subset \bigcup_{G \text{ in } C} \{X : P_G \neq G\},$$

and the number of gates in the circuit is S, we conclude that $\hat{P}(X) \neq C(X)$ for at most $2^n S/3^k$ possible circuit inputs X. Equivalently, there is a domain $D \subset \{0,1\}^n$ with $|D| \geq 2^n(1 - \frac{S}{3^k})$ and $\hat{P}(X) = C(X)$ for all $X \in D$. $\square$

Definition 9.1: Let $f : \{0,1\}^n \rightarrow Z_3$ and $D \subset \{0,1\}^n$. The degree of f in D, denoted by $deg_D(f)$, is the minimum degree of a polynomial $p(X)$ such that $p(X) = f(X)$ for all $X \in D$. $\square$

Lemma 9.3: For any $D \subset \{0,1\}^n$, if $deg_D(Parity) = h$, then for any $f : \{0,1\}^n \rightarrow Z_3$, we have $deg_D(f) \leq \frac{n}{2} + h$.

Proof: Every function $f : \{0,1\}^n \rightarrow Z_3$ can be expressed as a multilinear polynomial in $X = (x_1, ..., x_n) \in \{0,1\}^n$ over Z_3. If we let $z_i = 1 + x_i$, then we can rewrite any f as a multilinear polynomial in z_i. Therefore, it suffices to show that for any monomial $\prod_{i \in A} z_i$, where $A \subset \{1, ..., n\}$, we have

$$deg_D(\prod_{i \in A} z_i) \leq \frac{n}{2} + h.$$

Note that in Z_3, $2^m = 1$ if m is even and $2^m = 2$ if m is odd. Hence, $Parity(x_1, ..., x_n)$ can be written as $\prod_{i=1}^{n}(1 + x_i) - 1 = \prod_{i=1}^{n} z_i - 1$. Thus,

$$deg_D(\prod_{i=1}^{n} z_i) = deg_D(Parity) = h.$$ Now for a general monomial $\prod_{i \in A} z_i$, there are two cases to be considered.

1. If $|A| \leq n/2$, then clearly $deg_D(\prod_{i \in A} z_i) \leq n/2$.

2. If $|A| > n/2$, then $|A^c| \leq n/2$ and $\prod_{i \in A} z_i = \prod_{i \in A^c} z_i \cdot \prod_{i=1}^{n} z_i$, since $z_i^2 = 1$ in Z_3. Hence, $deg_D(\prod_{i \in A} z_i) \leq deg_D(\prod_{i \in A^c} z_i) + deg_D(\prod_{i=1}^{n} z_i) \leq n/2 + h$.

Hence, we have shown that $deg_D(f) \leq \frac{n}{2} + h$ for any function f. $\square$

Lemma 9.4: Let $D \subset \{0,1\}^n$. If every function $f : D \rightarrow Z_3$ has $deg_D(f) \leq \ell$, then $|D| \leq \sum_{j=0}^{\ell} \binom{n}{j}$.

Proof: The set of all functions $f : D \rightarrow Z_3$ forms a vector space over Z_3 of dimension $|D|$. Let M denote the set of all monomials of degree $\leq \ell$. By assumption, since every function f on D can be expressed as a polynomial of degree $\leq \ell$, M contains a basis of the vector space. But $|M| = \sum_{j=0}^{\ell} \binom{n}{j}$. Hence, the number of vectors in the basis, which is the dimension $|D|$, must be less than $\sum_{j=0}^{\ell} \binom{n}{j}$.   □

Theorem 9.2:   Any depth-d circuit of AND, OR, NOT, and MOD$_3$ gates requires at least $2^{\Omega(n^{1/2d})}$ gates to compute the n-variable Parity function.

Proof: Let C be a depth-d circuit of AND, OR, NOT, and MOD$_3$ gates with size $S = 2^{o(n^{1/2d})}$. Suppose C computes Parity. From the result in Theorem 9.1, we can choose k such that $2^n(1 - \frac{S}{3^k}) = 2^n(1 - o(1))$ and $(2k)^d = o(\sqrt{n})$; moreover, $deg_D(Parity) = o(\sqrt{n})$ for some $|D| = 2^n(1 - o(1))$. On the other hand, Lemma 9.3 and Lemma 9.4 imply that $|D| \leq \sum_{j=0}^{n/2 + o(\sqrt{n})} \binom{n}{j} = 2^{n-1} + o(2^n)$, which is a contradiction.   □

9.3   Simulating AC^0 Circuits by Depth-3 Threshold Circuits

In this section, we show that any function computable by a constant-depth polynomial-size AND-OR circuit (AC^0 function) can also be computed by a depth-3 threshold circuit of superpolynomial-size ($2^{O(\log^k n)}$). We use algebraic and probabilistic techniques similar to those in the previous section where we derived the lower bound results for AC^0 circuits. The idea is that for every AC^0 function, one can construct a random polynomial of low degree over the field Z_2, such that the random polynomial agrees with the AC^0 function with high probability. Using a standard argument by taking the majority function of independent copies of these random polynomials, we then show that for every input, the probability that the resulting Majority function does not agree with the AC^0 function is less than 2^{-n}. Hence, we conclude that there are polynomials such that the Majority function of them agrees with the AC^0 function for all inputs. We then show that any such function can be computed by a depth-3 threshold circuit (of superpolynomial size).

In the following discussion, all the arithmetic operations are defined over Z_2. Every Boolean function f can be considered as $f : \{0,1\}^n \rightarrow Z_2$. Also observe that in Z_2, we can express $\bigwedge_{i=1}^{n} x_i$ as $\prod_{i=1}^{n} x_i$.

Lemma 9.5: Given any $c > 0$, there are polynomials $Q_{OR}(y_1, y_2, ..., y_n)$ and $Q_{AND}(y_1, y_2, ..., y_n)$ of degree $O(\log n)$ defined over some probability distribution such that for every $(y_1, y_2, ..., y_n) \in \{0,1\}^n$,

$$Pr\{Q_{OR}(y_1, y_2, ..., y_n) \neq \bigvee(y_1, y_2, ..., y_n)\} = O(n^{-c}),$$

and

$$Pr\{Q_{AND}(y_1, y_2, ..., y_n) \neq \bigwedge(y_1, y_2, ..., y_n)\} = O(n^{-c}).$$

Proof: We first prove the result for the OR ($\bigvee$) function. Consider the polynomial $Q_{OR}(y_1, y_2, ..., y_n) = 1 - \prod_{i=1}^{k}(1 + \sum_{j=1}^{n} c_{ij} y_j)$, where each c_{ij} is chosen independently with $Pr\{c_{ij} = 1\} = Pr\{c_{ij} = 0\} = 1/2$. There are two cases to be considered.

1. If $\bigvee(y_1, y_2, ..., y_n) = 0$, then all $y_j = 0$. Thus, for every i, $1 + \sum_{j=1}^{n} c_{ij} y_j = 1$, and $Q_{OR}(y_1, y_2, ..., y_n) = 0$.

2. If $\bigvee(y_1, y_2, ..., y_n) = 1$, then some $y_j = 1$. For $i = 1, ..., k$, let $S_i = 1 + \sum_{j=1}^{n} c_{ij} y_j = 1 + \sum_{j: y_j = 1} c_{ij}$. It can be shown (see Exercise 9.1) that the S_i's are independent and uniformly distributed over Z_2. Therefore, $Pr\{Q_{OR} = 1 - \prod_{i=1}^{k} S_i = 0\} = Pr\{S_i = 1 \text{ for } i = 1, ..., k\} = 1/2^k$. In other words, $Pr\{Q_{OR}(y_1, y_2, ..., y_n) \neq \bigvee(y_1, y_2, ..., y_n)\} = 1/2^k$.

Now take $k = c \log n$. Then, Q_{OR} is a polynomial of degree $O(\log n)$ and $Pr\{Q_{OR} \neq OR\} = O(n^{-c})$.

For the AND ($\bigwedge$) function, $Q_{AND}(y_1, y_2, ..., y_n) = 1 - Q_{OR}(1 - y_1, 1 - y_2, ..., 1 - y_n)$, where Q_{OR} is distributed as before, satisfies the claim. $\square$

Lemma 9.6: Let $C(X) : \{0,1\}^n \rightarrow Z_2$ be computable by a polynomial-size depth-d AND-OR circuit. Given any $k > 0$, there is a polynomial $P_C(X)$ of degree $O(\log^d n)$ distributed over some probability distribution such that $Pr\{P_C(X) \neq C(X)\} = O(n^{-k})$ for every $X \in \{0,1\}^n$.

Proof: We prove this lemma by induction on the depth d. The basis case $d = 1$ has been proved in Lemma 9.5. For the induction step, let $C(X)$ be the output of a depth-d polynomial-size AND-OR circuit with input $X \in \{0,1\}^n$. First assume that the output gate of the circuit is an AND ($\wedge$) gate. Then, we can write $C(X) = \bigwedge_{i=1}^{p_n} c_i(X)$, where each $c_i(X)$ is computable by a depth-$(d-1)$ polynomial-size AND-OR circuit, and $p_n = O(n^l)$ for some constant l. By inductive hypothesis, corresponding to each $c_i(X)$, there is a random polynomial $P_{c_i}(X)$ of degree $O(\log^{(d-1)} n)$ such that $Pr\{P_{c_i}(X) \neq c_i(X)\} = O(n^{-(k+l)})$ for every $X \in \{0,1\}^n$. It follows that $Pr\{P_{c_i}(X) \neq c_i(X)$ for some $i\} = O(n^{-(k+l)} \cdot p_n) = O(n^{-k})$. Moreover, for the output AND gate of the circuit, there is a polynomial $P_{out}(c_1(X), ..., c_{p_n}(X))$ of degree $O(\log n)$ such that $Pr\{P_{out}(c_1(X), ..., c_{p_n}(X)) \neq \bigwedge_{i=1}^{p_n} c_i(X)\} = O(n^{-k})$. When substituting $P_{c_i}(X)$ for each $c_i(X)$ in the polynomial P_{out}, we obtain a polynomial $P_C(X)$ of degree $O(\log^d n)$; furthermore,

$$Pr\{P_C(X) \neq \bigwedge_{i=1}^{p_n} c_i(X)\} \leq Pr\{P_{c_i}(X) \neq c_i(X) \text{ for some } i\} +$$

$$Pr\{P_{out}(c_1(X), ..., c_{p_n}(X)) \neq \bigwedge_{i=1}^{p_n} c_i(X)\}$$

$$= O(n^{-k}).$$

If the output gate of the circuit is an OR gate, then one can use essentially the same arguments to show the result of Lemma 9.6. $\square$

Theorem 9.3: If $f(X) : \{0,1\}^n \to \{0,1\}$ is a function computable by a depth-d polynomial-size AND-OR circuit, then $f(X)$ can be computed by a depth-3 threshold circuit of size $n^{O(\log^d n)}$.

Proof: By Lemma 9.6, there is a random polynomial $P_f(X)$ (over Z_2) of degree $O(\log^d n)$ such that $Pr\{P_f(X) \neq f(X)\} = O(n^{-k})$ for every $X \in \{0,1\}^n$. Now take q_n independent copies of these polynomials, $P_f^{(i)}(X)$, $i = 1, ..., q_n$. Consider the Majority function of the values of these polynomials, i.e., $MAJ(P_f^{(1)}(X), P_f^{(2)}(X), ..., P_f^{(q_n)}(X))$. By standard argument using the Chernoff Bound, it is easy to see that for $q_n = O(n^c)$, we have

$$Pr\{MAJ(P_f^{(1)}(X), P_f^{(2)}(X), ..., P_f^{(q_n)}(X)) \neq f(X)\} < 2^{-n}.$$

Hence, there exist some (deterministic) polynomials $\hat{P}_f^{(i)}(X)$ of degree $O(\log^d n)$ such that $MAJ(\hat{P}_f^{(1)}(X), \hat{P}_f^{(2)}(X), ..., \hat{P}_f^{(q_n)}(X)) = f(X)$ for all $X \in \{0,1\}^n$.

Note that each $\hat{P}_f^{(i)}(X)$ can be computed by a depth-2 circuit in which the output gate is a Parity gate (addition in Z_2) and the gates in the first layer are AND gates of fan-in $O(\log^d n)$. There are $O(n^c)$ of these depth-2 circuits and each of them has size at most $n^{O(\log^d n)}$. We therefore obtain a depth-3 circuit of size $n^{O(\log^d n)}$ computing $f(X)$, where the first layer consists of AND gates, the second layer consists of Parity gates, and the output is a majority gate. Since a Parity gate can be expressed as a sum of linear threshold gates, the last two layers can be converted into two layers of threshold gates (see Section 3.2). Thus, we obtain a depth-3 threshold circuit of size $n^{O(\log^d n)}$, as in Theorem 9.3. $\square$

Exercises

9.1: [*A sum of independent identically distributed variables over a finite field*]
Let p be a prime. Let c_{ij}'s be independent random variables uniformly distributed over Z_p, i.e., $Pr\{c_{ij} = 0\} = Pr\{c_{ij} = 1\} = \cdots = Pr\{c_{ij} = p - 1\} = \frac{1}{p}$. Show that the random variables S_j's where $S_j = \sum_{i=1}^{t} c_{ij}$ are independent and uniformly distributed over Z_p.

9.2: [*The general case of Lemma 9.1*]
Let p be a prime and $y_i \in \{0,1\}$ for $i = 1, ..., t$. Assume that the arithmetic is defined over Z_p. For every integer $k > 0$, show the following:

(a) $NOT(y_i) = 1 - y_i$.

(b) $MOD_p(y_1, y_2, ..., y_t) = 1 - (\sum_{i=1}^{t} y_i)^{p-1}$.

(c) $\bigwedge(y_1, y_2, ..., y_t) = \prod_{i=1}^{t} y_i$.

(d) $\bigvee(y_1, y_2, ..., y_t) = 1 - \prod_{i=1}^{t}(1 - y_i)$.

(e) There is a polynomial $P_{OR}(y_1, y_2, ..., y_t)$ of degree $\leq (p-1)k$ such that $P_{OR} = OR$ over a domain $D \subset \{0,1\}^t$ with $|D| \geq 2^t(1 - \frac{1}{p^k})$.

(f) There is a polynomial $P_{AND}(y_1, y_2, ..., y_t)$ of degree $\leq (p-1)k$ such that $P_{AND} = AND$ over a domain $\tilde{D} \subset \{0,1\}^t$ with $|\tilde{D}| \geq 2^t(1 - \frac{1}{p^k})$.

9.3:   [*A complete function over a finite field*]

Let $\mathcal{F}$ be a finite field, $f(x_1, ..., x_n) : \{0,1\}^n \to \mathcal{F}$, and $D \subset \{0,1\}^n$. Let α be an element in $\mathcal{F}$ that is not 0 or 1 and let $y_i = (\alpha - 1)x_i + 1$. Recall that $deg_D(f)$ is the minimum degree of a polynomial $p(X)$ over $\mathcal{F}$ such that $p(X) = f(X)$ for all $X \in D$. Show that if $deg_D(\prod_{i=1}^n y_i) = h$, then for any $f : \{0,1\}^n \to \mathcal{F}$, we have $deg_D(f) \leq \frac{n}{2} + h$.

9.4:   [*AC^0 reductions between modulo functions*]

Let $\mathrm{mod}_{s,q}(X) = 1$ if the number of 1's in X is equal to s mod q, and $\mathrm{mod}_{s,q}(X) = 0$ otherwise. Note that $\mathrm{mod}_{0,q}(X)$ is the negation of $\mathrm{mod}_q(X)$. Let q be a divisor of r. Show that for $s = 0, ..., q-1$, $\mathrm{mod}_{s,q}(X)$ is AC^0 reducible to $\mathrm{mod}_r(X)$.

9.5:   [*Roots of unity*]

An element a in a finite field is said to be a qth root of unity if $a \neq 1$ and $a^q = 1$. Suppose $\mathcal{F}$ is a finite field that contains α as a qth root of unity. Let $X = (x_1, ..., x_n) \in D \subset \{0,1\}^n$ and $y_i = (\alpha - 1)x_i + 1$.

(a) If the number of 1's in X is equal to s mod q, show that the product $\prod_{i=1}^n y_i = \alpha^s$ in $\mathcal{F}$.

(b) Show that $\displaystyle\prod_{i=1}^n y_i = \sum_{s=0}^{q-1} \alpha^s \cdot \mathrm{mod}_{s,q}(X)$.

(c) Let $h = \max_{0 \leq s \leq q-1}\{deg_D(\mathrm{mod}_{s,q}(X))\}$. Show that for any $f : \{0,1\}^n \to \mathcal{F}$, we have $deg_D(f) \leq \frac{n}{2} + h$.

9.6:   [*Degrees of modulus functions*]

Suppose p and q are two distinct prime numbers. Then, a result from number theory states that there is a finite field $\mathcal{F}$ that contains a qth root of unity and Z_p as a subfield. Show that for any $f : \{0,1\}^n \to \mathcal{F}$, we have $deg_D(f) \leq \frac{n}{2} + h$, where $h = \max_{0 \leq s \leq q-1}\{deg_D(\mathrm{mod}_{s,q}(X))\}$.

9.7:   [*The general case of Theorem 9.2*]

Let p be a prime and $r > 1$ be any integer that is not a power of p. Use the preceding results and follow similar arguments in the chapter to show that any depth-d circuit C of AND, OR, NOT, and MOD$_p$ gates with input $X \in \{0,1\}^n$ computing $\mathrm{mod}_r(X)$ has size $2^{\Omega(n^{1/2d})}$.

9.8: [*The error incurred in computing Parity by AC^0 circuits*]
Show that the output of a depth-d AND-OR circuit of size $2^{\phi(n^{1/2d})}$ must differ from the Parity function on $2^{n-1} - o(2^n)$ input assignments, where n is the number of variables. Hence, an AC^0 circuit not only fails to compute the Parity function but will yield an error with probability close to $1/2$.

Bibliographic Notes

There are several known techniques in deriving the lower bounds on AC^0 circuits. It is independently shown in Furst, Saxe, and Sipser [36] and in Ajtai [4] that a constant-depth circuit of AND/OR gates computing the Parity function must require a superpolynomial number of gates. The lower bound has been improved from superpolynomial to exponential in Yao [148] and Håstad [50]. The techniques of Yao [149] and Håstad [50] primarily involve probabilistic arguments. On the other hand, Razborov [106] and Smolensky [131] use an algebraic approach that we have adopted in this chapter to derive the lower bound.

Allender [5], using much of the insights in Smolensky [131], shows that all AC^0 functions are computable by depth-3 threshold circuits of subexponential size. Yao [149] generalizes Allender's result and shows that functions in ACC are also computable by depth-3 threshold circuits of subexponential size. ACC is the class of functions computable by constant-depth polynomial-size circuits of AND/OR gates and MOD_m gates, where m is a fixed constant. It turns out that ACC is invariant even if we allow a fixed finite set of moduli for the MOD gates instead of a single modulus m. Beigel and Tarui [15] strengthen Yao's result and show that functions in ACC can be computed by depth-2 circuits of subexponential size, where the first layer consists of AND gates with polylogarithmic fan-in and the second layer consists of a symmetric gate.

Chapter 10

Lower Bounds via Communication Complexity

10.1   Introduction

This chapter concerns the node complexity of circuits with no restriction on the depth. For circuits of unrestricted depth and unbounded fan-in, deriving seemingly weak lower bounds, such as linear or polylogarithmic in the number of input variables, is considered challenging. The best-known lower bounds so far are still linear in the number of inputs. Using communication complexity concepts and techniques, we derive linear and almost-linear lower bounds on the size of circuits computing specific functions. This approach utilizes only basic features of the gates used, and hence, the bounds hold for general families of gates of which the symmetric and threshold gates are special cases. Thus, communication complexity arguments serve to generalize many of the best-known lower bounds and unify their proofs.

As in most of the previous chapters, we use the term *circuit size* to stand for the number of nodes of a circuit. Since we are dealing with various families of circuits, it is convenient to use the following shorthand notation. A circuit whose gates are all from a family of gates, say $\mathcal{G}$, is a $\mathcal{G}$-circuit. The $\mathcal{G}$-circuit complexity, $C_{\mathcal{G}}(f)$, of a Boolean function f is the size of the smallest $\mathcal{G}$-circuit that computes f. For example, $C_{\wedge,\vee}(f)$, $C_{\mathcal{GS}}(f)$, $C_{\mathcal{SYM}}(f)$, and $C_{\mathcal{TH}}(f)$ refer to the circuit complexity of the function f when the circuit comprises AND/OR/NOT, generalized symmetric, symmetric, and threshold gates, respectively.

In the next section, we define the notions of decomposition number and largest monochromatic rectangle of a function. These are simple attributes useful for analyzing the communication complexity of various functions.

In Section 10.3, we consider the family of polynomially rectangular gates. These gates, which include symmetric, generalized symmetric, and polynomially-bounded-weight threshold gates, compute functions with small decomposition numbers. We show that functions computed by small circuits of polynomially rectangular gates have small decomposition numbers. It follows that functions with high decomposition numbers require circuits of proportionally large size. We then use some effective techniques to derive lower bounds on the decomposition numbers and prove almost-linear lower bounds on the circuit complexity of several functions.

In Section 10.4, we strengthen the results for the family of triangular gates. These gates, which include all threshold gates, compute functions with large monochromatic rectangles. We show that any function computed by a small circuit of triangular gates contains a large monochromatic rectangle. Therefore, functions with only small monochromatic rectangles require circuits of proportionally large size.

We illustrate these results using the Inner Product Mod 2 (IP) function, the Comparison function (which were introduced in earlier chapters), and the Equality (EQ) function:

$$EQ(x_1, \ldots, x_{\frac{n}{2}}, y_1, \ldots, y_{\frac{n}{2}}) = \begin{cases} 1 & \text{if } x_i = y_i \text{ for all } 1 \le i \le \frac{n}{2}, \\ 0 & \text{otherwise.} \end{cases}$$

The bounds we derive imply:

1. Any realization of the n-variable EQ or IP by generalized symmetric gates requires $\Theta(n/\log n)$ gates. Namely, if the weights are bounded by n^k, then

$$\frac{1}{4(k+1)} \frac{n}{\log n} \le C_{\mathcal{GS}}(EQ), C_{\mathcal{GS}}(IP) \le \frac{\log 3}{2k} \frac{n}{\log n}.$$

2. Any realization of the n-variable EQ or IP by symmetric gates requires $\Omega(\frac{n}{\log n})$ gates:

$$C_{\mathcal{SYM}}(EQ), C_{\mathcal{SYM}}(IP) \ge \frac{n}{4 \log n}.$$

3. Any realization of the n-variable EQ or IP by AND/OR/NOT gates requires $\Theta(n)$ gates:

$$\frac{n}{2\log 3} \leq C_{\wedge,\vee}(\text{EQ}), C_{\wedge,\vee}(\text{IP}) \leq 2n.$$

4. Any realization of the n-variable IP by threshold gates requires $\Theta(n)$ gates:

$$\frac{1}{4}n \leq C_{\mathcal{TH}}(\text{IP}) \leq \frac{3}{4}n + 1.$$

Both upper and lower bounds apply to threshold circuits with exponential, as well as polynomially bounded, integer weights.

Note that the bounds in (1), (3), and (4) are tight up to a small multiplicative factor.

10.2 Communication Complexity Arguments

Let $f : \{0,1\}^n \to \{0,1\}$ be an n-variable Boolean function, and $Z = \{z_1, ..., z_n\}$ denote the set of variables of f. If $X \subset Z$ is a subset of variables, then an element of $\{0,1\}^{|X|}$ is a value assignment to the variables in X and is called an *X-input*.

Let $\{X,Y\}$ partition the set of variables (i.e., $X \cup Y = \{z_1, ..., z_n\}$ and $X \cap Y = \emptyset$). An X-input x together with a Y-input y correspond in an obvious way to an input that we call the *joint input* and will be denoted by (x,y). In the same way, the set of all inputs corresponds to the Cartesian product $\{0,1\}^{|X|} \times \{0,1\}^{|Y|}$. We can therefore associate with the function f and the partition $\{X,Y\}$ a matrix $M_{f,X,Y}$. We also refer to $M_{f,X,Y}$ as the function matrix of f under the partition $\{X,Y\}$. It has $2^{|X|}$ rows, each indexed by an X-input, $2^{|Y|}$ columns, each indexed by a Y-input, and

$$M_{f,X,Y}(x,y) = f(x,y).$$

An $\{X,Y\}$-*rectangle* is a Cartesian product $A \times B$, where A is a set of X-inputs and B is a set of Y-inputs. The *size* of the rectangle is $|A| \cdot |B|$, the number of inputs it contains. An $\{X,Y\}$-*decomposition* is a partition of $\{0,1\}^{|X|} \times \{0,1\}^{|Y|}$ into $\{X,Y\}$-rectangles. The *size* of the decomposition is the number of rectangles in the partition. A set of inputs is *f-constant* if f assigns the same value to all the elements in the set. An *f-constant*

$\{X,Y\}$-*decomposition* is an $\{X,Y\}$-decomposition whose rectangles are all f-constant.

The concept of rectangles plays a major role in the following communication complexity problem. As before, let f be an n-variable Boolean function and $\{X,Y\}$ a partition of the variables. A person P_X knows an X-input, a person P_Y knows a Y-input, and they communicate according to a predetermined protocol in order to find the value of f on their joint input. We are interested in $\hat{C}(f,X,Y)$, the minimum number of bits P_X and P_Y must transmit for the worst input.

The following results are well known:

1. Every protocol induces an $\{X,Y\}$-decomposition.

2. If the protocol always produces the correct answer, this decomposition is f-constant.

3. The number of bits required by the protocol for the worst input is at least the logarithm[1] of the size of the decomposition.

Let $\rho_{f,X,Y}$ be the smallest size of an f-constant $\{X,Y\}$-decomposition. Hence,

$$\hat{C}(f,X,Y) \geq \log \rho_{f,X,Y}. \tag{10.1}$$

It is shown in [3] that this bound is not far from being tight:

$$\hat{C}(f,X,Y) \leq \log^2 \rho_{f,X,Y}.$$

For that reason, several simple methods have been introduced to derive lower bounds on $\rho_{f,X,Y}$ for arbitrary f, X, and Y.

Largest f-constant rectangle   Let $L_{f,X,Y}$ be the size of the largest f-constant $\{X,Y\}$-rectangle. Clearly,

$$\rho_{f,X,Y} \geq \frac{2^n}{L_{f,X,Y}}.$$

Fooling set   An f-constant subset S of $\{0,1\}^X \times \{0,1\}^Y$ is an $\{X,Y\}$-*fooling set* if $(x_1,y_1),(x_2,y_2) \in S$ implies that either $f(x_1,y_2)$ or $f(x_2,y_1)$ differs from the common value of f over S. Let $F_{f,X,Y}$ be the

---

[1] All logarithms are to the base 2.

size of the largest $\{X, Y\}$-fooling set. An f-constant $\{X, Y\}$-rectangle contains at most one element of a given $\{X, Y\}$-fooling set. Hence,

$$\rho_{f,X,Y} \geq F_{f,X,Y}.$$

Rank   The matrix representing the indicator function of a rectangle has rank 1, and ranks are subadditive under matrix addition. Exercise 10.1 outlines a proof showing that

$$\rho_{f,X,Y} \geq 2\,\mathrm{rank}(M_{f,X,Y}) - 1.$$

In our applications, we can choose the most advantageous partition of the input variables. We therefore define the *decomposition number* of f,

$$\rho_f = \max\{\rho_{f,X,Y} : \{X, Y\} \text{ partitions } \{x_1, \ldots, x_n\}\},$$

to be the number of rectangles needed in the variable partition that yields the strongest bound in Eqn. (10.1). We use the methods above to derive lower bounds on the decomposition number of EQ, IP, and COMP.

Example 10.1:   We show that the decomposition numbers of EQ, IP, and COMP are larger than $2^{\frac{n}{2}}$. In the following, $X = \{x_1, \ldots, x_{n/2}\}$ and $Y = \{y_1, \cdots, y_{n/2}\}$. Every $\frac{n}{2}$-bit sequence corresponds in an obvious way to an X-input and to a Y-input. We can therefore talk about the joint input (x, x), where $x \in \{0, 1\}^{\frac{n}{2}}$.

Equality   The set $\{(x, x) : x \in \{0, 1\}^{\frac{n}{2}}\}$ is an $\{X, Y\}$-fooling set of size $2^{\frac{n}{2}}$, implying that $\rho_{\mathrm{EQ},X,Y} \geq 2^{\frac{n}{2}}$. In fact, $\rho_{\mathrm{EQ}} = \rho_{\mathrm{EQ},X,Y} = 2^{\frac{n}{2}+1}$.

Comparison   The same set serves as a fooling set here too; hence, $\rho_{\mathrm{COMP}} \geq 2^{\frac{n}{2}}$. In fact, $\rho_{\mathrm{COMP}} = \rho_{\mathrm{COMP},X,Y} = 2^{\frac{n}{2}+1} - 1$.

Inner Product Mod 2   It is shown in Chapter 8 that $M_{\mathrm{IP},X,Y}$ is a Hadamard matrix and, hence, has full rank over the reals. Thus, $\rho_{\mathrm{IP}} \geq 2^{\frac{n}{2}+1} - 1$.   □

10.3   Rectangular Gates

The last section was motivated by the notion that a function with a high decomposition number is 'complex'. To show that computing such a function requires many gates, we first show that the gates used are 'simple', i.e., they can be decomposed into a small number of rectangles.

A function f is r-*rectangular* for some integer r, if for every variable partition $\{X, Y\}$ there is an f-constant $\{X, Y\}$-decomposition consisting of at most r rectangles, namely, if

$$\rho_f \leq r.$$

Let $p : \mathcal{Z}^+ \to \mathcal{Z}$. A family $\mathcal{G}$ of functions is p-*rectangular* if for every $m \leq n$, all m-variable functions in $\mathcal{G}$ are $p(n)$-rectangular. The family is *polynomially rectangular* if it is p-rectangular for some polynomial p. These definitions apply to gates and families of gates via their underlying functions. The next lemma, its simple proof omitted, provides a basic tool for proving that a function is r-rectangular.

Lemma 10.1:   Let f be a Boolean function and let $\{X, Y\}$ partition the set of variables. If $f(x, y)$ can be expressed as $h(g_1(x), g_2(y))$, where $g_1 : \{0, 1\}^{|X|} \to \mathcal{Z}$ and $g_2 : \{0, 1\}^{|Y|} \to \mathcal{Z}$, then

$$\rho_{f,X,Y} \leq |g_1| \cdot |g_2|,$$

where $|g_i|$ is the size of the range of g_i.   □

To prove that a function is r-rectangular, we apply Lemma 10.1 to all possible partitions of the variables.

Example 10.2:   We show that the gate families mentioned in the introduction are polynomially rectangular. In the following, $\{X, Y\}$ is an arbitrary partition of the input variables $\{x_1, \ldots, x_n\}$.

AND/OR

$$\bigvee_{x_i \in X \cup Y} x_i = \left(\bigvee_{x_i \in X} x_i \right) \vee \left(\bigvee_{x_i \in Y} x_i \right),$$

hence, the lemma implies that every OR gate is 4-rectangular (in fact three rectangles are sufficient). The same holds for AND gates.

Symmetric gates

$$f(x,y) = h\left(\sum_{x_i \in X} x_i, \sum_{y_i \in Y} y_i\right),$$

hence,

$$\rho_{f,X,Y} \leq (|X| + 1) \cdot (|Y| + 1) \leq \left(\frac{n}{2} + 1\right)^2.$$

Generalized symmetric gates

$$f(x,y) = h\left(\sum_{x_i \in X} w_i x_i + \sum_{y_i \in Y} \tilde{w}_i y_i\right),$$

where the w_i's are integers bounded by some polynomial $p(n)$. The first sum attains at most $(|X| + 1) \cdot p(n)$ values and likewise for the second, and hence, f is $(\frac{n}{2} + 1)^2 \cdot p^2(n)$-rectangular. It follows that the family of generalized symmetric functions is polynomially rectangular.

Threshold gates

Since generalized symmetric functions are polynomially rectangular, it follows that the family of threshold functions with polynomially bounded integer weights is also polynomially rectangular. □

Theorem 10.1: Let $\mathcal{G}$ be a p-rectangular family of gates. If a $\mathcal{G}$-circuit consisting of k gates computes an n-variable function f, then

$$\rho_f \leq (p(n))^k.$$

Proof: We can label the gates in the circuit so that if $i < j$, then there is no directed path from the output of gate j to the input of gate i. Let g_j denote the function computed by gate j. We prove by induction on j that the vector-valued function $G_j = (g_1, g_2, \ldots, g_j)$ has $\rho_{G_j, X, Y} \leq (p(n))^j$ for all variable partitions $\{X, Y\}$.

The induction basis holds by definition; suppose it holds for j, and consider the $(j + 1)$st gate. Let $\{X, Y\}$ be a variable partition. There is a G_j-constant $\{X, Y\}$-decomposition consisting of at most $(p(n))^j$ rectangles. Let R be a rectangle in this decomposition. Over R, all of $g_1, \ldots, g_j$ are constant, and hence, the $(j + 1)$st gate coincides with a $p(n)$-rectangular function of the original variables. Therefore, R can be partitioned into $p(n)$ G_{j+1}-constant $\{X, Y\}$ rectangles, and the induction step follows.   □

Corollary 10.1:   Let $\mathcal{G}$ be a p-rectangular family of gates. For every n-variable function f,

$$C_{\mathcal{G}}(f) \geq \frac{\log \rho_f}{\log p(n)} \, . \qquad\qquad \square$$

We apply Corollary 10.1 to derive a lower bound on the number of gates needed to compute the Equality and the Inner Product Mod 2 functions.

Theorem 10.2:

(i) For circuits consisting of AND, OR, and NOT gates:

$$\frac{n}{2 \log 3} \leq C_{\wedge,\vee}(\text{EQ}), C_{\wedge,\vee}(\text{IP}) \leq 2n.$$

(ii) For circuits consisting of generalized symmetric gates:

$$C_{\mathcal{GS}}(\text{EQ}), C_{\mathcal{GS}}(\text{IP}) = \Theta\left(\frac{n}{\log n}\right).$$

More specifically, if the weights are bounded by n^k, then

$$\frac{1}{4(k+1)} \frac{n}{\log n} \leq C_{\mathcal{GS}}(\text{EQ}), C_{\mathcal{GS}}(\text{IP}) \leq \frac{\log 3}{2k} \frac{n}{\log n}.$$

(iii) For circuits consisting of symmetric gates:

$$C_{\mathcal{SYM}}(\text{EQ}), C_{\mathcal{SYM}}(\text{IP}) \geq \frac{n}{4 \log n}.$$

Proof: All six lower bounds follow from Corollary 10.1 as both EQ and IP have decomposition numbers of at least $2^{\frac{n}{2}}$. The upper bounds in (i) follow from a simple construction. Next we show how to achieve the upper bounds in (ii).

Let $m = 2\lfloor k \log n\rfloor$. Clearly, m-variable $\text{COMP}_{\frac{m}{2}}$ can be written as

$$\text{COMP}_{\frac{m}{2}}(x_1, \ldots, x_{\frac{m}{2}}, y_1, \ldots, y_{\frac{m}{2}}) = \text{sgn}\left(\sum_{i=1}^{m/2} 2^i (x_i - y_i)\right) \qquad (10.2)$$

and thus can be computed by a single threshold gate with weights of at most n^k. For $i = 1, \ldots, \lceil n/m \rceil$, let

$$x^i = x_{(i-1) \cdot m/2 + 1}, \ldots, x_{i \cdot m/2}$$

and

$$y^i = y_{(i-1) \cdot m/2 + 1}, \ldots, y_{i \cdot m/2}.$$

Then,

$$\mathrm{EQ}(x^i, y^i) = \mathrm{COMP}_{\frac{m}{2}}(x^i, y^i) + \mathrm{COMP}_{\frac{m}{2}}(y^i, x^i) - 1.$$

Hence, an m-variable EQ can be computed by a depth-2 threshold circuit with weights of at most n^k and where the top gate is just a weighted sum of the first-level outputs (without a threshold). Finally, observe that

$$\mathrm{EQ}(x_1, \ldots, x_{\frac{n}{2}}, y_1, \ldots, y_{\frac{n}{2}}) = \bigwedge_{i=1}^{\lceil n/m \rceil} \mathrm{EQ}(x^i, y^i).$$

Hence, the n-variable EQ function can be computed by a depth-2 threshold circuit of size $2\lceil n/(2k \log n) \rceil + 1$. When generalized symmetric gates are used instead of threshold gates, the number of gates can be reduced to $\lceil n/(2k \log n) \rceil + 1$.

When trying to meet the lower bound for IP, we cannot use threshold gates as we did for EQ. The next section shows that any threshold circuit for IP (even with exponential weights) has at least linear size. Yet, we can use the circuit structure applied to EQ. Every $(k \log n)$-variable function, in particular $\mathrm{IP}(x_1, \ldots, x_{k \log n/2}, y_1, \ldots, y_{k \log n/2})$, can be computed by a single generalized symmetric gate with weights of at most n^k. Use $\lceil n/k \log n \rceil$ generalized symmetric gates to compute the partial IPs, and then use a single (symmetric) gate to compute their parity.   □

In Chapter 6, it is shown that Comparison cannot be computed by a single threshold gate with polynomially bounded integer weights. It is, however, shown in Chapter 4 that it can be computed by a depth-2 polynomial-size threshold circuit with polynomially bounded integer weights $(\widehat{LT}_2)$. Next we establish a tight lower bound on the size of a generalized symmetric circuit computing Comparison and also derive a depth-3 threshold circuit with size that achieves the lower bound.

Theorem 10.3:

$$C_{\mathcal{GS}}(\mathrm{COMP}) = \Theta\left(\frac{n}{\log n}\right).$$

Proof: It is shown in Example 10.1 that $\rho_{\text{COMP}} = 2^{\frac{n}{2}+1}$. Hence, it follows from Corollary 10.1 that $C_{\mathcal{GS}}(\text{COMP}) = \Omega(\dfrac{n}{\log n})$. We next show that the lower bound can be met by a circuit in $\widehat{LT}_3$. It is not known whether the lower bound can be met by a circuit in $\widehat{LT}_2$.

Let $m = 2\lceil \log n \rceil$. For $i = 1, \ldots, \lceil n/m \rceil$, let

$$C_i = \text{sgn}\left(\sum_{j=(i-1)\cdot m/2+1}^{i\cdot m/2} 2^j (x_j - y_j) \right),$$

and

$$\tilde{C}_i = \text{sgn}\left(\sum_{j=(i-1)\cdot m/2+1}^{i\cdot m/2} 2^j (x_j - y_j) - 1 \right).$$

Note that both C_i and $\tilde{C}_i$ can be computed with threshold gates of polynomially bounded weights. Further,

$$C_i = 1 \text{ if } \sum_{j=(i-1)\cdot m/2+1}^{i\cdot m/2} 2^j x_j \geq \sum_{j=(i-1)\cdot m/2+1}^{i\cdot m/2} 2^j y_j,$$

and

$$\tilde{C}_i = 1 \text{ if } \sum_{j=(i-1)\cdot m/2+1}^{i\cdot m/2} 2^j x_j > \sum_{j=(i-1)\cdot m/2+1}^{i\cdot m/2} 2^j y_j.$$

Define Boolean expressions:

$$B_{\lceil n/m \rceil} = \tilde{C}_{\lceil n/m \rceil},$$

for $k = 2, \ldots, \lceil n/m \rceil - 1$

$$B_k = \tilde{C}_k \bigwedge_{j=k+1}^{\lceil n/m \rceil} C_j,$$

and

$$B_1 = \bigwedge_{j=1}^{\lceil n/m \rceil} C_j.$$

It is straightforward to see that

$$\text{COMP}_{\frac{n}{2}}(x_1, \ldots, x_{\frac{n}{2}}, y_1, \ldots, y_{\frac{n}{2}}) = \bigvee_{j=1}^{n} B_j.$$

The first layer of our circuit for the Comparison function has $O(n/\log n)$ gates computing the C_i and $\tilde{C}_i$. With these computed values as inputs, the second layer has $O(n/\log n)$ gates each computing the B_j. Finally the output gate computes the OR ($\bigvee$) of all the B_j's. The total number of gates is $O(n/\log n)$.   □

10.4   Triangular Gates

A Boolean matrix is *strictly triangular* if the entries in each row and column form a nondecreasing sequence. In a strictly triangular Boolean matrix, the sets of 1's and 0's resemble a (possibly truncated) triangle, and hence, the name. A matrix is *triangular* if its rows and columns can be permuted so that the resulting matrix is strictly triangular.

Lemma 10.2:   A Boolean matrix is triangular if and only if it does not contain any 2×2 rectangle of the form

$$
\begin{array}{cc} 0 & 1 \\ 1 & 0 \end{array} \quad \text{or} \quad \begin{array}{cc} 1 & 0 \\ 0 & 1 \end{array}
$$

(recall that a rectangle need not be contiguous).

Proof: By definition of a triangular matrix, one can permute the rows and columns so that the resulting matrix is strictly triangular. On the other hand, it is easy to verify that no 2×2 rectangles of the preceding form can be obtained by permuting the rows and columns of a strictly triangular matrix. Thus, a triangular matrix cannot contain 2×2 rectangles of the preceding form.

For the other direction, permute the rows so that the rows are arranged according to an increasing number of 1's in each row. Then, permute the columns so that the columns are arranged according to an increasing number of 1's in each column. The resulting matrix is strictly triangular for if in some column a 1 appears above a 0, then, as the number of 1's does not decrease with the rows, there must be another column where in the same locations a 0 appears above a 1, contradicting the noncontainment assumption.   □

Some properties of triangular matrices are apparent.

Corollary 10.2:

1. Every submatrix of a triangular matrix is triangular.

2. Every triangular matrix contains a constant rectangle whose size is at least $1/4$ the size of the given triangular matrix.

3. Every submatrix of a triangular matrix contains a constant rectangle whose size is at least $1/4$ the size of the given submatrix.

Proof: The first property follows directly from Lemma 10.2. To show the second property, permute the rows/columns until one obtains a strictly triangular matrix. Consider the midpoint (x, y). If the (x, y)th element of the matrix is 0, then the rectangle above and to the left of (x, y) is all 0; otherwise, the rectangle to the right and below (x, y) is all 1. The third property follows directly from the first two properties.   □

An n-variable Boolean function f is *triangular* if $M_{f,X,Y}$ is triangular for *all* $\{X, Y\}$-partitions of the variables. A family of functions is triangular if all the functions in the family are. The definition applies to gates and families of gates via the underlying functions.

Example 10.3:   Threshold gates (and in particular AND and OR gates) are triangular. We use Lemma 10.2 and show the result by contradiction. Let $f(X, Y) = \text{sgn}(\sum_{x_i \in X} u_i x_i + \sum_{y_i \in Y} v_i y_i)$ and suppose that it is not triangular. It follows from Lemma 10.2 that there exist $x, x' \in X$ and $y, y' \in Y$ such that $f(x, y) = f(x', y') = 1$ and that $f(x, y') = f(x', y) = 0$. Then, $\sum u_i x_i + \sum v_i y_i > \sum u_i x_i + \sum v_i y_i'$ while $\sum u_i x_i' + \sum v_i y_i < \sum u_i x_i' + \sum v_i y_i'$. This leads to an obvious contradiction.   □

Recall that $L_{f,X,Y}$ was defined to be the size of the largest f-constant $\{X, Y\}$-rectangle. Define L_f as

$$L_f = \min\{L_{f,X,Y} : \{X, Y\} \text{ partitions } \{x_1, \ldots, x_n\}\}.$$

Theorem 10.4:   If a circuit consisting of k triangular gates computes a function f, then

$$L_f \geq \frac{2^n}{4^k}.$$

Proof: As in Theorem 10.1, we label the gates in the circuit so that if $i < j$, then there is no directed path from the output of gate j to the input of gate i. Let g_j denote the function computed by gate j. We prove by induction on j that the vector-valued function $G_j = (g_1, g_2, \ldots, g_j)$ has $L_{G_j, X, Y} \geq \frac{2^n}{4^k}$ for all variable partitions $\{X, Y\}$.

The induction basis holds by property (2) of Corollary 10.2. Suppose it holds for j, and consider the $(j+1)$st gate. Let $\{X, Y\}$ be a variable partition. By induction hypothesis, there is a G_j-constant $\{X, Y\}$-rectangle R of size $2^n/4^k$. Over R, the outputs of the first j gates are fixed; hence, the input to the $(j+1)$st gate varies only with the original inputs. It follows that over R, the $(j+1)$st gate coincides with a triangular function whose inputs are the original inputs. By property (3), there must be a subrectangle of R of size $\geq |R|/4$ over which the $(j+1)$st gate has a constant output.   □

Corollary 10.3:   For every function f and every family $\mathcal{G}$ of triangular gates,

$$C_{\mathcal{G}}(f) \geq \frac{n - \log L_f}{2} \ .$$ □

Example 10.4:   Let $X = \{x_1, \ldots, x_{n/2}\}$ and $Y = \{y_1, \ldots, y_{n/2}\}$. Using Lemma 8.6 in Chapter 8 one can show that the largest IP-constant $\{X, Y\}$-rectangles are of size at most $2^{\frac{n}{2}}$ (see Exercise 10.5). Hence, it follows from Corollary 10.3 that

$$C_{\mathcal{TH}}(\mathrm{IP}) \geq \frac{n}{4} \ .$$ □

The bound on $C_{\mathcal{TH}}(\mathrm{IP})$ is tight up to a constant factor. A simple depth-3 threshold circuit with polynomially bounded integer weights can compute IP using $\frac{3}{4}n + 1$ gates (see Exercise 3.12 in Chapter 3). In a sense, the depth-3 circuit is also depth-optimal, because it is shown in Chapter 8 that every depth-2 threshold circuit for IP has exponential size if the weights at the second layer are polynomially bounded integers. It is not known whether there is a polynomial-size depth-2 threshold circuit for IP when exponential weights are allowed at the second layer.

10.5   Miscellaneous Applications

Here we will briefly discuss some applications of the results and techniques reviewed in the previous sections to threshold-circuit complexity, and circuit complexity with various gates.

10.5.1   Depth/Weight Trade-offs in Threshold Circuits

It is stated in Chapter 6 that any depth-d threshold circuit (with unrestricted weights) can be simulated by a depth-$(d+1)$ threshold circuit of polynomially bounded integer weights with only a polynomial factor increase in size. However, no upper or lower bounds have been shown for the degree of this polynomial factor.

One can realize the n-variable EQ using only three threshold gates with unrestricted weights in depth-2. Yet, Theorem 10.2 gave a lower bound of $\Omega(n/\log n)$ on the size of any threshold circuits (with polynomially bounded integer weights) for EQ. We therefore have:

Corollary 10.4:   There are n-variable Boolean functions whose threshold circuits (with polynomially bounded integer weights) have size $\Omega(n/\log n)$ times larger than the size of their depth-2 threshold circuits with unrestricted weights.   □

10.5.2   Weighted-Sum Gates

In our discussions, we often observed that the output gate of a given threshold circuit does not always require the *sgn* function usually associated with a threshold gate. A gate that computes a linear combination $\sum w_i x_i$ of its inputs (without taking a threshold) is a *weighted-sum gate*. No explicit function is known that requires superpolynomial size when realized by a depth-2 unrestricted-weight threshold circuit with a weighted-sum gate at the output. This is a weaker case of the more difficult open problem of proving that some given function requires superpolynomial size when computed by a depth-2 unrestricted-weight threshold circuit (with a threshold allowed in the output gate). We prove a partial result regarding weighted-sum gates in the context of the equality and other related functions.

As mentioned earlier, the n-variable EQ can be computed by a depth-2 circuit consisting of two threshold gates with exponential weights in the first layer and a weighted-sum gate in the second layer. We show that any circuit for EQ that consists of threshold gates with polynomially bounded integer weights at the first layer and of a weighted-sum gate at the second layer (possibly with unrestricted real weights) has exponential size.

Theorem 10.5:   Suppose that a depth-2 circuit comprising a weighted-sum gate (possibly with unrestricted real weights) at the output and $p(n)$-rectangular gates in the first layer computes the n-variable EQ. Then, the size of the circuit is at least $2^{\frac{n}{2}}/p(n)$.

Proof: Let $g_1, \ldots, g_k$ be the output functions of the k gates in the first layer of the circuit. Consider the 'natural' partition $X = \{x_1, \ldots, x_{n/2}\}$ and $Y = \{y_1, \ldots, y_{n/2}\}$ of the input variables. Note that for such a partition, the corresponding function matrix $M_{\text{EQ},X,Y} = I_{2^{n/2} \times 2^{n/2}}$. Since the output function is a weighted sum of g_i's, we have

$$M_{\text{EQ},X,Y} = \sum_{i=1}^{k} w_i M_{g_i,X,Y}.$$

By subadditivity of ranks,

$$\text{rank}(M_{\text{EQ},X,Y}) \leq \sum_{i=1}^{k} \text{rank}(M_{g_i,X,Y}).$$

But

$$\text{rank}(M_{\text{EQ},X,Y}) = 2^{\frac{n}{2}},$$

and for all $i \in \{1, \ldots, k\}$,

$$p(n) \geq \rho_{g_i,X,Y} \geq \text{rank}(M_{g_i,X,Y}).$$

Theorem 10.5 follows immediately from the preceding observations. $\square$

Corollary 10.5:   Suppose that a depth-2 circuit consisting of threshold gates with polynomially bounded integer weights in the first layer and a weighted-sum gate (possibly with exponential weights) at the output computes the n-variable EQ. Then, the size of the circuit is $\Omega(2^{\frac{n}{2}-\epsilon})$ for every $\epsilon > 0$.

Proof: Example 10.2 implies that any threshold gate with weights bounded by $p(n)$ is $(\frac{n}{2} + 1)^2 p^2(n)$-rectangular. $\square$

This result holds for any function f, (e.g., Comparison) for which $\text{rank}(M_{f,X,Y})$ is exponentially large for some partition $\{X, Y\}$ of the input variables.

10.5.3   Computing with MOD$_m$ Gates

We next consider computations of functions using depth-2 circuits in which the first layer has generalized symmetric gates and the output gate is a MOD$_m$ gate. Recall that a MOD$_m$ gate computes the following function:

$$\mathrm{MOD}_m(x_1, \ldots, x_n) = \begin{cases} 1 & \text{if } (\sum_{i=1}^{n} x_i) \not\equiv 0 \ (\mathrm{mod}\ m), \\ 0 & \text{otherwise.} \end{cases}$$

First we consider the case of MOD$_p$ gates, where p is a prime number. Similar arguments as in Theorem 10.5 apply, yielding the following.

Theorem 10.6:   Suppose that a depth-2 circuit comprising a MOD$_p$ gate at the output and $p(n)$-rectangular gates in the first layer, computes the n-variable EQ. Then, the size of the circuit is at least $2^{\frac{n}{2}}/p(n)$.

Proof: Let $g_1, \ldots, g_k$ be the output functions of the k gates in the first layer of the circuit. Consider again the 'natural' partition $X = \{x_1, \ldots, x_{n/2}\}$ and $Y = \{y_1, \ldots, y_{n/2}\}$ of the input variables. For such a partition, the corresponding function matrix $M_{\mathrm{EQ},X,Y} = I_{2^{n/2} \times 2^{n/2}}$. Since the output function is a MOD$_p$ gate, the sum of the function matrices, $M_{g_i,X,Y}$, modulo p must be a diagonal matrix with non-zero entries, i.e.,

$$(\sum_{i=1}^{k} M_{g_i,X,Y}) \bmod p = \mathrm{diag}(\lambda_1, \ldots, \lambda_{2^{n/2}}),$$

where $\lambda_i \neq 0$. The operations are defined over $GF(p)$, and hence, the rank arguments used in the proof of Theorem 10.5 hold (see Exercise 10.1). Then, using the subadditivity property of ranks we get:

$$\begin{aligned} \mathrm{rank}\ (\mathrm{diag}(\lambda_1, \ldots, \lambda_{2^{n/2}}) &= 2^{n/2} \\ &\leq \sum_{i=1}^{k} \mathrm{rank}(M_{g_i,X,Y}) \\ &\leq kp(n)\ . \end{aligned}$$

Hence, $k \geq 2^{n/2}/p(n)$.   □

If the output gate is a MOD$_m$ gate, where m is not necessarily prime, then the preceding arguments do not directly apply. However, a related concept of *variation rank* yields comparable results.

Definition 10.1:   Given a positive integer m, two integer matrices M_1 and M_2 will be said to be mod_m-equivalent if for all indices (i, j)

$$M_1(i, j) \equiv 0 \ (\text{mod } m) \iff M_2(i, j) \equiv 0 \ (\text{mod } m),$$

where $M_i(i, j)$ denotes the $(i, j)^{th}$ entry of matrix M_i.   □

Definition 10.2:   The variation rank of an integer matrix M with respect to a positive integer m is defined as

$$\text{var-rank}_{\text{mod}_m}(M) = \min\{\text{rank}(B) \ : \ B \text{ is mod}_m\text{-equivalent to } M\},$$

where the rank of a matrix is determined over $\mathcal{R}$.   □

Theorem 10.7:   Suppose that a depth-2 circuit comprising a MOD_m gate at the output and $p(n)$-rectangular gates in the first layer computes the n-variable EQ. Then, the size of the circuit is at least $\dfrac{2^{\frac{n}{2}}}{p(n) \log m}$.

Proof: Let $g_1, \ldots, g_k$ be the output functions of the k gates in the first layer of the circuit. Following the same arguments as in the proof of Theorem 10.6, consider the partition $X = \{x_1, \ldots, x_{n/2}\}$ and $Y = \{y_1, \ldots, y_{n/2}\}$ of the input variables. Under this partition, $M_{\text{EQ},X,Y} = I_{2^{n/2} \times 2^{n/2}}$, and it follows from the definition of a MOD_m gate that the output function of the circuit will be the EQ function if and only if $(\sum_{i=1}^{k} M_{g_i,X,Y})$ is mod_m-equivalent to $I_{2^{n/2} \times 2^{n/2}}$.

It can be shown (Exercise 10.6) that

$$\text{var-rank}_{\text{mod}_m}(I_{2^{n/2} \times 2^{n/2}}) \geq 2^{n/2} / \log m.$$

Thus,

$$kp(n) \geq \text{rank}(\sum_{i=1}^{k} M_{g_i,X,Y}) \geq 2^{n/2} / \log m \ .   □$$

Theorem 10.7 implies that if EQ is computed by a depth-2 circuit with generalized symmetric gates in the first layer and a MOD_m gate at the output, then exponentially many gates will be required.

If we define the variation rank of an n-variable Boolean function f as

$$\text{var-rank}_{\text{mod}_m}(f) = \max_{X,Y}\{\text{var-rank}_{\text{mod}_m}(M_{f,X,Y})\},$$

then we can state the following general result. The proof is similar to that of Theorem 10.7 and is omitted here.

Theorem 10.8:   Suppose that a depth-2 circuit comprising a MOD_m gate at the output and $p(n)$-rectangular gates in the first layer, computes an n-variable Boolean function f. Then, the size of the circuit is at least var-rank$_{\text{mod}_m}(f)/p(n)\log m$. $\square$

Exercises

10.1:   [*Decomposition number and rank*]
Let $\{X, Y\}$ be any partition of the input variables of a Boolean function f, and let $M_{f,X,Y}$ be the corresponding function matrix.

(a) Show that if all the $1'$s in $M_{f,X,Y}$ belong to a single f-constant rectangle, then rank$(M_{f,X,Y}) = 1$.

(b) Using the above observation and the subadditivity property of matrix-rank, show that $\rho_{f,X,Y} \geq 2\,\text{rank}(M_{f,X,Y}) - 1$, where $\rho_{f,X,Y}$ is the smallest size of an f-constant $\{X, Y\}$-decomposition. Note that if $M_{f,X,Y}$ is defined over any finite field $GF(p)$, then also the same result holds.

10.2:   [*Polynomially rectangular gates vs. generalized symmetric gates*]
Given $X = (x_1, \ldots, x_n) \in \{0, 1\}^n$, let the function msb (most significant bit) be defined as follows:

$$msb(X) = \begin{cases} k & \text{where } k \text{ is the largest index such that } x_k = 1, \\ 0 & \text{if } x_i = 0 \text{ for all } 1 \leq i \leq n. \end{cases}$$

If $I \subseteq \{1, \ldots, n\}$, then let $X_I = (x_{i_1}, \ldots, x_{i_l})$, where $i_j \in I$, and let

$$msb(X_I) = \begin{cases} k & \text{where } k \text{ is the largest index in } I \text{ such that } x_k = 1, \\ 0 & \text{if } x_i = 0 \text{ for all } i \in I. \end{cases}$$

For $X = (x_1, \ldots, x_n) \in \{0, 1\}^n$ and $Y = (y_1, \ldots, y_n) \in \{0, 1\}^n$, let $COMP_{msb}(X, Y)$ be defined as

$$COMP_{msb}(X, Y) = \begin{cases} 1 & \text{if } msb(X) \geq msb(Y), \\ 0 & \text{if } msb(X) < msb(Y). \end{cases}$$

(a) Show that $COMP_{msb} \in LT_1$.

(b) Let $I_1 \cup I_2 = J_1 \cup J_2 = \{1, \ldots, n\}$, where $I_1 \cap I_2 = J_1 \cap J_2 = \phi$, and let the inputs X and Y be correspondingly partitioned into X_{I_1}, Y_{J_1} and X_{I_2}, Y_{J_2}. Show that

$$COMP_{msb}(X,Y) = \begin{cases} 1 & \text{if } \max\{msb(X_{I_1}), msb(X_{I_2})\} \\ & \qquad \geq \max\{msb(Y_{J_1}), msb(Y_{J_2})\}, \\ 0 & \text{otherwise.} \end{cases}$$

Hence, conclude that $COMP_{msb}(X,Y)$ is a polynomially rectangular function.

(c) Assume that $COMP_{msb}(X,Y) = G(\sum_{i=1}^{n} \alpha_i x_i + \sum_{i=1}^{n} \beta_i y_i)$, where all $|\alpha_i|$ and $|\beta_i|$ are integers bounded by a polynomial $p(n)$. Show that for n sufficiently large there exist two X-inputs

$$Z = x_1, x_2, \ldots, x_{k-1}, 1, \overbrace{0, \ldots, 0}^{n-k}$$

$$Z' = x'_1, x'_2, \ldots, x'_{k-1}, 0, \overbrace{0, \ldots, 0}^{n-k}$$

such that $\sum_{i=1}^{n} \alpha_i z_i = \sum_{i=1}^{n} \alpha_i z'_i$. Next show that if

$$W = \overbrace{0, \ldots, 0}^{k-1}, 1, \overbrace{0, \ldots, 0}^{n-k}$$

then $COMP_{msb}(Z,W) = 1 \neq COMP_{msb}(Z',W)$ and

$$G(\sum_{i=1}^{n} \alpha_i z_i + \sum_{i=1}^{n} \beta_i w_i) = G(\sum_{i=1}^{n} \alpha_i z'_i + \sum_{i=1}^{n} \beta_i w_i).$$

Hence, conclude that $COMP_{msb}(X,Y)$ is not a generalized symmetric function.

10.3: [*Threshold functions and strictly triangular matrices*]
Let $f(X,Y) = \text{sgn}(\sum_{i=1}^{n} u_i x_i + \sum_{i=1}^{m} v_i y_i)$, where $X = \{x_1, \ldots, x_n\}$ and $Y = \{y_1, \ldots, y_m\}$ define a partitioning of the input variables, and let $M_{f,X,Y}$ be the matrix representation of f as induced by X and Y. Determine explicit permutations of the rows and columns of $M_{f,X,Y}$ which will make it strictly triangular. Note that Example 10.3 shows the existence of such permutations.

10.4:   [*Nontriangular functions*]

(a) Let $f(x, y, w, z) = (x \wedge y) \vee (w \wedge z)$. Show that if the input variables are partitioned as $X = \{x, y\}$ and $Y = \{w, z\}$ then the matrix representation $M_{f,X,Y}$ of f is triangular. Show, however, that the function f is not triangular.

(b) Let $f_p(x_{i_1}, x_{i_2}, \ldots, x_{i_k})$ be a projection of an n-variable $(k \leq n)$ function $f(x_1, x_2, \ldots, x_n)$, i.e., $f_p(\cdot)$ is derived from $f(\cdot)$ by assigning constants or variables (possibly repeated and complemented) to the variables in $f(\cdot)$. Show that if f_p is not a triangular function, then f is also not a triangular function. (*Hint:* Show that the matrix representation of any projection f_p is a rectangle of $M_{f,X,Y}$ for some partition $\{X, Y\}$ of the variables.)

10.5:   [*Largest IP-constant rectangle*]

In Chapter 8 (see Section 8.7.1) it is shown that in the $\{1, -1\}$ notation, the function matrix $M_{IP,X,Y}$ is a Hadamard matrix under the partition $X = \{x_1, \ldots, x_{n/2}\}$ and $Y = \{y_1, \ldots, y_{n/2}\}$. Moreover, Lemma 8.6 states that for any rectangle $A \times B$ of $M_{IP,X,Y}$, where $A, B \subseteq \{1, -1\}^{n/2}$,

$$\left| \sum_{(x,y) \in A \times B} M_{IP}(x, y) \right| \leq \sqrt{|A||B|2^{n/2}}.$$

Using the above result, show that the largest IP-constant $\{X, Y\}$-rectangles are of size at most $2^{\frac{n}{2}}$.

10.6:   [*Variation rank of the identity matrix*]

Let m be a natural number with the prime factorization $m = \prod_{i=1}^{r} p_i^{l_i}$. Show that

$$\text{var-rank}_{\text{mod}_m}(I_{N \times N}) = \frac{N}{\sum_{i=1}^{r} l_i}.$$

Hence, conclude that $\text{var-rank}_{\text{mod}_m}(I_{N \times N}) \geq N / \log m$.

10.7:   [*Differential dimension of analytic functions*]

Let S be a finite set of points in the n-dimensional complex vector space $\mathcal{C}^n$.

Let V denote the space of functions from S to $\mathcal{C}$. The differential dimension of an analytic function $g : \mathcal{C}^n \to \mathcal{C}$ over S is the dimension of the subspace of V spanned by the restrictions to S of g and all of its partial derivatives. In the above notation, the differential dimension of any function $b : S \to \mathcal{C}$ is the minimal differential dimension over S of any analytic function $g : \mathcal{C}^n \to \mathcal{C}$ that interpolates b. Let $S = \{1, \ldots, n\}$ and let $b : S \to \mathcal{C}$ for the rest of this exercise.

(a) If $b(i)$ equals the ith Fibonacci number, then show that b can be interpolated by an analytic function of the form $a_1 \exp^{\lambda_1 i} + a_2 \exp^{\lambda_1 i}$. Hence, show that the differential dimension of b is at most 2.

(b) If
$$b(i) = \begin{cases} 1 & \text{if } i \bmod 6 = 0, \\ 0 & \text{otherwise}, \end{cases}$$
then show that b can be interpolated by the following analytic function:
$$g(x) = \frac{1}{6}[1 + \exp^{\frac{2\pi j}{6} x} + \exp^{\frac{4\pi j}{6} x} + \exp^{\frac{6\pi j}{6} x} + \exp^{\frac{8\pi j}{6} x} + \exp^{\frac{10\pi j}{6} x}].$$

Hence, show that the differential dimension of b is at most 6.

10.8: *[Differential dimension of symmetric functions]*
The differential dimension of a Boolean function $f : \{0,1\}^n \to \{0,1\}$ is the minimal differential dimension over $S = \{0,1\}^n$ of any analytic function $g : \mathcal{C}^n \to \mathcal{C}$ that interpolates f.

(a) Using interpolation by exponentials, as outlined in Exercise 10.7, show that any function computed by a MOD_m gate has differential dimension at most m.

(b) Generalize the arguments for MOD_m gates to show that any function computed by a symmetric gate with m inputs has differential dimension at most $(m+1)$.

10.9:   [*Circuits and differential dimension*]

(a) Consider a circuit with k MOD$_6$ gates. Show that the output of the ith gate can be written as

$$\frac{1}{6}[1 + \exp^{\frac{2\pi j}{6}\sigma_i} + \exp^{\frac{4\pi j}{6}\sigma_i} + \exp^{\frac{6\pi j}{6}\sigma_i} + \exp^{\frac{8\pi j}{6}\sigma_i} + \exp^{\frac{10\pi j}{6}\sigma_i}],$$

where σ_i is the sum of the inputs to the ith gate. Next prove that the partial derivatives of this function (with respect to the input variables) can be written as polynomials in $y_i = \exp^{\frac{2\pi j}{6}\sigma_i}$. Hence, show that if a circuit consisting of k MOD$_6$ gates computes a Boolean function f, then the differential dimension of f is at most 6^k.

(b) Generalize the previous arguments to show that if a circuit consisting of k symmetric gates with maximum fan-in m computes a Boolean function f, then the differential dimension of f is at most $(m+1)^k$.

10.10:   [*Differential dimension and shifts*]
As before, let S be a finite set of points in $\mathcal{C}^n$, and let V denote the space of functions from S to $\mathcal{C}$. Given any analytic function $g : \mathcal{C}^n \to \mathcal{C}$ and any $v \in \mathcal{C}^n$, the *shifted* function g_v is defined as: $g_v(x) = g(x - v) \; \forall \, x \in \mathcal{C}^n$.

(a) Show that the subspace of V spanned by all the shifts of g restricted to S is contained within the subspace of V spanned by all the partial derivates of all orders of g restricted to S. (*Hint:* Taylor series expansion of $g_v(x)$ can be used to prove the result.)

(b) Using the above result show that if g interpolates a given Boolean function f, then the dimension of the space spanned by some of the shifts g_v, $v \in S = \{0,1\}^n$, gives a lower bound on the differential dimension of g.

(b) Show that the function $AND(x_1, \ldots, x_n)$ has differential dimension at least $(n+1)$.

10.11:   [*Communication complexity concepts and differential dimension*]

Given any Boolean function f, let $\{X, Y\}$ be any partition of the input variables, and let $M_{f,X,Y}$ be the corresponding function matrix.

(a) If an analytic function g interpolates f, then using the results in Exercise 10.10 show that the differential dimension of g is at least rank $(M_{f,X,Y})$. Thus, show that the differential dimension of a Boolean function f is $\Omega(r)$, where

$$r = \max\{\text{rank } (M_{f,X,Y}) \ : \ \{X,Y\} \text{ partitions the input variables}\}.$$

(b) Show that the differential dimension of $EQ(x_1,\ldots,x_{\frac{n}{2}},y_1,\ldots,y_{\frac{n}{2}})$ is at least $2^{n/2}$.

10.12:   *[Lower bounds using differential dimension]*

(a) Show that a circuit comprising MOD_m gates and computing the n-variable AND function has at least $\log_m(n)$ gates.

(a) Using the results of Exercise 10.9, show that if a circuit consisting of symmetric gates computes an n-variable Boolean function f, then $C_{SYM}(f)$ is $\Omega(\log(\text{differential dimension of } f)/\log n))$.

(c) Thus, show that $C_{SYM}(EQ)$ is $\Omega(n/\log n)$. This method illustrates an algebraic approach for proving lower bounds for circuits comprising symmetric gates.

Bibliographic Notes

$\Omega(\log n)$ lower bound on the size of threshold circuits computing the n-variable Parity function is shown in Wegener [140]. A matching upper bound was derived much earlier by Kautz [64]. The first-known linear lower bound on the size of unbounded fan-in circuits can be found in Hromkovic [60]. The circuit considered in [60] requires that each gate computes a commutative and associative function, and hence, is too restrictive to apply to symmetric or threshold circuits. Smolensky [132] establishes an $\Omega(n/\log n)$ lower bound on the size of circuits of symmetric gates computing the n-variable Equality function. Novel techniques such as analytic function interpolation of Boolean functions and the differential dimension are used in [132]; however, the techniques do not seem to apply to threshold circuits with unrestricted weights.

The general techniques in [132] are outlined in Exercises 10.7 to 10.12; the result in Exercise 10.11 relating the differential dimension of a Boolean function to the rank of its function matrix can be found in Roychowdhury, Siu, and Orlitsky [111]. Gröger and Turán [46] derive a linear lower bound ($n/4$) on the size of unrestricted-weight threshold circuits computing the n-variable IP. Combining the techniques in [46] and communication complexity arguments, Roychowdhury, Orlitsky, and Siu [111] obtain the lower bound results presented in this chapter. General classes of gates, such as the rectangular and triangular gates, are introduced in [111], and several of the matching upper bound results presented in this chapter are also derived in [111]. The variation rank of a matrix is defined in Krause and Waack [67], where the variation rank of an identity matrix (Exercise 10.6) is also determined. The result outlined in Exercise 10.2, which establishes the separation between polynomially rectangular and generalized symmetric gates, is from [38]. For other results relating communication complexity to circuit complexity, see Goldmann [39].

Chapter 11

Hopfield Network

11.1 Introduction

In this chapter, we study a model of a discrete time neural network proposed by Hopfield, and now generally referred to as the Hopfield network or the Hopfield model. The interesting properties of this model led to a resurgence of interest in the study of neural networks in the 1980s. Hopfield later extended his original discrete time model to a continuous time analog. Here, however, we focus only on the discrete time Hopfield network. Our goal is to derive several fundamental properties of the model and demonstrate its applications as well as its limitations.

Let us review the Hopfield model as introduced in Chapter 1. A Hopfield network can be viewed as a graph in which each node represents a neuron in the network. The order of the network is the number of neurons in the network. The connection strength between each pair of nodes (neurons) is represented by a weight associated with the edge connecting the corresponding pair of nodes. Formally, let H be a Hopfield network with n nodes denoted by $\{x_1, x_2, \ldots, x_n\}$. Let w_{ij} denote the weight associated with the edge (j, i) connecting node x_j to x_i, and t_i denote the threshold value attached to each node x_i. Let W be the $n \times n$ weight matrix whose $(i, j)^{th}$ entry is w_{ij}, and T be the n-dimensional threshold vector whose i^{th} entry is t_i.

Let $x_i(t) \in \{1, -1\}$, $t = 0, 1, 2, \ldots$, denote the state or output of the ith node at time t. The state of the network at time t is denoted by the vector $X(t) = [x_1(t)\ x_2(t)\ \cdots\ x_n(t)]^T$. In general, the state of the network at time

345

$t+1$ is a function of $\{W, T\}$ and the state $X(t)$ of the network at time t. The network is thus completely determined by the parameters $\{W, T\}$, the initial value of the states, $X(0)$, and the manner in which the nodes (neurons) are updated.

Each node is updated by computing a linear threshold function of the current states of the nodes in the network. If at time step t, node x_i is chosen to be updated, then at the next time step,

$$x_i(t + 1) = \mathrm{sgn}(P_i(t)) = \begin{cases} 1 & \text{if } P_i(t) > 0, \\ -1 & \text{if } P_i(t) < 0, \end{cases} \qquad (11.1)$$

where we define

$$P_i(t) = \sum_{j=1}^{n} w_{ij} x_j(t) - t_i.$$

Without loss of generality, we can assume that the w_{ij}'s and the t_i's are integers, and that $P_i(t) \neq 0$ for all t.

The next state of the network, $X(t + 1)$, is updated from the current state by evaluating Eqn. (11.1) for a prespecified subset of the nodes in the network. The sequence of states of the network $X(1)$, $X(2)$, $\ldots$, depends on the subset of nodes that is selected to be updated at each time step.

This chapter is organized as follows: The proof of convergence in the sequential mode of iteration together with the properties of other modes of iteration are presented in Section 11.2. In Section 11.3, the application of the Hopfield network to associative memory is addressed. In Section 11.4, we demonstrate the applications of the Hopfield network to two combinatorial optimization problems, namely, the Min-Cut Problem and the Traveling-Salesman Problem.

11.2   Properties of the State Space

A Hopfield network is said to be iterating in a sequential mode if at every iteration only a single node is updated (as in Eqn. (11.1)), while the states of the other nodes are left unchanged. A state vector $X \in \{1, -1\}^n$ is called a *stable state* of the network (with parameters $\{W, T\}$) if

$$X = \mathrm{sgn}(WX - T).$$

In the literature, a stable state is also sometimes referred to as a *fixed point* in the state space.

It follows from the definition that if at some time step t_o, the network has reached a stable state $X(t_o)$, then regardless of the mode of iteration, $X(t) = X(t_o)$ for all $t > t_o$, i.e., the state of the network will remain unchanged after t_o.

One of the important properties of the model is that when it iterates in a sequential mode, it always reaches a stable state (provided the weight matrix W is symmetric with non-negative diagonal entries and each node is updated sufficiently often). This convergence property of the Hopfield model has led to its various applications in associative memories and combinatorial optimization.

We first present a proof of this convergence property. The key observation is that in a sequential mode of operation, one can define an energy function for the network such that the energy function is nonincreasing at every iteration of the network. Since this energy function is bounded from below, after a finite number of iterations it must stop decreasing and converge to some value. As a result of the convergence of the energy function, we show that the state of the network must also converge.

Theorem 11.1:   Let H be a Hopfield network with weight matrix W and threshold vector T. If W is symmetric with non-negative diagonal elements, then when H is iterating in a sequential mode, and each node in H is updated sufficiently often, the network always converges to a stable state.   □

The assumption that each node in the network is updated sufficiently often is to exclude those degenerate cases where the iterations result in the updating of some of the nodes only finitely many times. For example, when the nodes are updated in a cyclic fashion: x_1, x_2, ..., x_n, x_1, x_2, ..., x_n, ... , the assumption is clearly satisfied. On the other hand, consider the uninteresting case in which the same node is updated at every (sequential) iteration. Then, clearly, the network will not change its state after the first iteration; but the resulting state may not be a stable state. A more general claim that takes into account the degenerate cases would be the following: The network will not change its state after a finite number of iterations. The proof of this more general result will also be evident from the proof below.

Proof: Define an energy function $E(t)$ as follows:

$$E(t) \;=\; 2X(t)^T T - X^T(t)WX(t) \quad \text{for} \ \ t = 0, 1, 2, \ldots \tag{11.2}$$

where the superscript T indicates the (vector or matrix) transpose. Let $\Delta E(t) = E(t+1) - E(t)$ be the difference in the energy function between consecutive steps of the (sequential) iteration. Let $\Delta X(t) = X(t+1) - X(t)$. Our first step is to show that $\Delta E(t) \le 0$ for all t. Let $\Delta x_i(t) = x_i(t+1) - x_i(t)$ and note from Eqn. (11.1) that for $i = 1, 2, \ldots, n$,

$$\Delta x_i(t) = \begin{cases} 2 & \text{if } x_i(t) = -1 \ \text{ and } \ \text{sgn}(P_i(t)) = 1, \\ 0 & \text{if } x_i(t) = \text{sgn}(P_i(t)), \\ -2 & \text{if } x_i(t) = 1 \ \text{ and } \ \text{sgn}(P_i(t)) = -1. \end{cases} \tag{11.3}$$

Since the network is iterating in a sequential mode, only one node (say, x_j) will be updated at every step, and $\Delta X(t) = [0, \ldots, \Delta x_j(t), \ldots, 0]^T$. Therefore, we obtain

$$\begin{aligned} \Delta E &= E(t+1) - E(t) \\ &= 2X(t+1)^T T - (X(t) + \Delta X(t))^T W (X(t) + \Delta X(t)) \\ &\quad - [2X(t)^T T - X^T(t)WX(t)] \\ &= 2\Delta X(t)^T T - \Delta X(t)^T W \Delta X(t) \\ &\quad - \Delta X(t)^T W X(t) - \Delta X(t)^T W^T X(t) \\ &= 2\Delta x_j(t) t_j - w_{jj}(\Delta x_j(t))^2 - \Delta x_j(t)\left(\sum_{k=1}^{n} w_{jk} x_k(t) + \sum_{k=1}^{n} w_{kj} x_k(t)\right) \\ &= -w_{jj}(\Delta x_j(t))^2 - 2 P_j \Delta x_j(t). \end{aligned} \tag{11.4}$$

The last line of Eqn. (11.4) follows from the assumption that W is symmetric. From Eqn. (11.3), we have for every j, $P_j \Delta x_j \ge 0$. Together with the assumption that $w_{jj} \ge 0$, we have $\Delta E \le 0$. Since

$$\max |E(t)| \le 2 \sum_{i=1}^{n} |t_i| + \sum_{i=1}^{n} \sum_{j=1}^{n} |w_{ij}|,$$

the energy function E is bounded from below, and hence, cannot decrease indefinitely. Therefore, after a finite number of iterations, E must converge to some value or, equivalently, $\Delta E = 0$.

It remains to be shown that the state of the network converges as a result of the convergence of $E(t)$. Recall (see Eqn. (11.1)) that for $j = 1, \ldots, n$, $P_j(t) \neq 0$ for all t. Since $P_j \Delta x_j$ and w_{jj} are both ≥ 0, and $\Delta E = 0$, from Eqn. (11.4) we must have $\Delta x_j = 0$ for $j = 1, \ldots, n$. Hence, the state of the network remains unchanged after the energy function has converged.   □

Remark 11.1:   When the Hopfield network iterates in a sequential mode, the proof of Theorem 11.1 shows that the energy function decreases until it reaches a local minimum. In other words, let E_{X_o} be the energy corresponding to a stable state X_o, and let a state $\tilde{X}$ differ from X_o by only one coordinate. Then, the energy corresponding to $\tilde{X}$, $E_{\tilde{X}}$ will be $\geq E_{X_o}$, provided $w_{ii} = 0$ (see Exercise 11.2). The state, however, may not converge to a global minimum of the energy function. This fact appears to be a major limitation in the applications of Hopfield networks. We shall discuss this issue in more detail in Section 11.4.   □

Recall from Section 1.2.4 that a Hopfield network is said to iterate in the parallel mode if at every iteration all the nodes are chosen to be updated. In matrix notation, each iteration of the parallel mode computes

$$X(t+1) = \text{sgn}(WX(t) - T),$$

where for vectors, the sgn function applies to every coordinate. Whereas there are many different modes of sequential iteration (depending on which node is selected to be updated at every iteration), it follows from the definition that there is only one mode of parallel iteration. In cases other than the sequential or the parallel mode, the network is said to be iterating in a hybrid mode.

Theorem 11.1 shows a convergence property of the Hopfield network iterating in a sequential mode. Do similar results hold for other modes of iteration? Recall that a Hopfield network is said to enter a limit cycle of length m if after a finite number of iterations the corresponding updating algorithm results in a sequence of states $X_1, \ldots, X_m, X_1, \ldots, X_m, \ldots$ repeating indefinitely, where the states $\{X_1, \ldots, X_m\}$ are all distinct. It turns out that when the network is iterating in the parallel mode, it need not converge. On the other hand, it can be shown that the Hopfield network always converges to a stable state or enters a limit cycle of length 2 in the parallel mode of iteration, provided the weight matrix W is symmetric.

The main idea in the proof of this result is also to construct a nonincreasing energy function associated with the parallel mode; however, the energy function for the parallel mode is different from the one used in the sequential mode, i.e., the one defined in Eqn. (11.2).

In the following, we present an alternative proof of this convergence result via a reduction approach. Instead of defining a new energy function, we simulate a network iterating in the parallel mode by another network iterating in a sequential mode. This simulation leads to a correspondence between the states of the two networks so that the state of one network can be derived from the state of the other. Because of this correspondence, the steady-state behavior of the network iterating in the parallel mode can be inferred from the convergence property of the other network.

Let H be a Hopfield network of order n with symmetric weight matrix W, threshold vector T, and nodes labeled as $X = \{x_1, \ldots, x_n\}$. Construct another network $\hat{H}$ of order $2n$ with weight matrix $\hat{W}$, threshold vector $\hat{T}$ and nodes labeled as $\hat{X} = \{\hat{x}_1, \ldots, \hat{x}_{2n}\}$, where

$$\hat{W} = \begin{bmatrix} 0 & W \\ W & 0 \end{bmatrix},$$

and

$$\hat{T} = \begin{bmatrix} T \\ T \end{bmatrix}. \tag{11.5}$$

First observe that the graph corresponding to the network $\hat{H}$ is *bipartite*: Consider a partition of the nodes into two sets,

$$N_1 = \{\hat{x}_1, \ldots, \hat{x}_n\}$$

and

$$N_2 = \{\hat{x}_{n+1}, \ldots, \hat{x}_{2n}\}.$$

It is clear from the structure of the matrix $\hat{W}$ that there is no connection between any two nodes in N_1 (or in N_2), since the weight on any edge connecting any two nodes in N_1 (or in N_2) is zero. In other words, every node in N_1 is only connected to nodes in N_2, and vice versa. Moreover, the weight matrix $\hat{W}$ of $\hat{H}$ is also symmetric. Furthermore, observe that the connection weight w_{ij} between x_i and x_j of H is the same as the connection weight between $\hat{x}_i$ in N_1 and $\hat{x}_{j+n}$ in N_2.

The following lemma establishes the correspondence between the states of H and $\hat{H}$.

Lemma 11.1: Let

$$N_1(t) = [\hat{x}_1(t) \cdots \hat{x}_n(t)],$$

and

$$N_2(t) = [\hat{x}_{n+1}(t) \cdots \hat{x}_{2n}(t)],$$

i.e., the state of $\hat{H}$ is $\hat{X}(t) = [N_1(t)\ N_2(t)]^T$, for $t = 0, 1, 2, \ldots$. Let the initial states of H and $\hat{H}$ be $X(0) = X_0$ and $\hat{X}(0) = [0\ X_0]^T$, respectively. Suppose that H iterates in the parallel mode, and $\hat{H}$ iterates in a sequential mode that cycles through the nodes in the following order: $(\hat{x}_1, \ldots, \hat{x}_{2n}, \hat{x}_1, \ldots, \hat{x}_{2n}, \ldots)$. Then, the states of $\hat{H}$ simulate the states in H as follows:

$$X(2t) = N_2(2nt),$$

and

$$X(2t+1) = N_1(n(2t+1)).$$

Proof: First observe the following fact: Since the nodes in N_1 (respectively, N_2) are not connected to each other, a cycle of sequentially updating all the nodes in N_1 (respectively, N_2) will result in the same state as a parallel iteration on N_1 (respectively, N_2).

If the state of N_2 is the same as some initial state of H, it follows that (see Eqn. (11.5)) performing a parallel iteration on N_1 will result in a state of N_1 equivalent to the state of H after a parallel iteration on the initial state of H. Because of this fact and $X(0) = N_2(0)$, we have $X(1) = N_1(n)$. Similarly, performing a parallel iteration on N_2 is equivalent to performing a parallel iteration on H again. Thus, combining with the preceding fact, we have $X(2) = N_2(n)$. One can now update the nodes of N_1 and N_2 in an alternate fashion. It follows easily by induction on t that for $t = 0, 1, 2, \ldots$, $X(2t) = N_2(2nt)$ and $X(2t+1) = N_1(n(2t+1))$. ☐

The following theorem establishes the behavior of Hopfield networks in the parallel mode of iteration.

Theorem 11.2: Let H be a Hopfield network with symmetric weight matrix W and threshold vector T. If H iterates in the parallel mode, then the network always enters a limit cycle of length at most 2.

Proof: We use the same notations as in Lemma 11.1. Given the Hopfield network H, construct a corresponding network $\hat{H}$ with parameters $\{\hat{W}, \hat{T}\}$ as described in the statement of Lemma 11.1. Since $\hat{H}$ is iterating in a sequential mode and $\hat{W}$ is symmetric with zero diagonal elements, Theorem 11.1 implies that $\hat{H}$ will converge to some stable state $\hat{X}_f$ after a finite number of iterations. Let N_{1f} and N_{2f} correspond to the first half and the second half of nodes in $\hat{X}_f$, respectively. It follows from Lemma 11.1 that the state of H will assume values $X(2t) = N_{1f}$ and $X(2t+1) = N_{2f}$ for all t larger than some finite t_f. Thus, if $N_{1f} \neq N_{2f}$, then the network H enters a limit cycle of length 2. If $N_{1f} = N_{2f}$, then the network H reaches a stable state. □

Note that the weight matrix in Theorem 11.2 can have negative diagonal elements, whereas the assumption of non-negative diagonal elements in the weight matrix is critical in the proof of Theorem 11.1. In fact, if the weight matrix has some negative diagonal elements, then it is easy to construct a simple network that does not converge to a stable state in a sequential mode of iteration (see Exercise 11.9). As mentioned before, instead of the reduction approach presented here, and one could prove Theorem 11.2 via the concept of an energy function (as in the proof of Theorem 11.1). A proof using such an approach is outlined in Exercise 11.3.

Both Theorem 11.1 and Theorem 11.2 assume that the weight matrix of the network is symmetric. Are there other matrix structures that would yield similar results on the state space of the network? In fact, when the weight matrix is antisymmetric, the network also exhibits interesting properties.

Theorem 11.3: Let H be a Hopfield network with an *antisymmetric* weight matrix W (i.e., $W^T = -W$) and with threshold vector $T = 0$. If H iterates in the parallel mode, then it always enters a limit cycle of length at most 4.

Proof: The theorem can be derived via a reduction approach. The proof is left as an exercise. □

The preceding theorems indicate some interesting properties of the state space of the Hopfield network when the weight matrix has special structures and the mode of iteration is restricted. If these restrictions are relaxed, the network could possibly have very long limit cycles. In fact, the length of the cycle could be exponential in the order of the network if some of the restrictions are relaxed. The preceding theorems guarantee that the network would either converge to a stable state (Theorem 11.1) or enter a limit cycle (Theorems 11.2 and 11.3) after a finite number of iterations. It is interesting to determine how long it will take for the network to reach these states from an arbitrary initial state. These issues will be addressed next.

11.2.1   On the Transient Period

The number of iterations for a Hopfield network to reach a stable state or a limit cycle is called the *transient period*; in general, it will depend on the mode of iteration and the initial state of the network. We shall derive an upper bound on the transient period in the worst-case scenario. Since the transient period is closely related to the energy function associated with the network, we first determine how long it will take for the energy function to converge.

Lemma 11.2:   Let $E(t)$ be the energy function as defined in Eqn. (11.2), where the elements in the weight matrix W and the threshold vector T are integers. Assume also that the Hopfield network is iterating in a sequential mode, so that by Theorem 11.1, $\Delta E(t) \leq 0$. Let $0 \leq t_1 < t_2 < \cdots < t_m$ to be the time steps where $\Delta E(t) < 0$. In other words, for $i = 0, 1, \ldots, m - 1$, t_{i+1} is the least integer $> t_i$ such that $E(t_{i+1}) < E(t_i)$ (where $t_o = 0$), and t_m is the least integer such that for all $t > t_m$, $E(t) = E(t_m)$. Then, the number of decrements (i.e., m) in the energy function before converging to a local minimum is bounded from above as

$$m \leq \sum_{i=1}^{n} |t_i| + \frac{1}{2} \sum_{i=1}^{n} \sum_{j=1}^{n} |w_{ij}|.$$

Proof: From the definition of m,

$$m \cdot \min|\Delta E(t_i)| \;\; \leq \;\; 2 \max|E(t)|. \tag{11.6}$$

From the expression in Eqn. (11.2), we have

$$\max|E(t)| \leq 2 \sum_{i=1}^{n} |t_i| + \sum_{i=1}^{n} \sum_{j=1}^{n} |w_{ij}|.$$

Since by assumption the elements of the weight matrix and the thresholds are integers and $P_j \neq 0$, we have $|P_j| \geq 1$. Further, from Eqn. (11.3), if $\Delta x \neq 0$, then $|\Delta x| = 2$. Again from Eqn. (11.4), we have

$$\min|\Delta E(t_i)| \geq 4.$$

Thus, Eqn. (11.6) becomes

$$m \leq \frac{2 \max|E(t)|}{\min|\Delta E(t_i)|} \leq \sum_{i=1}^{n} |t_i| + \frac{1}{2} \sum_{i=1}^{n} \sum_{j=1}^{n} |w_{ij}| . \qquad \square$$

With Lemma 11.2, we are now ready to derive an upper bound on the transient period of a Hopfield network. For the case of sequential iteration, we exclude those degenerate cases in which the network converges to a state that is not stable (see the remarks following Theorem 11.1). It is not difficult to construct a pathological case in which the sequential iteration can result in a transient period of any length, if there is no restriction on how we update the nodes.

A typical and interesting case of sequential iteration would be a cyclic mode of iteration in which each of the n nodes is updated sequentially in every n iterations (n is the order of the network): $x_{11}, x_{12}, \ldots, x_{1n}, \ x_{21}, x_{22}, \ldots, x_{2n}, \ldots$ where $\{x_{i1}, x_{i2}, \ldots, x_{in}\} = \{x_1, x_2, \ldots, x_n\}$ for every cycle $i = 1, 2, \ldots$. In a cyclic iteration, it is not necessary that the sequence of updated nodes be the same for every cycle.

Theorem 11.4:   Let H be a Hopfield network of order n with symmetric weight matrix W, $w_{ii} \geq 0$, and threshold vector T, iterating in a cyclic mode. Then, the state of the network always converges to a stable state with a transient period bounded from above by

$$n \cdot \left(\sum_{i=1}^{n} |t_i| + \frac{1}{2} \sum_{i=1}^{n} \sum_{j=1}^{n} |w_{ij}| \right).$$

Proof: At the end of a cycle of iteration, either (i) the state remains the same as in the last cycle of iteration, or (ii) some of the nodes change their states. In case (i), the network has converged to a stable state. In case (ii), there will be a decrement in the energy function, i.e., $\Delta E < 0$. Since every cycle consists of n sequential iterations, the transient period must be $\leq n$ $\times$ (the number of decrements in the energy function before converging to a local minimum). By Lemma 11.2, this is bounded from above by

$$n \cdot (\sum_{i=1}^{n} |t_i| + \frac{1}{2} \sum_{i=1}^{n} \sum_{j=1}^{n} |w_{ij}|). \qquad \square$$

We can also derive an upper bound on the transient period for the case of parallel iteration, using the correspondence between the parallel mode and the sequential mode established in Lemma 11.1.

Theorem 11.5:   Let H be a Hopfield network of order n with symmetric weight matrix W and threshold vector T iterating in the parallel mode. Then, the state of the network always reaches a limit cycle with a transient period P_{tr} bounded from above by

$$2 \sum_{i=1}^{n} |t_i| + \sum_{i=1}^{n} \sum_{j=1}^{n} |w_{ij}|.$$

Proof: Given the Hopfield network H, construct the corresponding network $\hat{H}$ with parameters $\{\hat{W}, \hat{T}\}$ as in the setup of Lemma 11.1. By Theorem 11.4, it is easy to see from the structure of $\hat{W}$ and $\hat{T}$ (Eqn. (11.5)) that the network $\hat{H}$ converges with a transient period bounded from above by

$$\hat{P}_{tr} = n \cdot (\sum_{i=1}^{2n} |\hat{t}_i| + \frac{1}{2} \sum_{i=1}^{2n} \sum_{j=1}^{2n} |\hat{w}_{ij}|) = n \cdot (2 \sum_{i=1}^{n} |t_i| + \sum_{i=1}^{n} \sum_{j=1}^{n} |w_{ij}|).$$

From the correspondence established in Lemma 11.1, every n sequential updates in $\hat{H}$ are equivalent to a parallel iteration in H. Thus, the transient period of H must be bounded from above by $\hat{P}_{tr}/n$, i.e., by

$$2 \sum_{i=1}^{n} |t_i| + \sum_{i=1}^{n} \sum_{j=1}^{n} |w_{ij}| . \qquad \square$$

Remark 11.2:   It follows from the results in Theorem 11.4 and 11.5 that when the parameters w_{ij} and t_i are small integers, say bounded by a fixed constant, then the transient period in the sequential (respectively, parallel) mode of iteration will grow at a rate at most cubic (respectively, quadratic) in the order of the network (i.e., $O(n^3)$ for a cyclic mode and $O(n^2)$ for the parallel mode). In general, if the weights are polynomially bounded integers, then the transient period is also polynomially bounded.   □

11.2.2   Hopfield Networks with Exponential Transient Periods

Since the results of Theorems 11.4 and 11.5 only yield upper bounds on the transient period, it is not clear whether the transient period can grow exponentially large with respect to the order of the network, when there is no restriction on the magnitudes of the parameters. In fact, for some applications, the elements in the weight matrix have special structures but can have arbitrarily large magnitudes; however, the transient period of the network is only linear $(O(n))$ for sequential iteration and is bounded from above by a fixed constant for parallel iteration [120]. In the following discussions, we present an explicit construction of Hopfield networks whose transient periods are exponentially large (in the number of nodes) for both parallel and sequential modes of iteration.

Consider an n-node network with zero thresholds and, as in Section 11.1, let $x_i(t)$ denote the output of the ith node at time t, and let $X(t) = [x_1(t), \dots, x_n(t)]$ denote the state of the network at time t. As defined in Section 11.1, let $P_i(t)$ be the weighted sum at gate i (the threshold $t_i = 0$ in the network). Then,

$$P_i(t) = \sum_{j=1}^{n} w_{ij} x_i(t), \qquad (11.7)$$

where w_{ij} is the weight associated with the edge connecting node j to node i.

Let $Y(t) = [P_1(t) \cdots P_n(t)]^T$ be the weighted-sum vector at time t. Then, we can rewrite Eqn. (11.7) in matrix form:

$$Y(t) = W \cdot X(t),$$

and in the parallel mode of iteration:

$$X(t+1) = \text{sgn}(Y(t)).$$

The restriction of the networks to zero thresholds provides for symmetry properties that help simplify some proofs.

Lemma 11.3:   If for some t and t',

$$X(t) = -X(t'),$$

then for all $i \geq 0$,

$$X(t+i) = -X(t'+i),$$

and

$$Y(t+i) = -Y(t'+i).$$

Proof: We prove this result by induction on i. It is true for $i = 0$, and let it be true for some other $i > 0$.

$$X(t+i) = -X(t'+i)$$

implies that

$$Y(t+i) = W \cdot X(t+i) = W \cdot (-X(t'+i)) = -W \cdot X(t'+i) = -Y(t'+i).$$

Hence, $X(t+i+1) = \text{sgn}(Y(t+i)) = \text{sgn}(-Y(t'+i)) = -\text{sgn}(Y(t'+i)) = -X(t'+i+1)$.   □

We want to construct a weight matrix such that when the network is initialized to some state $X(0)$ (by setting the output of each gate to $+1$ or -1), it will cycle through all 2^n states before returning to $X(0)$. For a network with n nodes, let W_n be the $n \times n$ matrix defined by

$$W_n(i,j) = \begin{cases} -\frac{3}{2^{j-1}} & \text{for } 1 \leq i < j \leq n, \\ (j-1) - \frac{1}{2^{j-1}} & \text{for } 1 \leq i = j \leq n, \\ (-1)^{i-j-1} & \text{for } 1 \leq j < i \leq n. \end{cases}$$

A few examples of the weight matrices and the corresponding cycle of states are given next.

For a single node:

$$W_1 = \begin{bmatrix} -1 \end{bmatrix}$$

t	$x_1(t)$	$P_1(t)$
1	1	-1.00
2	-1	1.00 .

For two nodes:

$$W_2 = \begin{bmatrix} -1 & -\frac{3}{2} \\ 1 & \frac{1}{2} \end{bmatrix}$$

t	$x_1(t)$	$P_1(t)$	$x_2(t)$	$P_2(t)$
1	1	-2.50	1	1.50
2	-1	-0.50	1	-0.50
3	-1	2.50	-1	-1.50
4	1	0.50	-1	0.50 .

And, for three nodes:

$$W_3 = \begin{bmatrix} -1 & -\frac{3}{2} & -\frac{3}{4} \\ 1 & \frac{1}{2} & -\frac{3}{4} \\ -1 & 1 & 1\frac{3}{4} \end{bmatrix}$$

t	$x_1(t)$	$P_1(t)$	$x_2(t)$	$P_2(t)$	$x_3(t)$	$P_3(t)$
1	1	-3.25	1	0.75	1	1.75
2	-1	-1.25	1	-1.25	1	3.75
3	-1	1.75	-1	-2.25	1	1.75
4	1	-0.25	-1	-0.25	1	-0.25
5	-1	3.25	-1	-0.75	-1	-1.75
6	1	1.25	-1	1.25	-1	-3.75
7	1	-1.75	1	2.25	-1	-1.75
8	-1	0.25	1	0.25	-1	0.25 .

We note that for four nodes, the weight matrix is:

$$W_4 = \begin{bmatrix} -1 & -\frac{3}{2} & -\frac{3}{4} & -\frac{3}{8} \\ 1 & \frac{1}{2} & -\frac{3}{4} & -\frac{3}{8} \\ -1 & 1 & 1\frac{3}{4} & -\frac{3}{8} \\ 1 & -1 & 1 & 2\frac{7}{8} \end{bmatrix} .$$

In these matrices, W_n can be obtained by augmenting W_{n-1} with an additional row and an additional column. This holds for all n and will be used in the proof of the following theorem.

Theorem 11.6: Let n be a positive integer and let W_n be defined as above. If

$$X(0) = [1 \cdots 1]^T,$$

i.e., if we initialize the output of each of the n gates to $+1$, then:

1. $X(0), \ldots, X(2^n - 1)$ are all different.

   (Hence the first 2^n states contain all possible states.)

2. For $i = 1, \ldots, n$,
   $$x_i(2^n - 1) = (-1)^{n-i-1}.$$

   (The 2^nth state is $[-1\ 1\ -1 \cdots -1]^T$ for odd n and $[1\ -1\ 1 \cdots -1]^T$ for even n.)

3. For $i = 1, \ldots, n$,
   $$P_i(2^n - 1) = \frac{1}{2^{n-1}}.$$

   (At the last state, the weighted sum of each gate is $\frac{1}{2^{n-1}}$; hence, the next state is $[1\ 1 \cdots 1]^T$.)

4. For $t = 1, \ldots, 2^n - 2$, and for $i = 1, \ldots, n$,

   $$\text{either} \quad P_i(t) \geq \frac{3}{2^{n-1}} \quad \text{or} \quad P_i(t) \leq -\frac{1}{2^{n-1}}.$$

   (At all steps but the last, the weighted sum at each node is either $\geq \frac{3}{2^{n-1}}$ or $\leq -\frac{1}{2^{n-1}}$.)

Proof: We prove the four statements in Theorem 11.6 by induction on n. The four statements clearly hold for $n = 1$. Assuming they hold for some n, we prove that they hold for $n + 1$. We begin by relating the first 2^n states (out of 2^{n+1}) to the corresponding states in the n-node network. We use the symbols $X^n(t)$ and $Y^n(t)$ to denote, respectively, the state vector and the weighted-sum vector of the network with n nodes.

We first prove the following claim:

Claim 1: For all $t \in \{0, \ldots, 2^n - 1\}$,

$$X^{n+1}(t) = [x_1(t) \cdots x_n(t)\ 1]^T. \tag{11.8}$$

That is, if two networks, one consisting of n nodes and weight matrix W_n, and the other consisting of $(n+1)$ nodes and weight matrix W_{n+1}, are both initialized to the all $+1$ state, then in each of the first 2^n steps, the states of nodes $1, \ldots, n$ in the $(n+1)$-node network equal the corresponding states in the n-node network, and the state of the node $n+1$ in the $(n+1)$-node network is $+1$.

This claim can be proved by induction on t. Eqn. (11.8) holds for the initial state vector $X(0) = [1 \cdots 1]^T$. Assuming it holds at step $t < 2^n - 1$, we show that it holds at step $t+1$. For $i = 1, \ldots, n$, the weighted sum of node i is

$$
\begin{aligned}
P_i^{n+1}(t) &= \sum_{j=1}^{n+1} W_{n+1}(i,j) \cdot x_j^{n+1}(t) \\
&= \sum_{j=1}^{n} W_n(i,j) \cdot x_j^n(t) + W_{n+1}(i,n+1) \cdot x_{n+1}^{n+1}(t) \\
&= P_i^n(t) + (-\frac{3}{2^n}) \cdot 1.
\end{aligned}
$$

Since $t < 2^n - 1$, Statement (4) of the theorem's induction hypothesis implies that one of the following holds:

1. Either $P_i^n(t) \geq \dfrac{3}{2^{n-1}}$, therefore

$$
P_i^{n+1}(t) \geq \frac{3}{2^{n-1}} - \frac{3}{2^n} = \frac{3}{2^n}, \tag{11.9}
$$

2. or $P_i^n(t) \leq -\frac{1}{2^{n-1}}$, therefore

$$
P_i^{n+1}(t) \leq -\frac{1}{2^{n-1}} - \frac{3}{2^n} \leq -\frac{5}{2^n}. \tag{11.10}
$$

Hence, for $i = 1, \ldots, n$,

$$
x_i^{n+1}(t+1) = x_i^n(t+1).
$$

For the $(n+1)$st node,

$$
P_{n+1}^{n+1}(t) = \sum_{j=1}^{n+1} W_{n+1}(n+1,j) \cdot x_j^{n+1}(t)
$$

$$= \sum_{j=1}^{n} (-1)^{n-j} \cdot x_j^n(t) + (n - \frac{1}{2^n}) \cdot 1$$

$$\geq 2 - \frac{1}{2^n}$$

$$\geq \frac{3}{2^n}, \qquad\qquad (11.11)$$

where the first inequality uses Statement (1) of the theorem's induction hypothesis, implying that $X(t) \neq X(2^n - 1)$ (for $0 \leq t \leq (2^n - 2)$) and Statement (2), implying that, therefore, $x_j^n(t) \neq (-1)^{n-j-1}$ for some $j \in \{1, \ldots, n\}$. This shows that

$$x_{n+1}^{n+1}(t+1) = +1,$$

completing the proof of Claim 1.

Next we determine the 2^nth state vector for the $(n+1)$-node network, by proving the following claim:

Claim 2:

$$Y^{n+1}(2^n - 1) = \left[-\frac{1}{2^n} \cdots -\frac{1}{2^n} \right]^T,$$

hence

$$X^{n+1}(2^n) = [-1 \cdots -1]^T.$$

We begin with the first n nodes. For $i = 1, \ldots, n$,

$$P_i^{n+1}(2^n - 1) = \sum_{j=1}^{n+1} W_{n+1}(i,j) \cdot x_j^{n+1}(2^n - 1)$$

$$= P_i^n(2^n - 1) + (-\frac{3}{2^n}) \cdot 1$$

$$= \frac{1}{2^{n-1}} - \frac{3}{2^n}$$

$$= -\frac{1}{2^n}.$$

For the last node we have

$$P_{n+1}^{n+1}(2^n - 1) = \sum_{j=1}^{n+1} W_{n+1}(n+1,j) \cdot x_j^{n+1}(2^n - 1)$$

$$= \sum_{j=1}^{n} W_n(n+1,j) \cdot x_j^n(2^n - 1)$$

$$+W_{n+1}(n+1, n+1) \cdot x_{n+1}^{n+1}(2^n - 1)$$

$$= \sum_{i=1}^{n} (-1)^{n-j}(-1)^{n-j-1} + (n - \frac{1}{2^n})$$

$$= -\frac{1}{2^n},$$

proving Claim 2.

We can now combine the results of the two claims to prove Statements (1), (2), (3), and (4) of the theorem's induction step.

Statement 1:   Claim 1 shows that $X^{n+1}(0), \ldots, X^{n+1}(2^n - 1)$ are all different and each ends with $+1$. Claim 2 shows that $X^{n+1}(2^n) = [-1 \cdots - 1]^T = -X^{n+1}(0)$; hence, Lemma 11.3 says that for $t = 2^n, \ldots, 2^{n+1} - 1$,

$$X^{n+1}(t) = -X^{n+1}(t - 2^n).$$

Therefore, $X^{n+1}(2^n), \ldots, X^{n+1}(2^{n+1} - 1)$ are all different. Furthermore, each ends with -1, and hence, differs from any of the first 2^n state vectors.

Statement 2:   Claim 1 and Statement (2) show that for $i = 1, \ldots, n$,

$$x_i^{n+1}(2^n - 1) = x_i^n(2^n - 1) = (-1)^{n-i-1},$$

and

$$x_{n+1}^{n+1}(2^n - 1) = 1 = (-1)^{n-(n+1)-1}.$$

Hence, Lemma 11.3 implies that for $i = 1, \ldots, n+1$,

$$x_i^{n+1}(2^{n+1} - 1) = -x_i^{n+1}(2^n - 1) = (-1)^{(n+1)-i-1}.$$

Statement 3:   Follows from Claim 2 and Lemma 11.3.

Statement 4:   For $t = 1, \ldots, 2^n - 2$, the statement follows from Eqns. (11.9), (11.10), and (11.11). For $t = 2^n, \ldots, 2^{n+1} - 1$, the statement follows from the same inequalities and Lemma 11.3. For $t = 2^n - 1$, the statement follows from Claim 2.   □

Corollary 11.1:   The n-node network with weight matrix W_n, initialized to any state, will keep cycling through all 2^n states.

Proof: It suffices to prove the result when the network is initialized to the all $+1$ state. For that case, the theorem says that the network goes through all 2^n steps, and that at the $(2^n - 1)$th step, the weight vector is $Y^n(2^n - 1) = [\frac{1}{2^{n-1}} \cdots \frac{1}{2^{n-1}}]^T$. Hence, $X^n(2^n) = [1 \cdots 1]^T$ and the cycle repeats.   □

As a result of Corollary 11.1 and the correspondence between the parallel mode and the cyclic mode established in Lemma 11.1, we can state the following.

Corollary 11.2:   There exists a Hopfield network of order n that has a transient period of $\Omega(n2^{n/2})$ in a cyclic mode of iteration.   □

11.3   Associative Memory

The potential applications of the Hopfield network as an associative memory device has led to considerable interest in the model. The basic goal of an associative memory can be informally described as follows: Store a set of s patterns, $M = \{X_1, \ldots, X_s\}$, such that when a new pattern $\tilde{X}$ is presented as the input, the associative memory should output a pattern X_k in M that is 'closest' to $\tilde{X}$ among all the patterns in M.

In general, depending on the applications, there are many different measures of 'closeness' or distances between two patterns. For example, a pattern could correspond to a set of pixel values on a digital image; the distance between two images (patterns) could be taken as the number of pixels different between them. Then, the associative memory problem can also be viewed as a problem of image reconstruction. In our framework, the patterns will be binary vectors, i.e., the set $M \subset \{1, -1\}^n$. The distance between two patterns will be defined as the number of coordinates that are different in the two patterns, or the *Hamming distance*.

A motivation behind using the Hopfield model as an associative memory is due to the convergence property of the model. The idea is that the set of stable states should correspond to the set of stored vectors M in the memory. When a pattern $\tilde{X}$ is applied to the network as the initial state, the network should converge to a stable state in M that is closest to $\tilde{X}$.

This concept is appealing due to its similarity with the way the brain retrieves information from its memory. In a conventional digital computer memory, each piece of information is stored one after another in a list with a different address. A piece of information is retrieved by giving its address in the memory. However, a small change in the address possibly due to some corruption by noise may result in the retrieval of information completely different from the one desired. In addition, the full address has to be given for accessing such memory. On the other hand, unlike conventional computer memories, the information in the Hopfield model of associative memory is stored as the set of weights distributed over the networks. The network is potentially capable of recalling the complete description of a pattern given only a partial or corrupted description of that pattern. Furthermore, each connection between the nodes contains traces of the information in all the stored patterns. Because of this nonlocalized or distributed property, a slight damage in the network itself is unlikely to change its characteristics. These fault-tolerance properties of the Hopfield model have made it attractive as an associative memory device.

Strictly speaking, the associative memory we described should be referred to as an *auto-associative* memory, whereupon presentation of a corrupted version of pattern X should result in a recall of X itself. There is another kind of associative memory in which an input pattern X is associated with an output pattern X' that is different from X; such memory is called *hetero-associative* memory. Here we are only concerned with the study of auto-associative memory and, for convenience, only the term associative memory will be used.

In this section, we address different issues that arise in the application of the Hopfield model as an associative memory. Our aim is to study the properties of this model – in particular, the relationship between its parameters and the stored patterns – and to demonstrate the limitations of the model.

11.3.1   Properties of the Stored Patterns

Suppose $M = \{X_1, X_2, \ldots, X_s\}$ is a set of patterns to be stored. In order for the Hopfield model to be useful as an associative memory, it is natural to require that if any X_i in M is presented as input to the network, then the output should recall X_i exactly. With this condition in mind, a fundamental question is: Can an arbitrary set of patterns be stored in the network?

The answer to this question is trivially true, provided there are no restrictions on the parameters (i.e., the weight matrix W and the threshold vector T) of the network. To see this, consider W to be the identity matrix and $T = 0$. Then, any pattern X will be recalled exactly. However, every $X \in \{1, -1\}^n$ not in M is also stored as a *spurious* memory. This is clearly undesirable. Of course, we would like to construct or program the network so that only the patterns in the set M are stored as memories. As we show later on, this is not always possible. In fact, we demonstrate in the next section that a popular method for storing a set of patterns, the *outer-product rule*, could result in an exponential number (in s) of spurious memories.

A fundamental problem is to determine what sets of patterns can be stored as memories in the network. Given any set M of patterns, we would like to store the patterns as stable states in the network. We next show that some conditions on the patterns need to be satisfied.

Theorem 11.7:   Let H be a Hopfield network with a zero-diagonal weight matrix W, (i.e., H has no self-loops as a graph). If M is a set of stable states in H, then no two distinct vectors in M can differ in only one coordinate.

Proof: Let $X = (x_1, x_2, \ldots, x_n)$ and $\tilde{X} = (\tilde{x}_1, \tilde{x}_2, \ldots, \tilde{x}_n)$ be two distinct stable states in H. Suppose X and $\tilde{X}$ differ only in the kth coordinate, i.e., $x_k \neq \tilde{x}_k$ and $x_j = \tilde{x}_j$ for $j \neq k$. It follows from the definition of a stable state that

$$x_k = \text{sgn}(\sum_{j=1}^{n} w_{kj} x_j - t_k). \tag{11.12}$$

Similarly,

$$\tilde{x}_k = \text{sgn}(\sum_{j=1}^{n} w_{kj} \tilde{x}_j - t_k). \tag{11.13}$$

Since by assumption, $w_{kk} = 0$ and $x_j = \tilde{x}_j$ for $j \neq k$, we can rewrite Eqn. (11.12) as

$$x_k = \text{sgn}(\sum_{j=1}^{n} w_{kj} \tilde{x}_j - t_k) = \tilde{x}_k,$$

which contradicts the fact that $x_k \neq \tilde{x}_k$.   □

Note that the assumption of Theorem 11.7 does not require the weight matrix to be symmetric. The only assumption is that the network is free of self-loops, which is a condition imposed on the connectivity of the network. In fact, we can derive a more general result than Theorem 11.7 relating the connectivity of the network to the properties of the stable states. For this purpose, we need to introduce additional terminologies.

Definition 11.1:   A set S of nodes in a graph is *independent* if no two nodes in S are connected to each other.   □

Theorem 11.8:   Let I be the index set of an independent set S of nodes in a Hopfield network H, i.e., $\{x_i : i \in I\} = S$, and if $i, j \in I$, then $w_{ij} = w_{ji} = 0$. Let $X = (x_1, x_2, \ldots, x_n)$ and $\tilde{X} = (\tilde{x}_1, \tilde{x}_2, \ldots, \tilde{x}_n)$ be two distinct stable states in H. Then, there exists a $j \notin I$ such that $x_j \neq \tilde{x}_j$.

Proof: We prove this theorem by contradiction. Suppose that

$$x_j = \tilde{x}_j \quad \text{for all} \quad j \notin I. \tag{11.14}$$

Since $X \neq \tilde{X}$, there exists an $i \in I$ such that $x_i \neq \tilde{x}_i$. Now both X and $\tilde{X}$ are stable, thus by definition

$$x_i = \text{sgn}(\sum_{j=1}^{n} w_{ij} x_j - t_i). \tag{11.15}$$

Similarly,

$$\tilde{x}_i = \text{sgn}(\sum_{j=1}^{n} w_{ij} \tilde{x}_j - t_i). \tag{11.16}$$

Since $w_{ij} = 0$ for all $i \in I$ and $j \in I$, by using Eqn. (11.14), Eqn. (11.15) can be rewritten as

$$\begin{aligned} x_i &= \text{sgn}(\sum_{j \notin I} w_{ij} x_j - t_i) \\ &= \text{sgn}(\sum_{j \notin I} w_{ij} \tilde{x}_j - t_i) \\ &= \text{sgn}(\sum_{j=1}^{n} w_{ij} \tilde{x}_j - t_i), \end{aligned}$$

which contradicts the fact that $x_i \neq \tilde{x}_i$.   □

Let the neighborhood of a node i be the set of all nodes j, such that $w_{ij} \neq 0$. Then, the preceding two theorems can be interpreted as follows: If any two stable states are identical in the neighborhood of a node i, then they must also be identical at node i. Notice that Theorem 11.7 is a special case of Theorem 11.8, since a single node without any self-loop is an independent set by itself.

To demonstrate an application of Theorem 11.8, consider the following case. Let $\hat{M} = \{X_1, X_2, \ldots, X_n\} \subset \{1, -1\}^n$, where $X_1 = [-1\ 1\ 1 \cdots 1]^T$, $X_2 = [1\ -1\ 1 \cdots 1]$, and in general, X_i is the vector that consists of all 1's, except at the ith coordinate where it is -1. It is easy to construct a completely connected network that has $\hat{M}$ as the set of stable states and no vector $\notin M$ is a stable state (see Exercise 11.5). The following corollary of Theorem 11.8 shows that any network that contains M as the set of stable states must be completely connected.

Corollary 11.3:   If $\hat{M}$ is a subset of the stable states in a Hopfield network H with weight matrix W (where $w_{ii} = 0$), then for $i \neq j$, either $w_{ij} \neq 0$ or $w_{ji} \neq 0$. Equivalently, H must be completely connected.

Proof: If $w_{ij} = w_{ji} = 0$ for some $i \neq j$, then nodes x_i and x_j together form an independent set. By Theorem 11.8, X_i and X_j cannot both be stable states.   □

11.3.2   On the Number of Spurious Memories

As we have already hinted at in the preceding section, an interesting issue concerning the use of the Hopfield network as an associative memory is: How should one program the network? Programming of a network can be defined as follows: Consider the set $M = \{X_1, X_2, \ldots, X_s\}$ that consists of s vectors in $\{1, -1\}^n$. Construct a network $N = \{W, T\}$ such that $M \subseteq M_N$, where M_N is the set of all possible stable states in the network N. Hopfield [58] suggested computing W by the *outer-product* method (which is a Hebb-type rule [52]). In particular,

$$W = \sum_{i=1}^{s}(X_i X_i^T - I_n), \tag{11.17}$$

where I_n is the $n \times n$ identity matrix. Using this method, T is chosen to be the all-zero vector. Note that if the X_i's are orthogonal, then

$$W X_1 = (n - s)X_1.$$

Thus, if $n > s$, then

$$X_i = \operatorname{sgn}(W X_i),$$

and every one of the X_i's is a stable state. Hence, a natural question is: Are there any other (spurious) stable states? Namely, what can be said about the number of stable states that are not in M?

In Exercise 11.10, a method for establishing an upper bound on the number of spurious memories is outlined. In particular, if an n-node Hopfield network is programmed using the outer-product rule involving s vectors (as described in Eqn. (11.17)), then the number of stable states is at most $2^{s \log(n+1)}$. We next show that in certain cases, the number of spurious memories (vectors which are in M_N but not in M) can indeed become exponentially large in s. The results hold for the following three cases, which cover all the possibilities for s:

1. $1 \leq s \leq \log n$.

2. $n - \log n \leq s < n$.

3. $s = 2^k$, where $0 \leq k < \log n$.

The main tool for proving the results is based on treating vectors over $\{1, -1\}^n$ as Boolean functions of $\log n$ variables. One approach is to use the polynomial representation of Boolean functions, which is described in Chapter 8. The first result (Theorem 11.9) provides an example in which the number of vectors, say s, to be stored is less than or equal to $\log n$, and the number of stable states in the resulting network is exponential in s. The proof holds for an arbitrary odd $s \leq \log n$. The idea in the proof is to represent vectors over $\{1, -1\}^n$ by polynomials. For example, consider vectors over $\{1, -1\}^8$, then any vector can be represented by a polynomial of three variables. The vectors that correspond to the three variables x_1, x_2, x_3 are

$$
\begin{aligned}
X_1 &= \begin{bmatrix} 1 & -1 & 1 & -1 & 1 & -1 & 1 & -1 \end{bmatrix}^T, \\
X_2 &= \begin{bmatrix} 1 & 1 & -1 & -1 & 1 & 1 & -1 & -1 \end{bmatrix}^T, \\
X_3 &= \begin{bmatrix} 1 & 1 & 1 & 1 & -1 & -1 & -1 & -1 \end{bmatrix}^T.
\end{aligned}
$$

Theorem 11.9:   Let $s \geq 1$ be odd and let $n = 2^k$, where $k \geq s$. Consider the s vectors $X_1, X_2, \ldots, X_s$ over $\{1, -1\}^n$ that correspond to the Boolean variables $x_1, \ldots, x_s$. Let W be the matrix that is computed by the outer-product method, i.e.,

$$
W = \sum_{i=1}^{s} (X_i X_i^T - I_n).
$$

Consider the vectors of the form

$$U_\beta = \text{sgn}(\sum_{i=1}^{s} \beta_i X_i) \,,$$

where $\beta = (\beta_1, \beta_2, \ldots, \beta_s)$. Then: (i) For all $\beta \in \{1, 0, -1\}^s$, such that the support of β (number of non-zero entries in β) is odd, the vector U_β is stable in the network $N = \{W, T\}$ (T is the all-zero vector). (ii) All the $\frac{3^s + 1}{2}$ U_β's that correspond to β with odd support are distinct.

Proof: The idea in the proof is to compute WU_β and to show that $\text{sgn}(WU_\beta) = U_\beta$. First, we prove that for β being the all-one vector and for all $i = 1, \ldots, s$, we have

$$X_i^T U_\beta = 2\binom{s-1}{\frac{s-1}{2}}.$$

Using the notation in Chapter 8, $X_1^T U_\beta = 2^s a_{1,0,\ldots 0}$. Namely, the inner product of U_β with X_1 is just the corresponding spectral coefficient multiplied by 2^s. Without loss of generality, we prove the result for $i = 1$. Notice that for the case where β is the all-one vector, the vector U_β corresponds to the *Majority* function of s variables. We have computed the spectral coefficients for *Majority* in Chapter 8; however, we present here an alternative (and perhaps more intuitive) approach for determining the linear spectral coefficients. To compute the spectral coefficient that corresponds to X_1, consider the vector

$$V = X_2 + X_3 + \cdots + X_s.$$

Clearly, V is zero in $2\binom{s-1}{\frac{s-1}{2}}$ entries. For example, for s=3, $X_2 + X_3$ is zero in four entries – see the example preceding the theorem. There is a symmetry in the values of the other entries, half of them being negative and half positive. Hence, in $X_1^T \text{sgn}(X_1 + V)$, the positive and negative entries cancel each other, and we get as a result $2\binom{s-1}{\frac{s-1}{2}}$. By a similar argument we get the result for general β with odd support w:

$$X_i^T U_\beta = \beta_i 2^{s-w+1}\binom{w-1}{\frac{w-1}{2}}.$$

Hence, for β with odd support w, we have

$$WU_\beta = 2^{s-w+1} \binom{w-1}{\frac{w-1}{2}} \sum_{i=1}^{s} \beta_i X_i - sU_\beta.$$

From the preceding equation, it follows that the sign of WU_β is the same as the sign of

$$\sum_{i=1}^{s} \beta_i X_i = U_\beta.$$

Thus, for all odd $s \geq 1$, and $\beta \in \{1, 0, -1\}^s$, such that the support of β is odd, we have $\operatorname{sgn}(WU_\beta) = U_\beta$. The U_β's are distinct because they all have distinct polynomial representations.   □

A natural question is whether such phenomena hold for other values of s. The next result (Theorem 11.10) shows that, indeed, for any $s = 2^k$, where $0 \leq k < \log n$, there are sets of s orthogonal vectors that result in an exponential number of spurious memories.

A note regarding the technique: In the foregoing theorem, the spurious memories are a nonlinear (threshold) function of the vectors that we want to store. In the results to follow, we use a different technique to prove that a state is stable: We prove that it is in the linear span of the vectors that were stored. By the following lemma, every state in the linear span of the stored vectors is also stable.

Lemma 11.4:   Let $1 \leq s < n$. Let $X_1, X_2, \ldots, X_s$ be a set of orthogonal vectors where $X_i \in \{1, -1\}^n$ for all $i = 1, \ldots, s$. Let

$$W = \sum_{i=1}^{s} (X_i X_i^T - I_n).$$

Then, a vector $V \in \{1, -1\}^n$ in the linear span of the X_i's corresponds to a stable state in $N = \{W, T\}$ with $T = 0$.

Proof: V is in the linear span; hence, there exist γ_i's such that

$$V = \sum_{i=1}^{s} \gamma_i X_i.$$

Thus, $WV = (n - s)V$. So if $s < n$, it follows that $\operatorname{sgn}(WV) = V$.   □

Theorem 11.10:   Let $n = 2^l$. For every $0 \le k < l$, there exists a set of 2^k orthogonal vectors $\{X_1, X_2, \ldots, X_{2^k}\}$ where $X_i \in \{1, -1\}^n$, such that when W is computed by the outer-product method, i.e.,

$$W = \sum_{i=1}^{2^k} (X_i X_i^T - I_n) ,$$

then the network $N = \{W, T\}$ with $T = 0$ has 2^{2^k} stable states.

Proof: The idea in the proof is to choose the vectors X_i's to be the basis functions of the Boolean functions with k variables (there are 2^k basis functions). For example, for $k = 2$, we consider the vectors that correspond to $1, x_1, x_2$, and $x_1 x_2$. Clearly, all Boolean functions of k variables (there are 2^{2^k} of those) are in the linear span of the basis functions. Hence, by Lemma 11.4, there are 2^{2^k} stable states in N. $\square$

In fact, one can give explicit constructions of networks (based on the preceding proof) in which there are exponentially many spurious memories. A construction of a set of such networks is outlined in Exercise 11.11.

The next interesting case is the one where s, the number of vectors to be stored, is very close to n, i.e., $n - \log n \le s < n$. It turns out that for $n - 1$, $n - 2$ and $n - 3$ orthogonal vectors, we can count exactly the number of vectors in the linear span. The next theorem presents results for these cases. Note that these results hold for any n for which there exist Hadamard matrices (not just for n being a power of 2, as in Theorem 11.10).

Theorem 11.11:   Let n be an integer such that a Hadamard matrix H_n of order n exists, then the number of vectors over $\{1, -1\}^n$ in the linear span of:

(i) $(n - 1)$ orthogonal vectors is

$$\binom{n}{\frac{n}{2}},$$

(ii) $(n - 2)$ orthogonal vectors is

$$\binom{\frac{n}{2}}{\frac{n}{4}}^2,$$

and .

(iii) $(n-3)$ orthogonal vectors is

$$\sum_{j=0}^{\frac{n}{4}} \binom{\frac{n}{4}}{j}^4.$$

Proof: Throughout the proof, we consider only vectors over $\{1,-1\}^n$. Counting the number of vectors in the linear span is the same as counting the number of vectors that are orthogonal to the null-space. Namely, we have to count the number of vectors over $\{1,-1\}^n$ that are orthogonal to a single vector (for (i)), two vectors (for (ii)), and three vectors (for (iii)).

Without loss of generality, assume that the single vector is the all-one vector. Clearly, the vectors that are orthogonal to this vector are those that consist of $\frac{n}{2}$ 1's and $\frac{n}{2}$ -1's. Hence, we get (i), i.e., the number of vectors in the linear span of any $(n-1)$ orthogonal vectors is $\binom{n}{\frac{n}{2}}$.

For (ii), we consider, without loss of generality, the all-one vector to be denoted by U_1, and a vector in which half of the entries are 1 and the other half are -1 to be denoted by U_2. Let $N^{++}(U,V)$ be the number of entries in which both U and V are 1. Similarly, let $N^{+-}(U,V)$ be the number of entries in which there is a 1 in U and a -1 in V, and let N^{-+} and N^{--} denote the other two cases. Assume that V is orthogonal to both U_1 and U_2; then we have:

$$N^{++}(U_2,V) + N^{+-}(U_2,V) = \frac{n}{2},$$

$$N^{++}(U_2,V) + N^{-+}(U_2,V) = \frac{n}{2},$$

$$N^{++}(U_2,V) + N^{--}(U_2,V) = \frac{n}{2},$$

and

$$N^{+-}(U_2,V) + N^{-+}(U_2,V) = \frac{n}{2}.$$

From the above equations, we can get a necessary and sufficient condition for V to be orthogonal to both U_1 and U_2:

$$N^{++}(U_2,V) = N^{--}(U_2,V) = N^{+-}(U_2,V) = N^{-+}(U_2,V) = \frac{n}{4}.$$

Hence, there are $\left(\frac{\frac{n}{2}}{\frac{n}{4}}\right)^2$ vectors that are orthogonal to both U_1 and U_2.

For (iii) we consider three orthogonal vectors, U_1, U_2, and U_3. We choose, without loss of generality, U_1 and U_2 as in (ii). From (ii) we know that there is a unique canonical form for U_3 in which the first quarter is 1, the second is -1, the third quarter is 1, and the fourth quarter is -1. Consider a vector V that is orthogonal to U_1, U_2, and U_3. Let j be the number of 1's in the first quarter of V. By a similar argument as in (ii), the number of -1's in the second quarter is j, the number of -1's in the third quarter is j, and the number of 1's in the fourth quarter is j. This is a necessary and sufficient condition for a vector to be orthogonal to U_1, U_2, and U_3. Hence, the number of vectors in a linear span of $(n-3)$ orthogonal vectors is

$$\sum_{j=0}^{\frac{n}{4}} \left(\frac{\frac{n}{4}}{j}\right)^4 .$$ □

So by Lemma 11.4, the number of stable states is indeed exponential for the cases of $n-1$, $n-2$, and $n-3$ orthogonal vectors. An important remark is that those are the only stable states, i.e., there are no other stable states besides those in the linear span.

The foregoing approach for counting vectors in the linear span does not work for $n-4$: we no longer have a canonical form (this phenomenon is related to the question of whether for each n divisible by 4, there exists any Hadamard matrix of order n [76]). Hence, we assume that n is a power of 2 (i.e., we consider the Sylvester-type Hadamard matrix) and we would like to count the number of 'high-frequency' Boolean functions – functions that have a zero constant term and zero linear terms in the polynomial representation. Counting the number of 'high-frequency' Boolean functions is the same as counting the number of vectors in the linear span of $n - \log n$ orthogonal vectors (those that correspond to higher order terms in the polynomial representation). In the following theorem, we provide a lower bound on this number.

Theorem 11.12: Let $n = 2^s$. Consider the set of vectors $S = \{1, X_1, \ldots, X_s\}$ over $\{1, -1\}^n$ that corresponds to the Boolean functions defined by $1, x_1, \ldots, x_s$. The number of vectors in $\{1, -1\}^n$ that are orthogonal to the vectors in S is at least $2^{\frac{n}{4}}$.

Proof: The number of Boolean functions (vectors over $\{1,-1\}^n$) that are functions only of the first $s-2$ variables is $2^{2^{s-2}} = 2^{\frac{n}{4}}$. Let $f(x_1,\ldots,x_s)$ be one of these functions, namely, f is a function only of the first $(s-2)$ variables. Let

$$g(x_1,\ldots,x_s) = x_{s-1} \oplus x_s \oplus f(x_1,\ldots,x_s) \, ,$$

where $\oplus$ is exclusive OR. The polynomial representation of g is

$$g(X) = x_{s-1}x_s f(X).$$

Since f is dependent only on the first $s-2$ variables, in the polynomial representation of $g(X)$, we have $a_\alpha = 0$ for all α of weight (number of 1's) less than or equal to 1. In other words, the polynomial representation of g has neither constant nor linear terms. In the language of vectors: We have exhibited a set of $2^{\frac{n}{4}}$ vectors that is orthogonal to S.   □

Theorem 11.12 shows that there exist $(n - \log n)$ orthogonal vectors (each vector assumes values from $\{1,-1\}^n$), such that at least $2^{\frac{n}{4}}$ other vectors lie in the linear span of the chosen vectors. Hence, it follows from Lemma 11.4 that there exists a set of vectors comprising $(n - \log n)$ vectors, such that a Hopfield network programmed using the outer-product rule will lead to exponentially many spurious stable states.

Thus, we have proved that the number of spurious memories resulting from using the outer-product method is in many cases exponentially large. Although the results are for some specific sets of vectors, it is not hard to see that they provide evidence for a more general phenomenon. In particular, Theorem 11.11 holds for any set of $n - 1$, $n - 2$, and $n - 3$ orthogonal vectors. Also, any set of orthogonal vectors that contains in its linear span a subset of one of the 'bad' sets of vectors (those that result in many spurious memories) will also be 'bad'. Exercise 11.12 outlines an alternative rule for deriving Hopfield networks, when the given vectors are not necessarily orthogonal. It is shown that the alternative technique might also lead to networks with many spurious stable states.

11.4   Combinatorial Optimization

In this section, we shall study the application of a Hopfield network to combinatorial optimization problems. In general, a combinatorial optimization problem consists of a cost function that we want to minimize (or maximize) over a finite but exponentially large set of possible solutions. An example of practical interest is the following:

Definition 11.2:   Traveling-Salesman Problem (TSP): Suppose we are given n cities and the distances d_{ij} between city c_i and city c_j. A salesman wants to make a closed tour that visits each city once and return to its starting point. The goal is to find a closed tour of minimum length among all possible choices.   □

For convenience, we shall consider a symmetric distance metric, i.e., $d_{ij} = d_{ji}$. Here, the cost function is the length of the possible closed tour. Since the number of solutions is finite, an obvious way of finding the optimal solution is to search through all possible cases. However, this approach is only feasible when the number of cities, n, is small. To see this, note that there are all together $(n-1)!/2$ possible solutions to this problem. Hence, an exhaustive search would require an astronomically large number of computations, even for moderately large n.

It is natural to ask whether there is an efficient way of finding the optimal solution to this problem. More precisely, we want to find an efficient algorithm that would compute the optimal solution of TSP for arbitrary inputs in a time that is proportional to a polynomial ($O(n^c)$) in the size of the inputs. Unfortunately, despite considerable research over several decades, no efficient algorithm for solving the TSP has been found. In fact, the TSP belongs to the class of NP-complete problems for which no efficient algorithm is known. A precise definition of NP-complete problems is beyond the scope of this book. For our purpose, it suffices to say that an NP-complete problem is computationally difficult, and it is a common conjecture that any algorithm that correctly finds an optimal solution to an NP-complete problem will require in the worst case an exponential amount of time.

Since it is unlikely that an efficient (polynomial-time) algorithm will exist for NP-complete problems, many researchers have spent their efforts devising efficient algorithms that will result in good, though not necessarily optimal, solutions. One successful approach to attacking these computationally difficult problems is to use the class of *local-search* algorithms.

The class of local-search algorithms can roughly be characterized as follows: Start at some initial feasible solution X_o and search for a better solu-

tion in its neighborhood. If an improved solution exists, repeat the search
procedure from the new solution; otherwise, the algorithm will terminate
and output X_o as a local optimum. We have seen in Section 11.2 that the
state space of a Hopfield network when iterating in a sequential mode will
always converge to a local minimum of an energy function:

$$E = 2X^TT - X^TWX = 2\sum_i x_it_i - \sum_{i,j} w_{ij}x_ix_j, \qquad (11.18)$$

where $X \in \{1, -1\}^n$, W is a symmetric matrix, and $w_{ii} \geq 0$. Thus, the
sequential iteration can be viewed as performing a local search of a minimum
of the corresponding energy function. Note that the energy E is a quadratic
function of the state variables X. Therefore, we could use a Hopfield network
to perform a local search for any combinatorial optimization problem that
can be formulated as the minimization of a quadratic cost function in the
form Eqn. (11.18). Next, we examine several specific NP-complete problems
and demonstrate how they can be formulated in terms of the Hopfield model.
We first point out some simple observations.

Lemma 11.5:   Any quadratic form in $X \in \{1, -1\}^n$

$$Q(X) = \sum_{ij} a_{ij}x_ix_j + \sum_i b_ix_i$$

with $a_{ii} = 0$ can be put into the form of Eqn. (11.18).

Proof: Simply let $t_i = b_i/2$, and $w_{ij} = -(a_{ij} + a_{ji})/2$ for all i and j in
Eqn. (11.18). Then, clearly W is a symmetric matrix with $w_{ii} = 0$. $\square$

Remark 11.3:   Given any quadratic form in $X \in \{1, -1\}^n$, $Q(X) = \sum_{ij} a_{ij}x_ix_j + \sum_i b_ix_i$, with $a_{ii} \neq 0$, it can be expressed as

$$Q(X) = c + \sum_{ij,\ i\neq j} a_{ij}x_ix_j + \sum_i b_ix_i = c + Q'(X),$$

where $c = \sum_i a_{i,i}$. Note that $Q'(X)$ is a quadratic form in $X \in \{1, -1\}^n$,
where a_{ii}'s are all 0. Moreover, any X_0 that minimizes $Q'(X)$ also minimizes
$Q(X)$. Hence, to minimize a quadratic form, one can assume without loss of
generality that a_{ii}'s are all zero (i.e., the condition in Lemma 11.5 is always
satisfied). $\square$

The following lemma states that the minimization of a quadratic form $Q(X)$ subject to some linear constraints is equivalent to the minimization of another unconstrained quadratic form $\tilde{Q}(X)$.

Lemma 11.6: Suppose there exists an $X_o \in \{1, -1\}^n$ that satisfies

$$\sum_{j=1}^{n} p_{ij} x_j = q_i \quad \text{for} \quad i = 1, \ldots, n, \tag{11.19}$$

where p_{ij} and q_i are integer constants. Then, $\tilde{X} \in \{1, -1\}^n$ minimizes the quadratic form

$$Q(X) = \sum_{ij} a_{ij} x_i x_j + \sum_i b_i x_i$$

subject to the constraints Eqn. (11.19) if and only if $\tilde{X}$ minimizes the following quadratic form:

$$\tilde{Q}(X) = Q(X) + 2\lambda \sum_{i=1}^{n} (\sum_{j=1}^{n} p_{ij} x_j - q_i)^2,$$

where λ is a constant such that $|Q(X)| < \lambda$ for all $X \in \{1, -1\}^n$.

Proof: It suffices to show that $\tilde{X}$ minimizes $\tilde{Q}(X)$ only if $\tilde{X}$ satisfies the constraints in Eqn. (11.19). By assumption, there exists an X_o that satisfies the constraints in Eqn. (11.19). Thus, $\min \tilde{Q}(X) \leq \tilde{Q}(X_o) = Q(X_o)$. Since $|Q(X)| < \lambda$ for all $X \in \{1, -1\}^n$, any $\hat{X}$ that does not satisfy Eqn. (11.19) will result in $\tilde{Q}(\hat{X}) \geq Q(\hat{X}) + 2\lambda > Q(X_o)$. Hence, $\hat{X}$ cannot achieve the minimum of $\tilde{Q}(X)$. $\qquad \square$

11.4.1 Min-Cut and Related Problems

It can be shown that the minimization of a quadratic form over $\{1, -1\}^n$ (or $\{0, 1\}^n$) is equivalent to finding a minimum cut in an undirected graph. Since the energy function of a Hopfield network is also quadratic in its state variables, it follows that the sequential iteration in a Hopfield network is equivalent to a local search for a minimum cut in the corresponding undirected graph. Before establishing such an equivalence relation, we first give a formal definition of a *cut* and its capacity in an undirected graph.

Definition 11.3:   **(Cut and Capacity)** Let G be an undirected graph with vertex set $X = \{x_1, \ldots, x_n\}$. Let w_{ij} be the weight associated with an edge connecting x_i and x_j. A *cut* is a partition $(S, \overline{S})$ of the nodes in X into sets S and $\overline{S}$ such that $S \cup \overline{S} = X$ and $S \cap \overline{S} = \phi$. The *capacity* of the cut $C(S, \overline{S})$ is the sum of weights on the edges that connect the nodes in S to the nodes in $\overline{S}$, i.e., $\displaystyle\sum_{x_i \in S \; x_j \in \overline{S}} w_{ij}$.   □

Note that our definition allows either S or $\overline{S}$ to be empty. In that case, the partition is called a *null cut* and the associated capacity is defined to be zero.

Definition 11.4:   **(Minimum Cut)** A *minimum cut* in a graph is a cut that has the minimum capacity among all other cuts in the graph.   □

There could be many cuts in a graph that achieve the same minimum capacity. We are interested in finding any minimum cut. It turns out that finding a minimum cut of a graph is in general NP-complete [37]. Next we show that finding a minimum cut in a graph is equivalent to minimizing a quadratic form.

With every cut $(S, \overline{S})$ in a graph G of n nodes $\{x_1, \ldots, x_n\}$, we associate an $X \in \{1, -1\}^n$ such that the nodes in S correspond to the 1's in X and the nodes in $\overline{S}$ correspond to the -1's in X.

Theorem 11.13:   Let w_{ij} be the weight of the edge connecting x_i and x_j in G. Then, any $X_o \in \{1, -1\}^n$ corresponding to a minimum cut of G achieves the minimum of the following quadratic form:

$$Q_W(X) = -X^T W X = -\sum_{i=1}^{n} \sum_{j=1}^{n} w_{ij} x_i x_j. \qquad (11.20)$$

Proof: Let $X \in \{1, -1\}^n$ correspond to a cut $(S, \overline{S})$ in G. Notice that $x_i x_j = -1$ if and only if $x_i \neq x_j$, or equivalently, x_i and x_j belong to different partitioned sets in $(S, \overline{S})$. Thus, $Q_W(X)$ can be expressed as

$$
\begin{aligned}
Q_W(X) &= -\left(\sum_{x_i = x_j} w_{ij} - \sum_{x_i \neq x_j} w_{ij} \right) \\
&= -\left(\sum_{x_i = x_j} w_{ij} + \sum_{x_i \neq x_j} w_{ij} - 2 \sum_{x_i \neq x_j} w_{ij} \right)
\end{aligned}
$$

$$= -\sum_{i=1}^{n}\sum_{j=1}^{n} w_{ij} + 2 \sum_{x_i \neq x_j} w_{ij}$$

$$= -\sum_{i=1}^{n}\sum_{j=1}^{n} w_{ij} + 4C(S,\overline{S}),$$

where $C(S,\overline{S}) = \dfrac{1}{2} \sum_{x_i \neq x_j} w_{ij}$ is the capacity of the cut. Observe that the first term on the last line $-\sum_{i=1}^{n}\sum_{j=1}^{n} w_{ij}$ is a constant independent of X. Therefore, any X that corresponds to a minimum cut capacity $C(S,\overline{S})$ will also achieve the minimum of the quadratic form Q.   □

Theorem 11.13 implies that to find a minimum cut in an undirected graph G (with weight matrix W), one can use a Hopfield network H_G with energy function $Q_W(X)$ to perform a local search for the minimum cut. The stable states of H_G will correspond to a local minimum cut of G. In fact, the converse also holds.

Theorem 11.14:   Given a Hopfield network H with thresholds T and zero-diagonal symmetric weight matrix W, there exists an undirected graph G_H whose minimum cut corresponds to the global minimum of the energy function associated with H.

Proof: Construct a graph G_H of $(n+1)$ nodes with its connection matrix being

$$\tilde{W} = \begin{bmatrix} W & T \\ T^T & 0 \end{bmatrix}.$$

It is easy to see that in G_H, the edge connecting nodes x_i and x_j has weight w_{ij}, and the edge connecting node x_i and x_{n+1} has weight t_i. If we fix the state variable $x_{n+1} = -1$ and replace (in Theorem 11.13) W by $\tilde{W}$ and X by $\tilde{X} = (x_1,\ldots,x_n,-1)$, then

$$Q_{\tilde{W}}(\tilde{X}) = -\tilde{X}^T\tilde{W}\tilde{X} = -X^TWX + 2X^TT,$$

which is the energy function associated with H. By Theorem 11.13, the minimum cut of G_H corresponds to the (global) minimum of $Q_{\tilde{W}}$.   □.

The Min-Cut Problem has played an important role in many practical applications, for example, integrated-circuit layout design. One can consider an integrated circuit as a graph in which each node represents a circuit element and each edge with $w_{ij} = 1$ represents a wire segment connecting the circuit elements x_i and x_j ($w_{ij} = 0$ means no direct connection between x_i and x_j). Then, finding a minimum cut of the graph is equivalent to partitioning the circuit into two components such that the number of wires connecting one component to the other is minimized. In practical situations, one might want to partition the circuit into two components such that each component has the same number of circuit elements. This can easily be formulated as a Min-Cut Problem with a linear constraint. In our formulation, we consider the more general case where w_{ij} is allowed to assume any integer values.

Theorem 11.15: **(Constrained Min-Cut Problem)** Finding a minimum cut $(S, \overline{S})$ in an undirected graph G of n nodes (n is even) with the constraint that the cardinality $|S| = |\overline{S}| = n/2$ is equivalent to minimizing the following quadratic form over $\{1, -1\}^n$:

$$\hat{Q}_W(X) = -\sum_{i=1}^{n}\sum_{j=1}^{n} w_{ij}x_i x_j + 2\lambda(\sum_{i=1}^{n} x_i)^2,$$

where w_{ij} are integers and $\lambda > \sum_{i=1}^{n}\sum_{j=1}^{n} |w_{ij}|$.

Proof: As in Theorem 11.13, let $X \in \{1, -1\}^n$ correspond to a cut $(S, \overline{S})$ in G. The constraint that the two partitioned sets S and $\overline{S}$ have the same cardinality is the same as requiring that there are as many 1's as -1's in X, i.e.,

$$\sum_{i=1}^{n} x_i = 0. \tag{11.21}$$

By Theorem 11.13, finding a minimum cut is equivalent to minimizing the quadratic form

$$Q_W(X) = -\sum_{i=1}^{n}\sum_{j=1}^{n} w_{ij}x_i x_j.$$

Thus, by Lemma 11.6, minimizing $Q_W(X)$ with the constraint Eqn. (11.21) is equivalent to minimizing the quadratic form

$$\hat{Q}_W(X) = Q_W(X) + 2\lambda(\sum_{i=1}^{n} x_i)^2 = -\sum_{i=1}^{n}\sum_{j=1}^{n} w_{ij}x_i x_j + 2\lambda(\sum_{i=1}^{n} x_i)^2,$$

where λ is any constant $> \sum_{i=1}^{n}\sum_{j=1}^{n}|w_{ij}|.$ $\qquad\square$

Very often, in circuit layout design applications, the weights of the graph are non-negative. In this case, when there is no constraint on the size of the partitioned sets, finding a minimum cut can be solved efficiently (in polynomial time) by standard network flow techniques. However, the constrained Min-Cut Problem remains NP-complete [37] even if all the non-zero weights of the graph are assumed to be 1 (such as in the preceding example in which each edge represents a wire segment).

11.4.2   The Traveling-Salesman Problem

Hopfield and Tank [59] originally formulated the Traveling-Salesman Problem (TSP) as the minimization of a quadratic form. The idea is to choose the weights w_{ij}'s and thresholds t_i's in a way such that the (global) minimum of E in Eqn. (11.18) would correspond to a minimum-length valid tour. To see this, let $x_{ij} = 1$ if city c_i is visited at the jth stop on the tour, and $x_{ij} = -1$ otherwise, for $i, j = 1, 2, \ldots, n$. For convenience, let $z_{ij} = (x_{ij}+1)/2$ so that

$$z_{ij} = \begin{cases} 1 & \text{if } x_{ij} = 1, \\ 0 & \text{if } x_{ij} = -1. \end{cases} \tag{11.22}$$

Note that there are n^2 nodes in the network. The total length of the tour is

$$L = \frac{1}{2}\sum_{i\neq j}^{n}\sum_{k=1}^{n} d_{ij}z_{ik}(z_{j,k-1} + z_{j,k+1}), \tag{11.23}$$

where $d_{ij} = d_{ji}$ is the distance between cities i and j, and $z_{j,n+1} = z_{j,1}$ and $z_{j,0} = z_{j,n}$, subject to the constraints:

$$\sum_{k=1}^{n} z_{ik} = 1 \quad \text{for } i = 1, \ldots, n, \tag{11.24}$$

$$\sum_{i=1}^{n} z_{ik} = 1 \quad \text{for } k = 1, \ldots, n. \tag{11.25}$$

Eqn. (11.24) implies that every city c_i is visited exactly once on the tour. Eqn. (11.25) implies that for every stop k on the tour, there is exactly one city being visited. To see that Eqn. (11.23) represents the length of a closed tour, simply notice that $z_{ik}(z_{j,k-1} + z_{j,k+1})$ is 1 if and only if the c_i and c_j are consecutive cities to be visited on the tour route, and the summation is then equal to the sum of the distances between those cities.

Clearly, we can express the above equations as the minimization of a single quadratic function of the state variables x_{ij}, such that L in Eqn. (11.23) is minimized and the linear constraints in Eqns. (11.24) and (11.25) are satisfied. In particular, it follows from Lemma 11.6 that a state $\tilde{X} \in \{1, -1\}$ corresponds to a minimum-length valid tour if and only if $\tilde{X}$ achieves the minimum of

$$\tilde{L} = \frac{1}{2} \sum_{i \neq j}^{n} \sum_{k=1}^{n} d_{ij} z_{ik}(z_{j,k-1} + z_{j,k+1}) + \lambda \left[\sum_{i=1}^{n} (\sum_{k=1}^{n} z_{ik} - 1)^2 + \sum_{k=1}^{n} (\sum_{i=1}^{n} z_{ik} - 1)^2 \right],$$

where λ can be any constant $> \sum_{i \neq j}^{n} d_{ij}$. $\square$

Exercises

11.1:   [*Convergence for hybrid mode of iteration*]
Show that the Hopfield network converges for any hybrid mode of operation, if W is symmetric, positive semi-definite (i.e., $\mathbf{x}^T W \mathbf{x} \geq 0$ for all $\mathbf{x}$), and the threshold vector $T = 0$.

11.2:   [*Local minimum of the energy function*]
Let W and T be the symmetric weight matrix and thresholds, respectively, associated with a Hopfield network operating in a sequential mode. Let E_{X_0} be the energy corresponding to a stable state X_0, and let a state $\tilde{X}$ differ from X_0 by only one coordinate. Suppose all diagonal elements of W are zero, i.e., $w_{ii} = 0$. Show that the energy corresponding to $\tilde{X}$, $E_{\tilde{X}}$ will be $\geq E_{X_0}$.

11.3:   [*Limit cycle in parallel mode*]
Consider a Hopfield network of n nodes operating in the parallel mode with threshold vector $T = 0$ and symmetric weight matrix W. Let $X(t)$ be the state of the network at time t. For X and $Y \in \{1, -1\}^n$, define the functions

$$E(X) = - \sum_{i=1}^{n} |\sum_{j=1}^{n} w_{ij} x_j|,$$

and

$$F(Y, X) = - \sum_{i=1}^{n} \sum_{j=1}^{n} y_i w_{ij} x_j.$$

(a) Show that $E(X(t)) = F(X(t+1), X(t))$.

(b) Show that $E(X) \leq F(Y, X)$ for all $Y, X \in \{1, -1\}^n$.

(c) Show that $E(X(t)) \leq E(X(t-1))$.

(d) Show that equality in (c) implies that $x_i(t+1) = x_i(t-1)$ for all i such that $(WX(t))_i \neq 0$. Conclude that the network converges to a limit cycle of length at most 2. Why can't we assume $(WX(t))_i \neq 0$ for all i in this case?

11.4: *[Limit cycle in case of antisymmetric weight matrix]*
Show that if a Hopfield network with antisymmetric weight matrix W (i.e., $W^T = -W$) and with threshold vector $T = 0$ iterates in the parallel mode, it must enter a limit cycle of length at most 4.

11.5: *[An n-node network with exactly n stable states]*
Let $\hat{M} = \{X_1, X_2, \ldots, X_n\} \subset \{1, -1\}^n$, where $X_1 = [-1, 1, 1, \ldots, 1]^T$, $X_2 = [1, -1, 1, \ldots, 1]^T$, and in general X_i is the vector that consists of all 1's, except at the ith coordinate where it is -1. Construct a completely connected network for which the set of stable states equals $\hat{M}$ (i.e., any vector $\notin \hat{M}$ is not a stable state of the network).

11.6: *[Networks with exactly two stable states]*

(a) Let n be an odd integer, $\mathbf{u} \in \{1, -1\}^n$ (a column vector), and $W = \mathbf{u}\mathbf{u}^T$ be the weight matrix of a Hopfield network with threshold vector $T = 0$. Show that $\mathbf{u}$ and $-\mathbf{u}$ are the only two stable states in this Hopfield network. Moreover, show that for *any* mode of iteration (assuming each node is updated infinitely often), the network will converge to $\mathbf{u}$ or $-\mathbf{u}$ from any initial condition.

(b) Consider a network with zero threshold values, where the weight matrix W has the following elements: $w_{ii} = 0$ and $w_{ij} = c$ for all $i \neq j$, where $c > 0$. Show that the corresponding network has *only* two stable states: $[1 \cdots 1]^T$ and $[-1 \cdots -1]^T$.

11.7: [*Independent set and number of stable states*]
Let H be a Hopfield network of order n, and S be the maximum size of an independent set associated with the graph of H. Show that the total number of stable states of H is bounded above by 2^{n-S}.

11.8: [*An upper bound on the number of different Hopfield networks*]
Give an effective upper bound on the number of different Hopfield networks with n nodes. (*Hint:* Apply the result on the number of linear threshold functions of n inputs.)

11.9: [*Weight matrix with negative diagonal elements*]
Consider a Hopfield network of two nodes with $x_1(t+1) = \mathrm{sgn}(-2x_1(t) + x_2(t))$ and $x_2(t+1) = \mathrm{sgn}(x_1(t) - 2x_2(t))$. This network has a symmetric matrix with negative diagonal elements, where $w_{11} = w_{22} = -2$ and $w_{12} = w_{21} = 1$. Show that given any initial state, the network will not converge to a stable state in any sequential mode.

11.10: [*An upper bound on the number of spurious memories*]
Let the vectors in the set $M = \{X_1, X_2, \ldots, X_s\}$, $X_i \in \{1, -1\}^n$, be programmed as stable states of a network by the outer-product rule, i.e., a Hopfield network H is obtained with $W = \sum_{i=1}^{s}(X_i X_i^T - I_n)$, and $T = 0$.

(a) Consider a modified network $H' = \{W', \mathbf{0}\}$, such that $W' = \sum_{i=1}^{s} X_i X_i^T$. Show that if $Y \in \{1, -1\}^n$ is a stable state of the original network H, then it is also a stable state of the modified network H'.

(b) Show that if $Y \in \{1, -1\}^n$ is a stable state of the network H', then it can be expressed as $Y = \mathrm{sgn}(\sum_{i=1}^{s} p_i X_i)$, where $p_i = X_i^T Y$. Hence, show that the total number of stable states in H' is $\leq C(n, s)$ (see Section 2.7).

(c) Using parts (a) and (b), show that the number of spurious memories in H (i.e., the network generated by the outer-product rule) is $\leq 2^{s \log(n+1)}$.

11.11:   [*Networks with exponentially many spurious memories*]
Consider the Sylvester-type Hadamard matrix H_n $(n = 2^k)$, which is recursively defined as follows:

$$H_1 = [1] \quad \text{and} \quad H_n = \begin{bmatrix} H_{\frac{n}{2}} & H_{\frac{n}{2}} \\ H_{\frac{n}{2}} & -H_{\frac{n}{2}} \end{bmatrix} \quad (n \geq 2).$$

(a) Consider the network obtained by programming the first $\frac{n}{2}$ columns
    of H_n as stable states via the outer-product method. Show that the
    weight matrix of the network is

$$W = \begin{bmatrix} 0 & \frac{n}{2}I_{\frac{n}{2}} \\ \frac{n}{2}I_{\frac{n}{2}} & 0 \end{bmatrix},$$

where $I_{\frac{n}{2}}$ is an $\frac{n}{2} \times \frac{n}{2}$ identity matrix. Show that the network consists
of $\frac{n}{2}$ pairs of nodes, each connected with an edge of weight $\frac{n}{2}$. Hence,
show that the network has exponentially many spurious memories by
proving that the total number of stable states in the network is $2^{\frac{n}{2}}$.

(b) In general, consider the network obtained by programming the first
    $s = \dfrac{n}{2^k}$ $(1 \leq k < \log n)$ columns of H_n as stable states via the outer-
    product method. Show that the network comprises s disjoint com-
    ponents, where each component is a completely connected network
    (without any self-loops) of 2^k nodes such that each edge has weight s.
    Hence, show that the total number stable states in the network is 2^s.

11.12:   [*A spectral scheme for programming Hopfield networks*]
Consider a set of vectors $\{X_1, \ldots, X_s\}$, where $X_i \in \{1, -1\}^n$ and $s \leq (n-1)$.
Let $Y = [X_1 \;\cdots\; X_s]$ be the matrix defined by the vectors, and let

$$W = Y(Y^T Y)^{-1} Y^T$$

be the projection matrix.

(a) Show that $W X_i = X_i$ for $i = 1, \ldots, s$. Hence, show that every X_i is a
    stable state in a network with weight matrix W and threshold $\mathbf{0}$.
    Since X_i's are the eigenvectors of W, this method is referred to as the
    'spectral scheme' for programming Hopfield networks.

(b) Show that in the network defined by $\{W, \mathbf{0}\}$, every vector belonging to
    the linear span of $\{X_1, \ldots, X_s\}$ is also a stable state of the network.

Bibliographic Notes

The Hopfield model is proposed in a seminal paper of Hopfield [58], which discusses the conditions under which the network is guaranteed to converge and suggests the application of the model for solving combinatorial optimization problems. It is shown in Goles, Fogelman, and Pellegrin [42, 43] that the Hopfield network always converges to a stable state or enters a limit cycle of length 2 in the parallel mode of iteration, provided the weight matrix W is symmetric. Theorem 11.3 is proved in [42] using an energy function. Our proof is taken from Bruck and Goodman [19], where the result is rederived using a reduction approach. It is shown in [42] that in the case of parallel iteration, the transient period can be exponentially large in the order of the network. The construction in Section 11.2.2 showing an exponentially long $\Omega(2^n)$ limit cycle in an n-node Hopfield network is due to Orlitsky [93]. Theorems 11.7 and 11.8 can be found in Bruck and Sanz [21]. The results in Section 11.3.2 on the number of spurious memories in Hopfield networks are from Bruck and Roychowdhury [20]. Analytical techniques, such as the polynomial representation of Boolean functions, are applied in [20] to derive bounds on the number of stable states in Hopfield networks. Moreover, the explicit constructions of Hopfield networks with exponentially many spurious memories derived in [20] (also presented in Section 11.3.2) provide direct evidence for the results in [1, 78, 136], where such a phenomenon is suggested based on probabilistic arguments.

Lemma 11.6 is derived in Picard and Ratliff [103]. It is also shown in [103] that the minimization of a quadratic form over $\{1, -1\}^n$ (or $\{0, 1\}^n$) is equivalent to finding a minimum cut in an undirected graph. The fact that the sequential iteration in a Hopfield network is equivalent to a local search for a minimum cut in the corresponding undirected graph is observed in [19]. For results regarding the statistical properties of Hopfield networks, see Venkatesh [137] and the references therein.

Glossary

$O(g(n))$ A non-negative quantity (dependent on n) that is bounded above by $cg(n)$, for some constant $c > 0$, as $n \to \infty$.

$\Omega(g(n))$ A non-negative quantity (dependent on n) that is bounded below by $cg(n)$, for some constant $c > 0$, as $n \to \infty$.

$\Theta(g(n))$ A non-negative quantity (dependent on n) that is bounded above and below by $c_2 g(n)$ and $c_1 g(n)$, respectively, for some constants $c_2 > c_1 > 0$, as $n \to \infty$.

$\bigwedge(x_1, \ldots, x_n)$ AND function of n variables, also represented as $x_1 \wedge \cdots \wedge x_n$.

$\bigvee(x_1, \ldots, x_n)$ OR function of n variables, also represented as $x_1 \vee \cdots \vee x_n$.

EQ $EQ(X, Y) = (\overline{x_1 \oplus y_1}) \wedge \cdots \wedge (\overline{x_n \oplus y_n})$.

IP $IP(X, Y) = (x_1 \wedge y_1) \oplus \cdots \oplus (x_n \wedge y_n)$.

$COMP_n$ $COMP_n(X, Y) = \mathrm{sgn}(\sum_{i=0}^{n-1} 2^i (x_i - y_i))$.

CQ $CQ(X) = (x_1 \wedge x_2) \oplus (x_1 \wedge x_3) \oplus \cdots \oplus (x_{n-1} \wedge x_n)$.

EX_k^n $EX_k^n(X)$ is 1 if $\displaystyle\sum_{i=1}^{n} x_i = k$, and is 0 otherwise.

MAJ_n $MAJ_n(X)$ is 1 if $\displaystyle\sum_{i=1}^{n} x_i \geq \frac{n}{2}$, and is 0 otherwise.

PAR_n The Parity function of n variables, also represented as $x_1 \oplus \cdots \oplus x_n$ or $XOR(x_1, \ldots, x_n)$.

MOD_m For $(x_1, \ldots, x_n) \in \{0, 1\}^n$, $\mathrm{MOD}_m(x_1, \ldots, x_n)$ is 1 if $(\displaystyle\sum_{i=1}^{n} x_i) \not\equiv 0$ (mod m), and is 0 otherwise.

$DIV_k(X/Y)$ The quotient of two n-bit integers, X and Y, truncated to $(n + k)$ bits, where the fractional part consists of k bits.

AND-OR circuit A Boolean circuit of unbounded fan-in AND/OR gates with inverters.

AC^0	The class of Boolean functions computable by constant-depth polynomial-size AND-OR circuits.
ACC	The class of Boolean functions computable by constant-depth polynomial-size circuits of AND/OR gates and MOD_m gates, where m is a fixed constant.
TC^0	The class of Boolean functions computable by constant-depth polynomial-size threshold circuits.
NC^1	The class of Boolean functions computable by $O(\log n)$-depth bounded fan-in AND-OR circuits.
LT_d	The class of Boolean functions computable by depth-d polynomial-size threshold circuits with unrestricted weights.
$\widehat{LT}_d$	The class of Boolean functions computable by depth-d polynomial-size threshold circuits with polynomially bounded integer weights.
$\widetilde{LT}_d$	The class of Boolean functions computable by depth-d polynomial-size threshold circuits where the weights at the output gate are polynomially bounded integers (with no restriction on the weights of the other gates).
$\widetilde{SP}$	The class of Boolean functions in which every function can be expressed as the sign of a sparse polynomial.
$\|f\|_{\mathcal{F}}$	The L_1 spectral norm of a Boolean function f.
X^α	$x_1^{\alpha_1} x_2^{\alpha_2} \ldots x_n^{\alpha_n}$, where $(x_1, \ldots, x_n) \in \{1, -1\}^n$ and $(\alpha_1, \ldots, \alpha_n) \in \{0, 1\}^n$.
$SD(f)$	The smallest degree of any multilinear polynomial $F(X)$ such that $f(X) = \mathrm{sgn}(F(X))$ and $F(X) \neq 0$ for all $X \in \{1, -1\}^n$, also referred to as the strong degree of f.
$deg_D(f)$	The minimum degree of any polynomial $p(X)$ such that $p(X) = f(X)$ for all $X \in D \subseteq \{0, 1\}^n$.
H_n	A Hadamard matrix of order n.
$I_{n \times n}$	An $n \times n$ identity matrix.
γ element	An element computing functions of the form $\gamma(\sum_{i=1}^m w_i x_i - t)$.
γ network	A feedforward network of γ elements.
$\mathcal{G}$ circuit	A circuit whose gates are all from a family of gates $\mathcal{G}$.
$C_{\mathcal{G}}(f)$	The size (number of gates) of the smallest $\mathcal{G}$ circuit that computes f.

Bibliography

[1] Y. S. Abu-Mostafa and J.-M. St. Jacus. Information Capacity of Hopfield Model. *IEEE Transactions on Information Theory*, 31(4):461–464, July 1985.

[2] D. H. Ackley, G. E. Hinton, and T. J. Sejnowski. A Learning Algorithm for Boltzmann Machines. *Cognitive Science*, 9:147–169, 1985.

[3] A. V. Aho, J. D. Ullman, and M. Yanakakis. On notions of information transfer in VLSI circuits. In *Proceedings of the 15th Annual ACM Symposium on Theory of Computing*, pages 133–139, 1983.

[4] M. Ajtai. $\sum_1^1$-formulae on finite structures. *Annals of Pure and Applied Logic*, 24:1–48, 1983.

[5] E. Allender. A note on the power of threshold circuits. In *Proceedings of the 30th Annual Symposium on Foundations of Computer Science*, pages 580–584, 1989.

[6] N. Alon and J. Bruck. Explicit Constructions of Depth-2 Majority Circuits for Comparison and Addition. *SIAM Journal on Discrete Mathematics*, 7(1):1–8, February 1994.

[7] I. Anderson. *Combinatorics of Finite Sets*. Oxford University Press, New York, 1987.

[8] J. Aspnes, R. Beigel, M. Furst, and S. Rudich. The Expressive Power of Voting Polynomials. In *Proceedings of the 23rd Annual ACM Symposium on Theory of Computing*, pages 286–291, 1991.

[9] K. A. Baker. A Generalization of Sperner's Lemma. *Journal of Combinatorial Theory*, 6:224–225, 1969.

[10] P. Baldi. Neural networks, orientations of the hypercube, and algebraic threshold functions. *IEEE Transactions on Information Theory*, 34(3):523–530, May 1988.

[11] P. Beame, E. Brisson, and R. Ladner. The Complexity of Computing Symmetric Functions Using Threshold Circuits. *Theoretical Computer Science*, 100(1):253–265, June 1992.

[12] P. W. Beame, S. A. Cook, and H. J. Hoover. Log depth circuits for division and related problems. *SIAM Journal on Computing*, 15(4):994–1003, November 1986.

[13] R. Beigel. Do extra threshold gates help? In *Proceedings of the 24th Annual ACM Symposium on Theory of Computing*, pages 450–454, 1992.

[14] R. Beigel, N. Reingold, and D. Speilman. PP is closed under intersection. In *Proceedings of the 23rd Annual ACM Symposium on Theory of Computing*, pages 1–9, 1991.

[15] R. Beigel and J. Tarui. On ACC. In *Proceedings of the 32nd Annual Symposium on Foundations of Computer Science*, pages 783–792, 1991.

[16] A. L. Blum and R. L. Rivest. Training a 3-Node Neural Network is NP-Complete. *Neural Networks*, 5(1):117–127, 1992.

[17] Y. Brandman, A. Orlitsky, and J. Hennessy. A Spectral Lower Bound Technique for the Size of Decision Trees and Two level AND/OR Circuits. *IEEE Transactions on Computers*, 39(2):282–287, February 1990.

[18] J. Bruck. Harmonic analysis of polynomial threshold functions. *SIAM Journal on Discrete Mathematics*, 3(2):168–177, May 1990.

[19] J. Bruck and J. W. Goodman. A Generalized Convergence Theorem for Neural Networks. *IEEE Transactions on Information Theory*, 34(5):1089–1092, September 1988.

[20] J. Bruck and V. P. Roychowdhury. On the Number of Spurious Memories in a Hopfield Network. *IEEE Transactions on Information Theory*, 36(2):393–397, March 1990.

[21] J. Bruck and J. Sanz. A study on neural networks. *International Journal of Intelligent Systems*, pages 59–75, 1988.

[22] J. Bruck and R. Smolensky. Polynomial Threshold Functions, AC^0 Functions, and Spectral Norms. *SIAM Journal on Computing*, 21(1):33–42, February 1992.

[23] R. C. Buck. Partition of Space. *American Mathematical Monthly*, 50:541–544, 1943.

[24] G. A. Carpenter and S. Grossberg. The Art of Adaptive Pattern Recognition. *Computer*, 21(3):77–88, March 1988.

[25] A. K. Chandra, S. Fortune, and R. Lipton. Unbounded fan-in circuits and associative functions. *Journal of Computer and System Sciences*, 30(2):222–234, April 1985.

[26] A. K. Chandra, L. Stockmeyer, and U. Vishkin. Constant depth reducibility. *SIAM Journal on Computing*, 13(2):423–439, May 1984.

[27] C. K. Chow. On The Characterization of Threshold Functions. *Proceedings of the 6th Symposium on Switching Circuit Theory and Logical Design*, pages 34–38, 1961.

[28] L. O. Chua and L. Yang. Cellular Neural Networks: Applications. *IEEE Transactions on Circuits and Systems*, 35(10):1273–1290, October 1988.

[29] L. O. Chua and L. Yang. Cellular Neural Networks: Theory. *IEEE Transactions on Circuits and Systems*, 35(10):1257–1272, October 1988.

[30] J. W. Cooley and J. W. Tukey. An Algorithm for the Machine Computation of Complex Fourier Series. *Mathematics of Computation*, 19(90):297–301, April 1965.

[31] T. M. Cover. Geometrical and Statistical Properties of Systems of Linear Inequalities with Applications in Pattern Recognition. *IEEE Transactions on Electronic Computers*, EC-14:326–334, June 1965.

[32] T. M. Cover. Capacity problems for linear machines. In L. Kanal, editor, *Pattern Recognition*, pages 283–289. Washington, Thompson Book Co., 1968.

[33] G. Cybenko. Approximations by superpositions of a sigmoidal function. *Mathematics of Control, Signals, Systems*, 2(4):303–314, 1989.

[34] B. DasGupta and G. Schnitger. The power of approximating: a comparison of activation functions. In S. J. Hanson, J. D. Cowan, and C. Lee Giles, editors, *Advances in Neural Information Processing Systems 5*, pages 615–622. San Mateo, CA: Morgan Kaufman, 1993.

[35] F. Fich. *Two Problems in Concrete Complexity: Cycle Detection and Parallel Prefix Computation*. PhD thesis, University of California, Berkeley, 1982.

[36] M. Furst, J. B. Saxe, and M. Sipser. Parity, Circuits and the Polynomial-Time Hierarchy. In *Proceedings of the 22nd Annual Symposium on Foundations of Computer Science*, pages 260–270, 1981. Also appears in *Mathematical Systems Theory*, 17(1):13-27, April 1984.

[37] M. Garey and D. Johnson. *Computers and Intractability: a guide to the theory of NP-completeness*. San Francisco: W. H. Freeman, 1979.

[38] M. Goldmann. Private communication, 1993.

[39] M. Goldmann. Communication complexity and lower bounds for threshold circuits. In V.P. Roychowdhury, K.-Y. Siu, and A. Orlitsky, editors, *Theoretical Advances in Neural Computation and Learning*. Boston: Kluwer Academic Publishers, 1994.

[40] M. Goldmann, J. Håstad, and A. Razborov. Majority Gates vs. General Weighted Threshold Gates. In *Proceedings of the 7th Annual Structure in Complexity Theory Conference*, pages 2–13, 1992.

[41] M. Goldmann and M. Karpinski. Simulating threshold circuits by majority circuits. In *Proceedings of the 25th Annual ACM Symposium on Theory of Computing*, pages 551–560, 1993.

[42] E. Goles. Antisymmetrical neural networks. *Discrete Applied Mathematics*, 13(1):97–100, January 1986.

[43] E. Goles, F. Fogelman, and D. Pellegrin. Decreasing energy functions as a tool for studying threshold networks. *Discrete Applied Mathematics*, 12(3):261–277, November 1985.

[44] S. W. Golomb. *Shift Register Sequences*. Laguna Hills, CA: Aegean Park Press, 1982.

[45] A. A. Gončar. On the rapidity of rational approximation of continuous functions with characteristic singularities. *Mat. Sbornik*, 2(4):561–568, 1967.

[46] H. D. Gröger and G. Turán. On linear decision trees computing Boolean functions. In J. L. Albert, B. Monien, and M. R. Artalejo, editors, *Proceedings of the 18th International Colloquium on Automata, Languages and Programming*, pages 707–718, 1991.

[47] S. Grossberg. Some Nonlinear Networks Capable of Learning a Spatial Pattern of Arbitrary Complexity. *Proceedings of the National Academy of Sciences*, pages 368–372, 1968.

[48] A. Hajnal, W. Maass, P. Pudlák, M. Szegedy, and G. Turán. Threshold circuits of bounded depth. In *Proceedings of the 28th Annual Symposium on Foundations of Computer Science*, pages 99–110, 1987. Also appears in *Journal of Computer and System Sciences*, 46(2):129-154, April 1993.

[49] M. H. Hassoun and J. Song. Adaptive Ho-Kashyap Rules for Perceptron Training. *IEEE Transactions on Neural Networks*, 3(1):51–61, January 1992.

[50] J. Håstad. Almost optimal lower bounds for small depth circuits. In *Proceedings of the 18th Annual ACM Symposium on Theory of Computing*, pages 6–20, 1986.

[51] J. Håstad. On the size of weights for threshold gates. *SIAM Journal on Discrete Mathematics*, 1994. In press.

[52] D. O. Hebb. *The organization of behavior: a neuropsychological theory*. New York: Wiley, 1949.

[53] T. Hededus and N. Megiddo. On the geometric separability of Boolean functions. Technical Report RJ 9147, IBM Research Division, December 1992.

[54] J. Hertz, A. Krogh, and R. G. Palmer. *Introduction to the Theory of Neural Computation*. Addison-Wesley, 1991.

[55] T. Hofmeister. Depth-efficient threshold circuits for arithmetic functions. In V.P. Roychowdhury, K.-Y. Siu, and A. Orlitsky, editors, *Theoretical Advances in Neural Computation and Learning.* Boston: Kluwer Academic Publishers, 1994.

[56] T. Hofmeister, W. Hohberg, and S. Köhling. Some Notes on Threshold Circuits, and Multiplication in Depth 4. *Information Processing Letters,* 39(4):219–225, 30 August 1991.

[57] T. Hofmeister and P. Pudlák. A proof that division is not in TC_2^0. Forschungs-bericht Nr. 447, 1992, Uni Dortmund.

[58] J. J. Hopfield. Neural networks and physical systems with emergent collective computational abilities. *Proceedings of the National Academy of Sciences,* 79:2554–2558, April 1982.

[59] J. J. Hopfield and D. W. Tank. Neural computation of decisions in optimization problems. *Biological Cybernetics,* 52:141–152, 1985.

[60] J. Hromkovic. Linear Lower Bounds on Unbounded Fan-in Boolean Circuits. *Information Processing Letters,* 21(2):71–74, 16 August 1985.

[61] S. Hu. *Threshold Logic.* Berkeley: University of California Press, 1965.

[62] R. Impagliazzo, R. Paturi, and M. Saks. Size-Depth Tradeoffs for Threshold Circuits. In *Proceedings of the 25th Annual ACM Symposium on Theory of Computing,* pages 541–550, 1993.

[63] S. Judd. On the complexity of loading shallow neural networks. *Journal of Complexity,* 4(3):177–192, September 1988.

[64] W. Kautz. The Realization of Symmetric Switching Functions with Linear-Input Logical Elements. *IRE Transactions on Electronic Computers,* EC-10(3):371–378, September 1961.

[65] V. M. Khrapchenko. Asymptotic Estimation of Addition Time of a Parallel Adder. *Problemy Kibernetiki,* 19:107–122, 1967.

[66] T. Kohonen. *Self-Organization and Associative Memory, 3rd ed.* Berlin, New York: Springer-Verlag, 1989.

[67] M. Krause and S. Waack. Variation ranks of communication matrices and lower bounds for depth two circuits having symmetric gates with unbounded fan-in. In *Proceedings of the 32nd Annual Symposium on Foundations of Computer Science,* pages 777–782, 1991.

[68] R. E. Ladner and M. J. Fischer. Parallel Prefix Computation. *Journal of ACM,* 27(4):831–838, October 1980.

[69] R. J. Lechner. Harmonic analysis of switching functions. In A. Mukhopad-hyay, editor, *Recent Developments in Switching Theory.* New York: Academic Press, 1971.

[70] P. M. Lewis and C. L. Coates. *Threshold Logic*. New York: Wiley, 1967.

[71] N. Linial, Y. Mansour, and N. Nisan. Constant Depth Circuits, Fourier Transforms, and Learnability. In *Proceedings of the 30th Annual Symposium on Foundations of Computer Science*, pages 574–579, 1989.

[72] D. Lubell. A Short Proof of Sperner's Lemma. *Journal of Combinatorial Theory*, 1:299, 1966.

[73] O. B. Lupanov. Circuits using threshold elements. *Soviet Physics Doklady*, 17(2):91–93, August 1971.

[74] O. B. Lupanov. On the synthesis of networks of threshold elements. *Problemy Kibernetiki*, 26:109–140, 1973.

[75] W. Maass, G. Schnitger, and E. Sontag. On the computational power of sigmoid versus Boolean threshold circuits. In *Proceedings of the 32nd Annual Symposium on Foundations of Computer Science*, pages 767–776, 1991.

[76] F. J. MacWilliams and N. J. A. Sloane. *The Theory of Error-Correcting Codes*. Amsterdam, NY: North-Holland, 1973.

[77] W. S. McCulloch and W. Pitts. A Logical Calculus of Ideas Immanent in Nervous Activity. *Bulletin of Mathematical Biophysics*, 5:115–133, 1943.

[78] R. J. McEliece, E. C. Posner, E. R. Rodemich, and S. S. Venkatesh. The capacity of Hopfield associative memory. *IEEE Transactions on Information Theory*, 33(4):461–482, July 1987.

[79] K. Mehlhorn and F. P. Preparata. Area-Time Optimal VLSI Integer Multiplier with Minimum Computation Time. *Information and Control*, 58(1–3):137–156, July–Sept. 1983.

[80] R. Minnick. Linear-Input Logic. *IRE Transactions on Electronic Computers*, EC-10(1):6–16, March 1961.

[81] M. Minsky and S. Papert. *Perceptrons: An Introduction to Computational Geometry*. Cambridge, MA: MIT Press, Expanded edition, 1988.

[82] D. E. Muller. Application of Boolean algebra to switching circuit design and to error detection. *Transactions of the IRE, Professional Group on Electronic Computers*, EC-3(3):6–12, 1954.

[83] D. E. Muller and F. P. Preparata. Bounds to Complexities of Networks for Sorting and for Switching. *Journal of ACM*, 22(2):195–201, April 1975.

[84] S. Muroga. The principle of majority decision logic elements and the complexity of their circuits. *International Conference on Information Processing, Paris, France*, June 1959.

[85] S. Muroga. Lower Bounds on the Number of Threshold Functions and a Maximum Weight. *IEEE Transactions on Electronic Computers*, EC-14(2):136 – 148, April 1965.

[86] S. Muroga. *Threshold Logic and its Applications*. NY: Wiley-Interscience, 1971.

[87] E. I. Nechiporuk. The synthesis of networks from threshold elements. *Problemy Kibernetiki*, 11:49–62, April 1964.

[88] D. J. Newman. Rational Approximation to $|x|$. *Michigan Math. Journal*, 11:11–14, 1964.

[89] I. Ninomiya. A theory of coordinate representation of switching functions. *Memoirs of the Faculty of Engineering, Nagoya University*, 10:175–190, 1958.

[90] A. M. Odlyzko. On subspaces spanned by random selections of ± 1 vectors. *Journal of Combinatorial Theory, Series A*, 47:124–133, 1988.

[91] Yu. P. Ofman. The Algorithmic Complexity of Discrete Functions. *Soviet Physics Doklady*, 7:589–591, 1963.

[92] P. E. O'Neil. Hyperplane cuts of an n-cube. *Discrete Mathematics*, 1(2):193–195, 1971.

[93] A. Orlitsky. unpublished manuscript, 1990.

[94] C. H. Papadimitriou and K. Steiglitz. *Combinatorial Optimization: Algorithms and Complexity*. Englewood Cliffs, NJ: Prentice Hall, 1982.

[95] I. Parberry and G. Schnitger. Parallel Computation with Threshold Functions. *Journal of Computer and System Sciences*, 36(3):278–302, June 1988.

[96] I. Parberry and G. Schnitger. Relating Boltzmann Machines to Conventional Models of Computation. *Neural Networks*, 2(1):59–67, 1989.

[97] M. Paterson, N. Pippenger, and U. Zwick. Faster circuits and shorter formulae for multiple addition, multiplication and symmetric Boolean functions. In *Proceedings of the 31st Annual Symposium on Foundations of Computer Science*, pages 642–650, 1990.

[98] M. Paterson and U. Zwick. Shallow Multiplication Circuits. In *Proceedings of the 10th IEEE Symposium on Computer Arithmetic*, pages 28–34, 1991.

[99] R. Paturi. On the Degree of Polynomials that Approximate Symmetric Boolean Functions. In *Proceedings of the 24th Annual ACM Symposium on Theory of Computing*, pages 468–474, 1992.

[100] R. Paturi and M. Saks. On Threshold Circuits for Parity. In *Proceedings of the 31st Annual Symposium on Foundations of Computer Science*, pages 397–404, 1990.

[101] U. Peled and B. Simeone. Polynomial-Time Algorithms for Regular Set-Covering and Threshold Synthesis. *Discrete Applied Mathematics*, 12(1):57–69, September 1985.

[102] P. P. Petrushev and V. A. Popov. *Rational approximation of real functions*. Cambridge, NY: Cambridge University Press, 1987.

[103] J. C. Picard and H. D. Ratliff. Minimum cuts and related problems. *Networks*, 5:357–370, 1974.

[104] N. Pippenger. The complexity of computations by networks. *IBM Journal of Research and Development*, 31(2):235–243, March 1987.

[105] P. Raghavan. Learning in Threshold Networks: A Computation Model and Applications. Technical Report RC 13859, IBM Research Division, July 1988.

[106] A. A. Razborov. Lower Bounds on the size of bounded-depth networks over the basis $\{\wedge, \oplus\}$. Tech. Rep. Acad. of Sciences, Moscow, 1986.

[107] N. P. Redkin. Synthesis of threshold circuits for certain classes of Boolean Functions. *Kibernetika*, 5:6–9, 1970.

[108] J. Reif. On Threshold Circuits and Polynomial Computation. In *Proceedings of 2nd Annual Structure in Complexity Theory Conference*, pages 118–123, 1987.

[109] F. Rosenblatt. *Principles of Neurodynamics*. Washington, DC: Spartan Books, 1962.

[110] G.-C. Rota. On the Foundations of Combinatorial Theory I. Theory of Möbius Functions. *Z. Wahrsch. Verw. Gebiete*, 2:340–368, 1964.

[111] V. P. Roychowdhury, A. Orlitsky, and K.-Y. Siu. Lower Bounds on Threshold and Related Circuits via Communication Complexity. *IEEE Transactions on Information Theory*, 40(2):467–474, March 1994.

[112] V. P. Roychowdhury, K.-Y. Siu, and T. Kailath. Classification of Linearly Non-Separable Patterns by Linear Threshold Elements. *IEEE Transactions on Neural Networks*, 1995. In press.

[113] V. P. Roychowdhury, K.-Y. Siu, and A. Orlitsky, editors. *Theoretical Advances in Neural Computation and Learning*. Boston: Kluwer Academic Publishers, 1994.

[114] V. P. Roychowdhury, K.-Y. Siu, A. Orlitsky, and T. Kailath. A Geometric Approach to Threshold Circuit Complexity. In *Proceedings of the Fourth Annual Workshop on Computational Learning Theory (COLT)*, pages 97–111, August 1991.

[115] V. P. Roychowdhury, K.-Y. Siu, A. Orlitsky, and T. Kailath. Vector Analysis of Threshold Functions. *Information and Computation*, 1995. In press.

[116] D. E. Rumelhart, J. L. McClelland, and the PDP Research Group. *Parallel Distributed Processing: Explorations in the Microstructure of Cognition, v. 1-2*. Cambridge, MA: MIT Press, 1986.

[117] L. Schläfli. *Gesammelte Mathematische Abhandlungen I.* Basel, Switzerland: Verlag Birkhäuser, 1950.

[118] A. Schonhage and V. Strassen. Schnelle Multiplikation großer Zahlen. *Computing*, 7:281–292, 1971.

[119] C. Shannon. The synthesis of two-terminal switching circuits. *Bell Syst. Tech. J.*, 28:59–98, 1949.

[120] Y. Shrivastava, S. Dasgupta, and S. M. Reddy. Guaranteed Convergence in a Class of Hopfield Networks. *IEEE Transactions on Neural Networks*, 3(6):951–961, November 1992.

[121] R. C. Singleton. A test for linear separability as applied to self-organizing machines. In M. C. Youtis, G. T. Jacobi, and G. D. Goldstein, editors, *Self-Organizing Systems.* Spartan Books, Washington D.C., 1962.

[122] K.-Y. Siu and J. Bruck. Neural Computation of Arithmetic Functions. *Proceedings of the IEEE*, 78(10):1669–1675, October 1990. Special Issue on Neural Networks.

[123] K.-Y. Siu and J. Bruck. On the Power of Threshold Circuits with Small Weights. *SIAM Journal on Discrete Mathematics*, 4(3):423–435, August 1991.

[124] K.-Y. Siu, J. Bruck, T. Kailath, and T. Hofmeister. Depth Efficient Neural Networks for Division and Related Problems. *IEEE Transactions on Information Theory*, 39(3):946–956, May 1993.

[125] K.-Y. Siu and V. P. Roychowdhury. On Optimal Depth Threshold Circuits for Multiplication and Related Problems. *SIAM Journal on Discrete Mathematics*, 7(2):284–292, May 1994.

[126] K.-Y. Siu, V. P. Roychowdhury, and T. Kailath. Computing with Almost Optimal Size Threshold Circuits. *IEEE International Symposium on Information Theory, Budapest, Hungary*, June 1991.

[127] K.-Y. Siu, V. P. Roychowdhury, and T. Kailath. Depth-Size Tradeoffs for Neural Computation. *IEEE Transactions on Computers*, 40(12):1402 – 1412, December 1991. Special Issue on Neural Networks.

[128] K.-Y. Siu, V. P. Roychowdhury, and T. Kailath. Computing with Almost Optimal Size Neural Networks. In S. J. Hanson, J. D. Cowan, and C. Lee Giles, editors, *Advances in Neural Information Processing Systems 5*, pages 19–26. San Mateo, CA: Morgan Kaufman, 1993.

[129] K.-Y. Siu, V. P. Roychowdhury, and T. Kailath. Rational Approximation Techniques for Analysis of Neural Networks. *IEEE Transactions on Information Theory*, 40(2):455–466, March 1994.

[130] K.-Y. Siu, V. P. Roychowdhury, and T. Kailath. Toward Massively Parallel Design of Multipliers. *Journal of Parallel and Distributed Computing*, 1995. In press.

[131] R. Smolensky. Algebraic methods in the theory of lower bounds for Boolean circuit complexity. In *Proceedings of the 19th Annual ACM Symposium on Theory of Computing*, pages 77–82, 1987.

[132] R. Smolensky. On interpolation by analytical functions with special properties and some weak lower bounds on the size of circuits with symmetric gates. In *Proceedings of the 31st Annual Symposium on Foundations of Computer Science*, pages 628–631, 1990.

[133] M. Snir. Depth-size tradeoffs for parallel prefix computation. *Journal of Algorithms*, 7(2):185–201, June 1986.

[134] J. Spencer. *Ten Lectures on the Probabilistic Method.* Philadelphia, PA: SIAM, 1987.

[135] M. Szegedy. *Algebraic Methods in Lower Bounds for Computational Models with Limited Communication.* PhD thesis, University of Chicago, 1989.

[136] F. Tanaka and S. F. Edwards. Analytic theory of the ground state properties of a spin glass. I. Ising spin glass. *Journal of Physics F (Metal Physics)*, 10(12):2769–2778, December 1980.

[137] S. S. Venkatesh. Connectivity Versus Capacity in the Hebb Rule. In V.P. Roychowdhury, K.-Y. Siu, and A. Orlitsky, editors, *Theoretical Advances in Neural Computation and Learning*. Boston: Kluwer Academic Publishers, 1994.

[138] A. Vergis, K. Steiglitz, and B. Dickinson. The Complexity of Analog Computation. *Mathematics and Computers in Simulation*, 28(2):91–113, April 1986.

[139] C. S. Wallace. A suggestion for a fast multiplier. *IEEE Transactions on Electronic Computers*, EC-13(1):14–17, 1964.

[140] I. Wegener. *The Complexity of Boolean Functions.* New York, John Wiley & Sons, 1987.

[141] I. Wegener. Optimal lower bounds on the depth of polynomial-size threshold circuits for some arithmetic functions. *Information Processing Letters*, 46(2):85–87, 17 May 1993.

[142] A. Weinberger and J. L. Smith. A One-Microsecond Adder Using One-Megacycle Circuitry. *IRE Transactions on Electronic Computers*, 5:67–73, 1956.

[143] P. Werbos. *Beyond Regression: New Tools for Prediction and Analysis in the Behavioral Sciences.* PhD thesis, Harvard University, 1974.

[144] B. Widrow and M. E. Hoff. Adaptive Switching Circuits. *1960 IRE Western Electric Show and Convention Record, Part 4*, pages 96–104, August 1960.

[145] B. Widrow and M. Lehr. 30 Years of Adaptive Neural Networks: Perceptron, Madaline, and BackPropagation. *Proceedings of the IEEE*, 78(9):1415–1442, September 1990.

[146] B. Widrow and S.D. Stearns. *Adaptive Signal Processing*. Englewood Cliffs, NJ: Prentice-Hall, 1985.

[147] R. O. Winder. *Threshold Logic*. PhD thesis, Princeton University, 1962.

[148] A. Yao. Separating the polynomial-time hierarchy by oracles. In *Proceedings of the 26th Annual Symposium Foundations of Computer Science*, pages 1–10, 1985.

[149] A. Yao. On ACC and Threshold Circuits. In *Proceedings of the 31st Annual Symposium on Foundations of Computer Science*, pages 619–627, 1990.

[150] Yu. A. Zuev. Asymptotics of the logarithm of the number of threshold functions of the algebra of logic. *Soviet Mathematics Doklady*, 39(3):512–513, 1989.

[151] T. Zaslavsky. Facing up to arrangements. *Mem. American Mathematical Society*, (154), 1975.

[152] E. I. Zolotarjov. Application of the elliptic functions to the problems on the functions of the least and most deviation from zero (Russian). *Zapiskah Rossijskoi Akad. Nauk.*, 1877.

Index